IUTAM Bookseries

Volume 40

The IUTAM Bookseries publishes the refereed proceedings of symposia organised by the International Union of Theoretical and Applied Mechanics (IUTAM).

Every two years the IUTAM General Assembly decides on the list of IUTAM Symposia. The Assembly calls upon the advice of the Symposia panels. Proposals for Symposia are made through the Assembly members, the Adhering Organizations, and the Affiliated Organizations, and are submitted online when a call is launched on the IUTAM website.

The IUTAM Symposia are reserved to invited participants. Those wishing to participate in an IUTAM Symposium are therefore advised to contact the Chairman of the Scientific Committee in due time in advance of the meeting. From 1996 to 2010, Kluwer Academic Publishers, now Springer, was the preferred publisher of the refereed proceedings of the IUTAM Symposia. Proceedings have also been published as special issues of appropriate journals. From 2018, this bookseries is again recommended by IUTAM for publication of Symposia proceedings.

Indexed in Ei Compendex and Scopus.

Dilmurat Azimov
Editor

Proceedings of the IUTAM Symposium on Optimal Guidance and Control for Autonomous Systems 2023

Springer

Editor
Dilmurat Azimov
Department of Mechanical Engineering
University of Hawai'i at Manoa
Honolulu, HI, USA

ISSN 1875-3507 ISSN 1875-3493 (electronic)
IUTAM Bookseries
ISBN 978-3-031-39302-0 ISBN 978-3-031-39303-7 (eBook)
https://doi.org/10.1007/978-3-031-39303-7

This Springer imprint is published by the registered company Springer Nature Switzerland AG
The registered company address is: Gewerbestrasse 11, 6330 Cham, Switzerland

Paper in this product is recyclable.

Contents

Part I
Guidance, Navigation and Control, and Sensor Technologies

Chapter 1
Force Estimation and Control Enhanced by a Force-Derivative Sensor

Yu-Sheng Lu and Liang-Hao Chen

Abstract The paper presents a scheme for force estimation and control of a motion system in contact with an environment. Especially, force-derivative sensing is used to facilitate force estimation and control. Regarding force estimation, this paper proposes an integrated force estimator (IFE) that fuses an extended state observer (ESO) and force-derivative sensing. The force-derivative sensing provides the derivative of force, referred to as yank. Meanwhile, the ESO receives positional information and estimates a force. The IFE fuses the yank signal with the force estimate produced by the ESO and provides an improved force estimate. Compared with previous research on yank, the proposed system does not require an additional force sensor to simultaneously obtain both yank and force information. The proposed IFE can provide a better force estimate than traditional ESOs and alleviates the drift problem encountered in the direct integration of yank. Concerning force control, derivative control can be used to offer damping due to the existence of force-derivative sensing. Experimental results on a platform with a linear motor are reported, and a comparative study is conducted in this work. The experimental results demonstrate the feasibility and advantage of the proposed scheme.

Keywords Motion system · Piezoelectric sensor · Force derivative · Force observation · Extended state observer · Sensor fusion

1.1 Introduction

Commercial motion systems, including rotary motor systems, linear motor systems, CNC control systems, and robotic arm systems, usually provide speed or position control loops. However, for some applications in contact with the environment, simple speed and position control loops are no longer sufficient, such as clamping force control of injection molding machines, edge trimming of robotic arms, grinding,

Y.-S. Lu (✉) · L.-H. Chen
National Taiwan Normal University, Taipei 106, Taiwan
e-mail: luys@ntnu.edu.tw

D. Azimov (ed.), *Proceedings of the IUTAM Symposium on Optimal Guidance and Control for Autonomous Systems 2023*, IUTAM Bookseries 40,
https://doi.org/10.1007/978-3-031-39303-7_1

polishing, and assembly applications. Although commercial motor servo systems usually also provide current control loops, users can make use of this to have the motor produce the desired torque/force, but the torque/force generated by the motor cannot represent the contact force between the system and the environment, so it needs further consideration to obtain contact force information. The strong demand for contact force information can also be inferred from the advent of many force sensors.

The ways to obtain force information can be roughly divided into two categories: one is to use physical force sensors, and the other is to use force observers. In using force sensors, although there have been many applications of sensing principles, there are two most widely used and effective force sensors: load cells with strain gauges and piezoelectric force sensors [1]. Compared with load cells with strain gauges, piezoelectric force sensors have the following benefits: high rigidity of piezoelectric crystals, small size, and large dynamic range [2]. Therefore, piezoelectric sensors are used for cutting force measurement of machine tools, where high stiffness is required, and vibration sensing which requires high bandwidth. However, piezoelectric force sensors are unable to measure static forces, which limits their application to motion control systems. Moreover, these piezoelectric force sensors are unable to measure the derivative of force, referred to as yank.

Concerning force observers using algorithmic approaches, physical force sensors are not used, but other sensors are required to measure the current, position, or acceleration of the system, and then the mathematical model of the system is used to estimate the force using algorithms. However, to estimate environmental forces using position information, the limited position sensing resolution makes it impossible to effectively observe high-frequency environmental forces, which is the major drawback of the force observers.

To combine the best features of these two approaches, this paper proposes a hybrid approach that enables the measurement of static forces and provides high dynamic bandwidth as well. In the study [3], a piezoelectric yank sensor has been developed. Based on this yank sensor, this paper intends to use an extended state observer (ESO) and combines it with the yank sensing to obtain a force estimate. More specifically, this paper proposes an integrated force estimator (IFE) for the combination of a yank sensor and an ESO to obtain better force estimation than traditional ESO. Compared with physical force sensors, the proposed method is expected to obtain static and dynamic force information without redesigning the structure of the sensor. Compared with the previous research on the yank sensor, the proposed method can eliminate the need for an additional force sensor while simultaneously obtaining information on both force and yank. More importantly, it measures force-derivative that can be used to implement the derivative control in a force controller for improved closed-loop damping. The proposed scheme is applied to a physical linear-motor system to investigate its applicability and feasibility, and experimental results are reported in this paper.

1.2 Plant Model and an ESO

Consider a plant, which is modeled as a second-order mass-damper motion system with one input and one output:

$$m\ddot{x} + b\dot{x} = ku + d + f \tag{1.1}$$

in which x is the position, m is the mass of the movable mass, b is the viscous damping coefficient, k is a constant gain, and u is the input to the plant; f is the environmental force, which represents the reaction force exerted by the external environment on the plant; d represents the disturbance force within the plant, including friction between mechanical moving parts and cogging torque in electric motors. Assume that d is known. The most direct way to estimate the environmental force is to rewrite (1.1) to obtain the environmental force, $f = m\ddot{x} + b\dot{x} - ku - d$; this idea has been used in the literature [4] to obtain contact force estimation. However, the realization of such a method requires an acceleration signal or the differentiation of the position signal twice, so either an additional accelerometer needs to be installed or the noise amplification caused by the two times of differentiation will be encountered. To avoid this implementation difficulty, the literature [5] used the so-called momentum observer to estimate the environmental force; in addition, many studies [6] used the concept of the disturbance estimator to propose a reaction force observer to avoid directly using acceleration signals, but these studies did not observe the state variables of the plant while estimating the environmental force. In order to observe the environmental force and state variables at the same time, the design of an ESO is the most commonly used method [7].

To design an ESO for estimating an environmental force. First, it is assumed that d is known, and $\dot{f} = 0$. Let f be regarded as a state variable, and the extended state equation is:

$$\begin{aligned} \dot{x}_1 &= x_2, \\ m\dot{x}_2 &= -bx_2 + ku + d + f, \\ \dot{f} &= 0, \end{aligned} \tag{1.2}$$

in which $x_1 = x$, and $x_2 = \dot{x}$. Design a linear ESO [7] as

$$\begin{aligned} \dot{\hat{x}}_1 &= \hat{x}_2 + \ell_{L1}\left(x_1 - \hat{x}_1\right), \\ \dot{\hat{x}}_2 &= m^{-1}\left(-b\hat{x}_2 + ku + d + \hat{f}_L\right) + \ell_{L2}\left(x_1 - \hat{x}_1\right), \\ \dot{\hat{f}}_L &= \ell_{L3}m\left(x_1 - \hat{x}_1\right), \end{aligned} \tag{1.3}$$

where $\hat{x}_1$, $\hat{x}_2$, and $\hat{f}_L$ are estimates of x_1, x_2, and f, respectively, and ℓ_{Li}, $i = 1, 2, 3$, are constant gains, which the designer can determine using the pole placement method. If all the poles of the linear ESO are set to $-\omega_L$, the observer gains are determined by:

$$\ell_{L1} = 3\omega_L - m^{-1}b, \quad \ell_{L2} = 3\omega_L^2 - m^{-1}b\ell_{L1}, \quad \ell_{L3} = \omega_L^3. \tag{1.4}$$

Large values of ω_L can be chosen to have fast observer dynamics.

1.3 Integrated Force Estimator (IFE)

Block diagram of the proposed scheme is shown in Fig. 1.1, which uses a piezoelectric transducer and a position sensor that commonly exists in motion systems. Piezoelectric transducer is connected to a transimpedance amplifier (TIA) to obtain the yank, y_p. Meanwhile, the linear ESO receives the position signal to produce the position and velocity estimates and also a force estimate, $\hat{f}_L$. The force estimate, $\hat{f}_L$, by the linear ESO is considered a preliminary force estimate. Because only the positional measurement is used, and the limited positional resolution deteriorates the quality of this force estimate. Finally, the IFE integrates y_p and $\hat{f}_L$ for improved force estimation. Overall speaking, the proposed system produces a yank signal and estimates of force, position, and velocity.

Direct integration of y_p with respect to time, denoted as $I\{y_p\}$, can produce a nice force estimate but subject to a drift problem. In order to avoid the drift problem and measure the static force, this paper proposes integrated force estimators (IFEs) to capture the high-frequency part of $I\{y_p\}$ and the low-frequency part of $\hat{f}_L$. Consider a first-order IFE (FO-IFE) described by:

$$\dot{\hat{f}}_{FO} = y_p - c\left(\hat{f}_{FO} - \hat{f}_L\right), \tag{1.5}$$

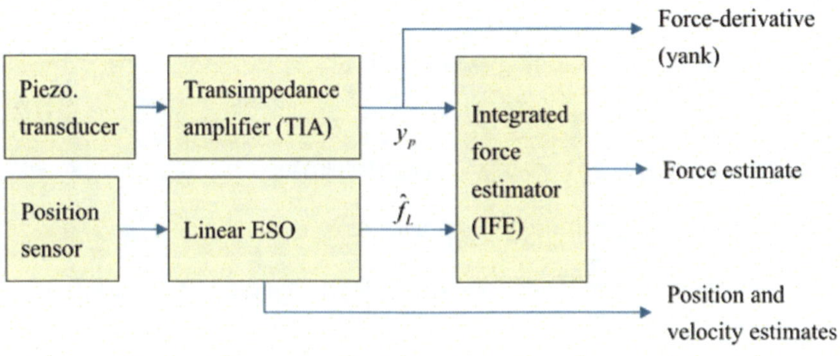

Fig. 1.1 Block diagram of the proposed scheme

in which $\hat{f}_{FO}$ is the force estimate by the FO-IFE, and c is a positive parameter. The FO-IFE can also be expressed as $\hat{f}_{FO}(t) = \int_0^t y_p(\tau)d\tau - c\int_0^t \left(\hat{f}_{FO}(\tau) - \hat{f}_L(\tau)\right)d\tau$. To explore the estimation dynamics of FO-IFE, let $y_p = y + n_p$, where $y = \dot{f}$, and n_p represents the measurement noise of force derivative. From (1.5) we can get:

$$\hat{f}_{FO}(s) = H_{FO}(s)f(s) + L_{FO}(s)\hat{f}_L(s) + \frac{1}{s+c}n_p(s), \tag{1.6}$$

in which $H_{FO}(s) = s/(s+c)$, $L_{FO}(s) = c/(s+c)$, and the function whose argument is s represents the Laplace transform of the function whose corresponding argument is t. From (1.6), the relationship between n_p and $\hat{f}_{FO}$ can be obtained: $\hat{f}_{FO}(s)/n_\mathrm{p}(s) = 1/(s+c)$, so FO-IFE can attenuate the influence of n_p on $\hat{f}_{FO}$ at low frequencies by c times, and can make the high-frequency components of n_p have almost no influence on $\hat{f}_{FO}$. Interestingly, $H_{FO}(s)$ is a high-pass filter, and $L_{FO}(s)$ is a low-pass filter, and $H_{FO}(s) + L_{FO}(s) = 1$, which means that $H_{FO}(s)$ and $L_{FO}(s)$ is a set of complementary filters. It can be seen from (1.6) that FO-IFE extracts the high-frequency part of $I\{y_p\}$, and then combines the low-frequency part of the linear ESO's force estimate, $\hat{f}_L$, to synthesize the force estimate, $\hat{f}_{FO}$. The FO-IFE captures the integrated high-frequency part of $I\{y_p\}$ to avoid the drift problem, and captures the low-frequency part of $\hat{f}_L$ to reduce high-frequency noise and measure static force. This illustrates the benefits of integrating the force differential sensor and the linear ESO.

FO-IFE gives $\hat{f}_{FO}(s)/n_\mathrm{p}(s) = 1/(s+c)$, which means that FO-IFE cannot completely eliminate the influence of low-frequency n_p on $\hat{f}_{FO}$. To further reduce the impact of low-frequency n_p on force estimation, this paper proposes a second-order integrated force estimator (SO-IFE) design described by:

$$\dot{\hat{f}}_{SO} = y_p - c_2\left(\hat{f}_{SO} - \hat{f}_L\right) - c_1\int_0^t \left(\hat{f}_{SO}(\tau) - \hat{f}_L(\tau)\right)d\tau, \tag{1.7}$$

where c_2 and c_1 are adjustable positive parameters. From (1.7) we can get:

$$\hat{f}_{SO}(s) = H_{SO}(s)f(s) + L_{SO}(s)\hat{f}_L(s) + \frac{s}{s^2+c_2s+c_1}n_p(s), \tag{1.8}$$

in which $H_{SO}(s) = \frac{s^2}{s^2+c_2s+c_1}$, and $L_{SO} = \frac{c_2s+c_1}{s^2+c_2s+c_1}$. Similarly, $H_{SO}(s)$ is a high-pass filter, and $L_{SO}(s)$ is a low-pass filter, and $H_{SO}(s) + L_{SO}(s) = 1$, meaning that $H_{SO}(s)$ and $L_{SO}(s)$ are a set of complementary filters. From (1.8), we can get the relationship between n_p and $\hat{f}_{SO}$: $\hat{f}_{SO}(s)/n_p(s) = s/\left(s^2+c_2s+c_1\right)$, so SO-IFE makes high-frequency n_p almost have no influence on $\hat{f}_{SO}$, and further completely eliminates the influence of static n_p on force estimation. Assuming that n_p is the output offset of the force yank sensor and almost remains constant, then $\hat{f}_{SO}$ does not subject to the drift problem in force estimation.

1.4 Experimental Study

1.4.1 Experimental Setup and System Identification

This paper validates the proposed method using a linear motor system. Figure 1.2 shows a photograph of the experimental platform. A driver (SGDV-1R6A05B, YASKAWA) operating in current mode drives a linear motor (SGLGW-40A253CP, YASKAWA), which has a rated thrust of 93 N and a rated speed of 2 m/s. The linear scale on the platform is RGS20-S optical ruler (Renishaw), and the corresponding read head is RGH24-B30L00A (Renishaw), giving a minimum resolution of 0.078 μm. As for the piezoelectric transducer, charge type KF24 (MMF) is used, and its charge sensitivity is 349.89 pC/N. The amplifier matched with this piezoelectric transducer is a TIA circuit, whose output corresponds to yank. In addition, the load cell using strain gauges has a measurement range of ±100 N. The load cell is installed on the environment and can be in direct contact with the piezoelectric transducer, so both the load cell and the piezoelectric transducer are subject to the same magnitude of force. Note that the load cell is used for a comparative study, and the proposed method does not require this load cell.

A TI TMS320C6713 DSP with Xilinx XCV50 FPGA forms a DSP/FPGA real-time controller, in which FPGA stands for field-programmable gate array, and DSP

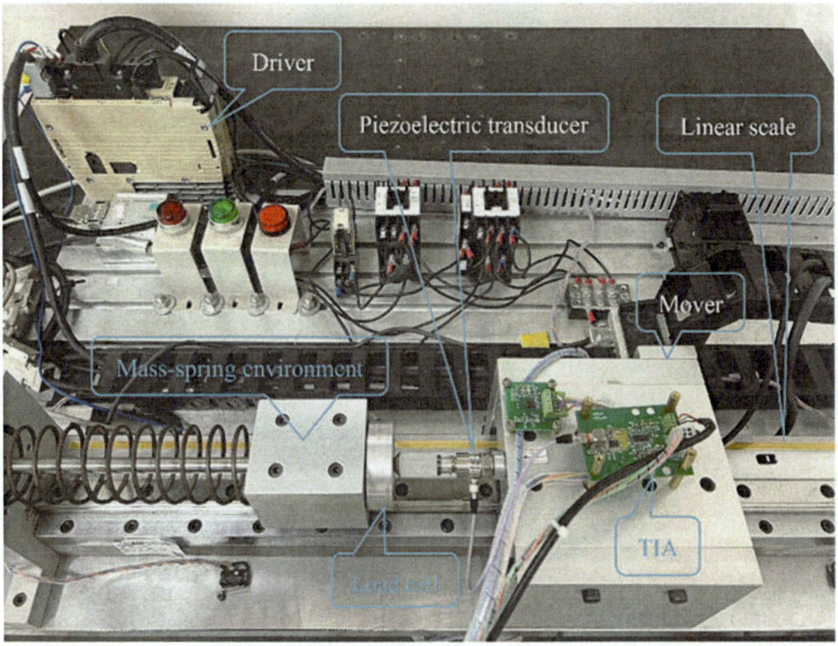

Fig. 1.2 Photo of the experimental system

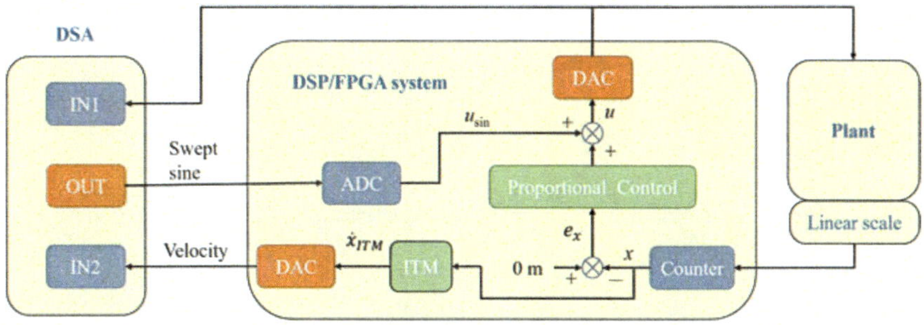

Fig. 1.3 Block diagram for measuring the frequency response of the plant

stands for digital signal processor. All input–output peripherals are handled by the FPGA that also requests an external interrupt to the DSP. After getting the interrupt request, the DSP starts an interrupt service routine for the ESO, IFE, and force controller. The FPGA clock is divided to produce the interrupt request signal to the DSP, giving a sampling frequency of 10,986.33 Hz. Figure 1.3 shows the block diagram for measuring the frequency response of the plant, in which ADC and DAC respectively stand for analog-to-digital and digital-to-analog converters. Regarding system identification, a dynamic signal analyzer, DSA (PXI-8106, NI), was used to generate a 4 – Vpp swept sine signal, and its frequency increased from 1 to 100 Hz to stimulate the plant, making the linear motor's mover move. The FPGA decodes the optical encoder's output signal and counts. The DSP reads back the counter value from the FPGA and converts it into a position. In addition, FPGA uses ITM (inverse time method, also known as fixed position method) to estimate the mover's speed, which is denoted as $\dot{x}_{ITM}$. Figure 1.3 shows the structure of the frequency response of the measurement system. As shown in Fig. 1.3, a proportional controller is used to roughly control the mover at the midpoint of the stroke to avoid the mover hitting the limit position during the measurement process.

As shown in Fig. 1.3, the DSP converts the speed estimate, $\dot{x}_{ITM}$, into a proportional voltage through the DAC, and sends it to the input IN2 of the DSA. Since IN1 of the DSA receives the input voltage of the plant at the same time, the DSA can measure the frequency response of the plant, as shown in Fig. 1.4. Figure 1.4 also shows the result of curve fitting, giving a system model: $\dot{x}_{ITM}/u = 2.325/(s + 11.29)$. It can be seen from (1.1) that $\dot{x}/u = k/(ms + b)$. From k =93 [N]/8 [V]= 11.625 N/V, it can be known that the mass of the mover is 4.99 kg, and the damping coefficient is 56.45 N/(m/s).

1.4.2 *Implementation of the ESO*

This paper uses ESO for low-frequency force estimation, but assumes that the internal disturbance force, d, of the motion system is known. In order to identify d, which

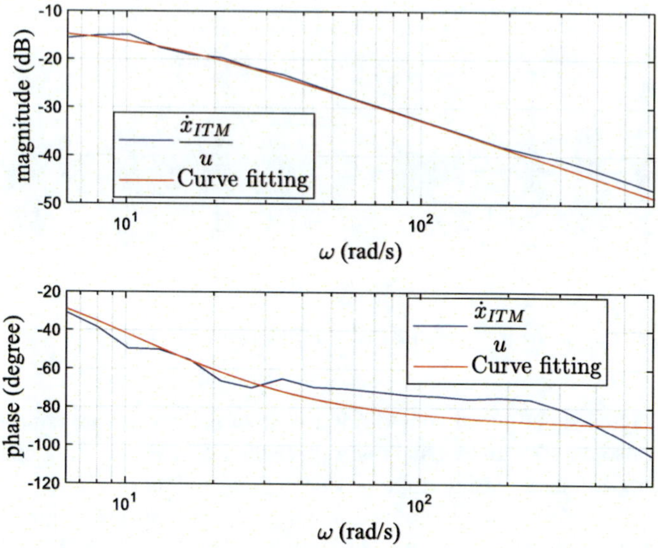

Fig. 1.4 Frequency response of the plant

is mainly friction force, d needs to be modeled. The modeling method is to use a traditional disturbance observer (DOB) to estimate d in the absence of environmental external forces ($f = 0$). The General Kinetic Friction (GKF) model is then fitted, and finally, the LuGre friction model that includes the GKF model is used. The DOB uses a first-order Q-filter with a cutoff frequency of 500 Hz. The experimental method adopts the method in the study [8], and a PD controller is used to track a ramp position command, so that the mover can almost maintain a constant speed. Because the friction force of the mover is usually different at different positions and/or speeds, the dynamic friction force ($\hat{d}_{ITM}$) is estimated through the DOB during the movement of the mover, and the DSP records $\hat{d}_{ITM}$, the current position (x), and speed of the mover ($\dot{x}_{ITM}$) when the mover is almost at a constant speed. In addition, the DSP also enables the PD controller to receive ramp position commands of different slopes, so that the DSP can record the friction force at different positions and speeds.

After DSP collects the data, the curve-fitting tool of MATLAB is used to fit according to the GKF model and to obtain the model parameters at different positions. The friction force ($\hat{d}_f$) represented by the GKF model [8] is:

$$\hat{d}_f(x, \dot{x}) = g(x, \dot{x})\mathrm{sgn}(\dot{x}) + b_f \dot{x} \tag{1.9}$$

in which b_f is the viscous damping coefficient, and $g(x, \dot{x})$ is given by:

$$g(x, \dot{x}) = d_C(x) + (d_S(x) - d_C(x)) \exp\left(-\left|\frac{\dot{x}}{\dot{x}_{str}}\right|^2\right) \tag{1.10}$$

Here, d_C is Coulomb force, d_S is stiction force, and $\dot{x}_{str}$ is Stribeck velocity. When the mover is at different positions, the corresponding model parameters are also different. Figure 1.5 shows $\hat{d}_{ITM}$ and $\hat{d}_f$ at different speeds when the mover is at the position of 19 cm, which shows that the original data ($\hat{d}_{ITM}$) and the fitting result ($\hat{d}_f$) are similar. To obtain better friction estimation, the LuGre model is then introduced [9]:

$$\hat{d}_L(x, \dot{x}) = \alpha_0 z + \alpha_1 \dot{z} + b_f \dot{x}, \quad \dot{z} = \dot{x} - \frac{\alpha_0 |\dot{x}|}{g(x, \dot{x})} z, \tag{1.11}$$

where $\hat{d}_L$ is the output of this model, z is the state variable of this model, and α_1 and α_0 are model parameters. In this study, $\alpha_1 = 100$, and $\alpha_0 = -10^{-5}$. Figure 1.6 shows the verification results of the friction force under position control. It can be seen that compared with $\hat{d}_f$, it is seen that $\hat{d}_L$ is closer to $\hat{d}_{ITM}$ minus the environmental force, so the LuGre model is used in this paper.

This paper applies the friction estimation of LuGre model to the linear ESO. To find an optimal parameter for the linear ESO, define a performance index, $ISE = \int_0^T \delta^2 dt$, where T is the length of the experiment, and $\delta = L^2\{\hat{f}_L\} - f_{ld}$. Here, $\hat{f}_L$ denotes the force estimate by the ESO, and $L^2\{\cdot\}$ represents a zero-phase low-pass filter with a cutoff frequency of 20 Hz. The DSP is used to control the position of the mover. During the position control, the linear ESO produces $\hat{f}_L$, which is however not used for feedback control. For the linear ESO, many experiments were conducted, and it was found that $\omega_L = 400\pi$ can make the performance index better. Hence, the linear ESO with $\omega_L = 400\pi$ is used in the following experiments.

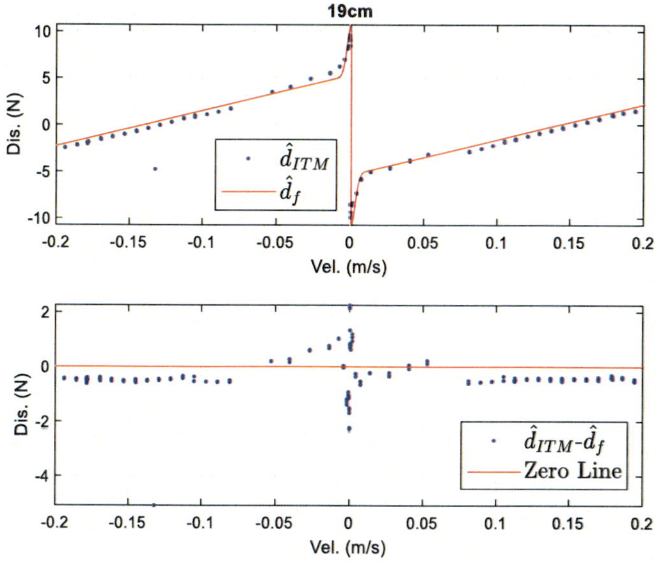

Fig. 1.5 Disturbance estimate when the mover is at the position of 19 cm

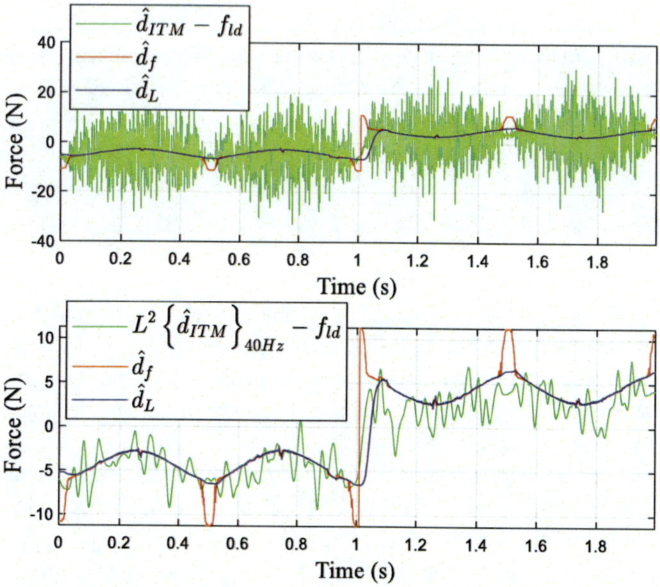

Fig. 1.6 Friction model verification under position control

1.4.3 Design and Implementation of Integrated Force Estimator (IFE)

IFEs, including FO- and SO-IFE, are to be designed and implemented. The position control of the linear motor platform is first performed, and the FO- and SO-IFE estimate the force at the same time, which is not used for feedback control but only used to select the parameters of FO- and SO-IFE. In tuning the SO-IFE, let $c_2 = 2\sqrt{c_1}$. Taking ISE as the performance index, several experiments were carried out to obtain the better IFE parameter. The results are $c = 0.5$ for the FO-IFE, and $c_2 = 2\sqrt{c_1} = 2 \times 2$ for the SO-IFE. Figure 1.7 shows the force measurement and estimation results under position control, where $I\{\cdot\}$ represents the integration with respect to time. It can be seen from Fig. 1.7 that $I\{y_p\}$ drifts, resulting in an error in force estimation. Although the estimation by the linear ESO has no drift phenomenon, its high-frequency components around 200 Hz are noticeably attenuated and out of phase compared with the load-cell measurement. Figure 1.8 shows the force measurement and IFEs' force estimation results under the same experiment. It can be seen that both FO- and SO-IFE can avoid the drift phenomenon, and the phase of their high-frequency components is roughly the same as the load-cell measurement. Therefore, both FO- and SO-IFEs have better force estimation than the direct yank integral and the ESO.

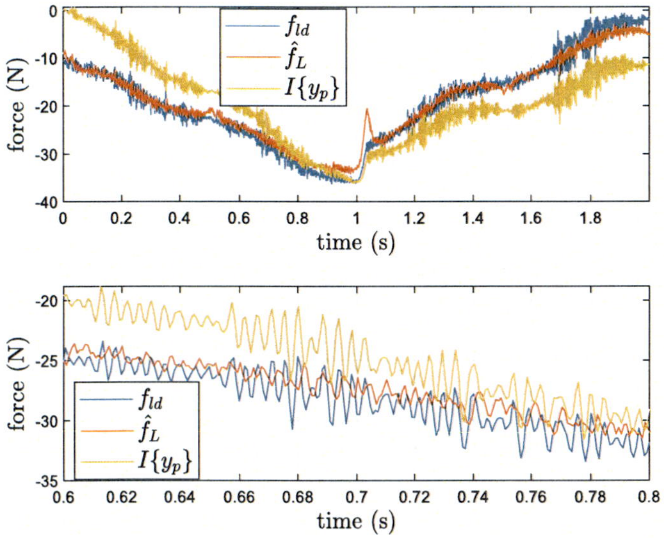

Fig. 1.7 Force measurement and estimation under position control

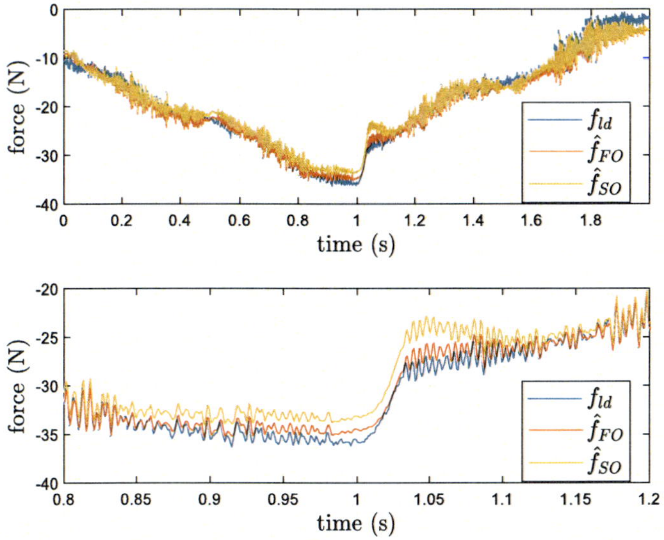

Fig. 1.8 Force measurement and IFE force estimation under position control

1.4.4 Force Control and Discussion on the IFEs' Performance

PI (Proportional Integral) force control was performed using force measurement and force estimates, and the FO- and SO-IFE performance was evaluated with force control results. Define the force error $e_f = f_{fb} - r$, where r is the force command, and f_{fb} is the feedback force signal, which can be f_{ld}, $\hat{f}_{FO}$, or $\hat{f}_{SO}$. The PI force control law is designed as $u = u_{PI}$:

$$u_{PI} = k^{-1}\left[K_{Pf}e_f + K_{If}\left(\int_0^t e_f(\tau)d\tau + h_0\right) - \hat{d}\right] \tag{1.12}$$

in which K_{Pf} and K_{If} are the proportional and integral control gains respectively, and the parameter h_0 represents the initial integral value. The purpose of adding this parameter to the PI force control law is to keep the control effort continuous and achieve bumpless transfer when the system is switched from position control to force control. Here, h_0 is specified as:

$$h_0 = K_{If}^{-1}\left(k\,u(0^-) + \hat{d}(0^+) - K_{Pf}e_f(0^+)\right) \tag{1.13}$$

The system performs position control at the beginning, and switches to force control when the mover reaches the position of 12 cm, in which $K_{Pf} = 0.3$, $K_{If} = 0.03$, and the force command is $r = -20$ N. Figure 1.9 shows the force and yank response under force control. It can be seen that there is still a drift phenomenon in $I\{y_p\}$, but it has less high-frequency noise than f_{ld}. On the other hand, $\hat{f}_L$ can gradually tend to the near low frequency content of f_{ld}, but at high frequencies (around 200 Hz) it is out of phase compared with the other two.

Keep the proportional gain unchanged, but increase the integral gain to $K_{If} = 0.09$. Figure 1.10 shows the force response, which contains the results of three individual experiments. Figure 1.10a–c show the experimental results of force control with the feedback of f_{ld}, $\hat{f}_{FO}$, and $\hat{f}_{SO}$, respectively. When controlling force by feeding back one signal, the other two force signals are also recorded, and three force signals are simultaneously presented in each plot. It can be seen from Fig. 1.10 that $\hat{f}_{FO}$ and $\hat{f}_{SO}$ have lower high-frequency noise than f_{ld}, and the drift phenomenon is not significant. Compared with Fig. 1.9, Fig. 1.10 shows that the increase of the integral gain reduces the rise time of the response and speeds up the dynamics of the closed-loop system, but the oscillations of the transient force responses become larger. In particular, the system response of the feedback of $\hat{f}_{FO}$ or $\hat{f}_{SO}$ is still stable, while the feedback system with f_{ld} becomes unstable. This shows that the quality of $\hat{f}_{FO}$ or $\hat{f}_{SO}$ produced by the IFE is higher than that of the load cell.

The yank signal can be obtained, and the differential control can be added to the PI control law to obtain a PID control law, which is:

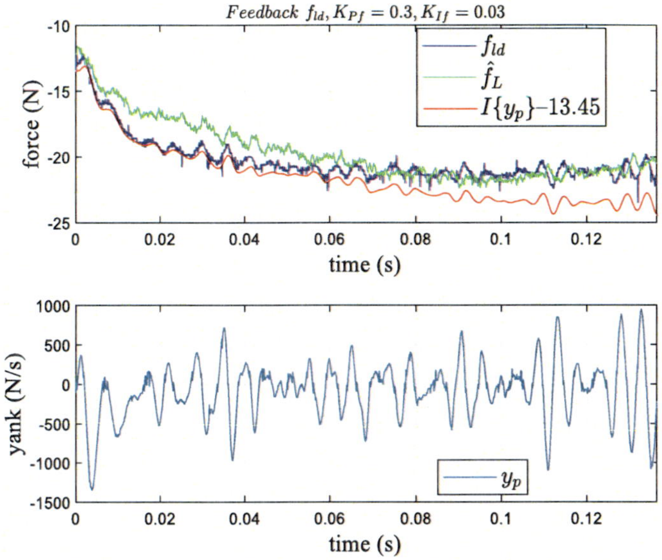

Fig. 1.9 Force and yank response under force control

$$u_{PID} = k^{-1}\left[K_{Pf}e_f + K_{Df}y_p + K_{If}\left(\int_0^t e_f(\tau)d\tau + h_0\right) - \hat{d}\right] \quad (1.14)$$

where K_{Df} is the differential control gain. Setting $K_{Df} = 0.0025$, Figs. 1.11 and 1.12 show the PID-controlled force responses with the feedback of $\hat{f}_{FO}$ and $\hat{f}_{SO}$, respectively. Compared with the PI control results shown in Fig. 1.10b and c, Figs. 1.11 and 1.12 show that the control with yank can effectively reduce the oscillations of the transient response and make the closed-loop system more stable. The experimental results of this PID control show the benefits of introducing the yank. The introduction of yank can increase the system damping, so that the bandwidth of the force control system can be further improved, and a higher performance force control system can be realized.

1.5 Conclusions

For a motion system in contact with the environment, yank and force information is necessary. Compared with directly differentiating the force signal, which can easily amplify noise, this paper uses a yank sensor and avoids amplifying high-frequency noise. As for obtaining force information, the proposed scheme uses the yang sensor, the ESO, and an integrated force estimator (IFE). The IFE integrates the high-frequency components of the yank integral and the low-frequency components of the ESO's force estimation to synthesize the final force estimation. The IFE can

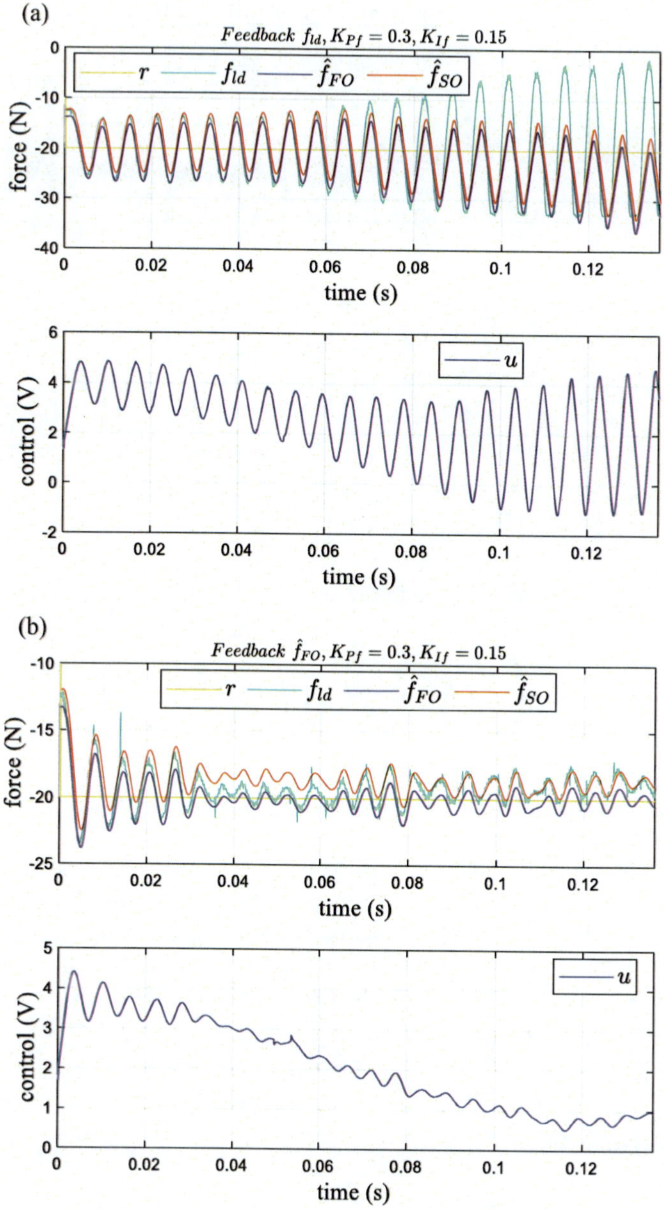

Fig. 1.10 Force responses with PI control by feeding back: **a** f_{ld}, **b** $\hat{f}_{FO}$, and **c** $\hat{f}_{SO}$

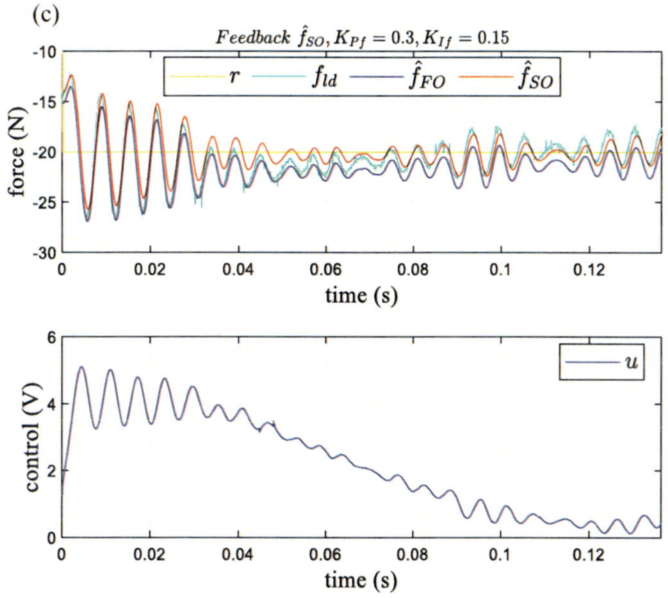

Fig. 1.10 (continued)

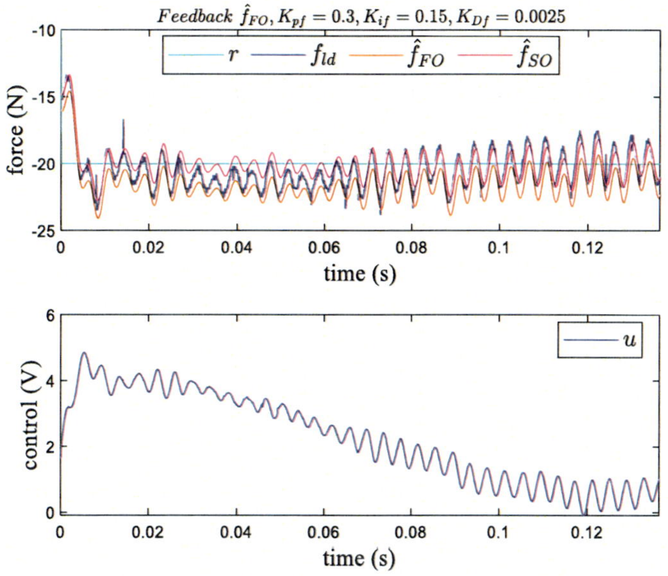

Fig. 1.11 PID-controlled force response with FO-IFE feedback

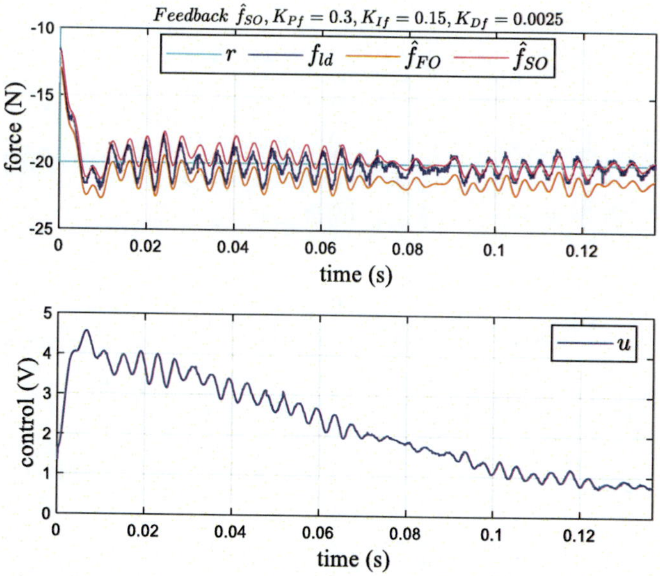

Fig. 1.12 PID-controlled force response with SO-IFE feedback

alleviate the drift problem of yank integral, reduce the negative impact of ESO's force estimation by position noise, and obtain high-quality force signals. The experimental results of PI control also show that the force signal quality of IFE is better than that of load cell. Yank and force information can be used to realize the PID force controller, and the experimental results confirm that the introduction of yank can increase the system damping and realize a high-bandwidth force control system.

References

1. Kleckers, T.: Force sensors for strain gauge and piezoelectric crystal-based mechatronic systems—a comparison. In: Proc. IEEE International Instrumentation and Measurement Technology Conference, pp. 2306–2308. Graz, Austria (2012)
2. Lu, Z., Chen, C. Y., Lin, W.: Force sensing and control in micromanipulation. IEEE Trans. Syst. Man Cybern. C: Appl. Rev. **36**(6), 713–724 (2006)
3. Lu, Y. S., Chen, L. H.: Towards measuring the rate of force with piezoelectric material. In: Proceedings on 2021 International Automatic Control Conference, 1004. CACS, Taiwan (2021)
4. Jeong, J.W., Chang, P.H., Park, K.B.: Sensorless and modeless estimation of external force using time delay estimation: application to impedance control. J. Mech. Sci. Technol. **25**(8), 2051–2051 (2011)
5. Popov D., Klimchik, A.: Real-time external contact force estimation and localization for collaborative robot. In: Proceedings on IEEE International Conference on Mechatronics, pp. 646–651. IEEE, Germany (2019).
6. Sariyildiz, E., Ohnishi, K.: An adaptive reaction force observer design. IEEE/ASME Trans. Mechatron. **20**(2), 750–760 (2015)

7. Sun, B., Gao, Z.: A DSP-based active disturbance rejection control design for a 1-kW H-bridge DC-DC power converter. IEEE Trans. Industrial Electron. **52**, 1271–1277 (2005)
8. Papadopoulos, E. G., Chasparis, G. C.: Analysis and model-based control of servomechanisms with friction. In: Proceedings on IEEE/RSJ International Conference on Intelligent Robots and Systems, pp. 2109–2114. IEEE, Switzerland (2002)
9. Kermani, M.R., Patel, R.V., Moallem, M.: Friction identification and compensation in robotic manipulators. IEEE Trans. Instrum. Meas. **56**(6), 2346–2353 (2007)

Chapter 2
Iterative Control Framework with Application to Guidance and Attitude Control of Spacecraft Rendezvous and Docking

Xun Liu, Hashem Ashrafiuon, and Sergey G. Nersesov

Abstract In contrast with traditional control techniques that require a priori verification of complex mathematical conditions, the Iterative Control Framework (ICF) developed previously by the authors, uses a novel control logic to guarantee the stability of a closed-loop system *without* a priori verification of Lyapunov-like conditions. The underlying idea is to reconfigure the control vector at each time step provided by a real-time computational routine running in the background. The control approach is applicable to a broad class of complex nonlinear systems but it is particularly suitable for systems inherently admitting only control actions of short duration, such as short- and long-range missiles, satellites, and spacecrafts. In this work, we apply the ICF to guidance and attitude control of spacecrafts. Our numerical results show the efficacy of the proposed control framework by attaining all docking criteria for Soyuz spacecraft which involve the alignment of linear and angular positions and velocities of the spacecraft and the target vehicle.

Keywords Iterative control · Non-monotonically decreasing functions · Thruster management of spacecrafts · Genetic algorithm

2.1 Introduction

Traditional control methods for nonlinear dynamical systems are predicated on the verification of mathematical conditions whose complexity is the main limitation of the applicability of those methods. The most basic of such conditions is verifying the existence of a Lyapunov function which is a positive definite function of the system

X. Liu · H. Ashrafiuon · S. G. Nersesov (✉)
Department of Mechanical Engineering, Villanova University, Villanova, PA 19085-1681, USA
e-mail: sergey.nersesov@villanova.edu

X. Liu
e-mail: xliu8@villanvoa.edu

H. Ashrafiuon
e-mail: hashem.ashrafiuon@villanova.edu

D. Azimov (ed.), *Proceedings of the IUTAM Symposium on Optimal Guidance and Control for Autonomous Systems 2023*, IUTAM Bookseries 40,
https://doi.org/10.1007/978-3-031-39303-7_2

state and whose value must strictly decrease with time along the trajectories of the closed-loop dynamical system for all initial conditions in some neighborhood of the desired equilibrium state [1–6].

Verification of Lyapunov conditions is complex and depends on the dimensionality of the dynamical system, nonlinearities in the dynamics, physical constraints, modeling uncertainties and disturbances, quantification of the system parameters, and many other factors. Rigorous mathematical verification of Lyapunov conditions becomes infinitely more difficult in real-world systems with high-dimensional, complex dynamics.

In [7], the authors developed a fundamentally new framework of control logic that can be readily applied to complex nonlinear systems. The approach, called the Iterative Control Framework (ICF), uses a novel control logic to guarantee the stability of any closed-loop system without a priori verification of complex mathematical conditions. Through ICF, control inputs are designed over a temporal period to satisfy mathematical conditions that are numerically verified in real-time. A simple numerical procedure is introduced to determine the instants of time when the control inputs are to be applied.

In this paper, we implement the ICF to guidance and attitude control of spacecrafts during the final stage of space docking. Since spacecrafts are equipped with impulsive thrusters, this is an ideal case study for ICF due to the short duration of actuation. Space docking is one of the most sophisticated maneuvers in space engineering when not only relative position and orientation must be aligned but also relative linear and rotational velocities of the spacecraft and the target vehicle (e.g., space station) must be decreased to almost zero, in order to guarantee successful docking. Furthermore, our control algorithm is augmented with a feature to avoid collision in close proximity of the two vehicles to secure their smooth engagement. We use docking criteria for the Soyuz spacecraft and our simulations show that all the criteria are met by the end of the simulation run.

2.2 Iterative Control Framework

In this section, we briefly review the main elements of the Iterative Control Framework (ICF) [7]. Consider a general form of a nonlinear dynamical system given by

$$\dot{\mathbf{x}}(t) = f(t, \mathbf{x}(t), \mathbf{u}(t)), \quad \mathbf{x}(t_0) = \mathbf{x}_0, \tag{2.1}$$

where $t \geq t_0$, $\mathbf{x}(t) \in \mathbb{R}^n$ is the vector of the system state and $\mathbf{u}(t) \in \mathbb{R}^m$ is the vector of the control input. Introduce a state-dependent positive-definite function $\mathcal{V}(\mathbf{x})$, $\mathbf{x} \in \mathbb{R}^n$. Note that the *only* requirement for the function $\mathcal{V}(\cdot)$ is that it is positive-definite with respect to the state of (2.1). Next, introduce a monotonically decreasing time-varying function $g(t)$, $t \geq t_0$, such that $g(t_0) = \mathcal{V}(\mathbf{x}_0)$ and either the limit of

$g(\cdot)$ is zero as $t \to \infty$ or its value reaches zero in finite time. For example, such function $g(\cdot)$ can have exponential or linear form.

Assume that for the control input vector $\mathbf{u}(\cdot) = [u_1(\cdot), \ldots, u_m(\cdot)]^{\mathrm{T}}$, each variable $u_i(\cdot)$ has its saturation limit, that is,

$$|u_i(t)| \leq u_{i\text{sat}}, \quad t \geq t_0, \quad i = 1, \ldots, m, \tag{2.2}$$

and let u_i^* be any practically acceptable value for the ith control input within the range $u_i^* \in [0, u_{i\text{sat}}]$, $i = 1, \ldots, m$. This practically acceptable value depends on the timing of the control action, the control objective for a specific application, and the safety considerations for the saturation limits. Next, define

$$\mathbf{u}^* \triangleq [u_1^*, \ldots, u_m^*]^{\mathrm{T}} \in \mathbb{R}^m, \tag{2.3}$$

and define a matrix $\mathbf{M} \in \mathbb{R}^{m \times m}$ given by $\mathbf{M} = \text{diag}[\alpha_1, \ldots, \alpha_m]$ where

$$\alpha_i = \begin{cases} 1, & \text{if } u_i \text{ is applied in the positive direction,} \\ -1, & \text{if } u_i \text{ is applied in the negative direction,} \\ 0, & \text{if } u_i \text{ is not applied.} \end{cases} \tag{2.4}$$

Thus,

$$\mathbf{u}_{\mathrm{T}} = \mathbf{M}\mathbf{u}^* = [\alpha_1 u_1^*, \ldots, \alpha_m u_m^*]^{\mathrm{T}}, \quad t \in [t^*, t^* + \Delta t], \tag{2.5}$$

defines the control input applied to the system (2.1), at the instant of time t^* with the duration Δt. Note that the values of t^* and α_i, $i = 1, \ldots, m$, are to be determined at each instant of application of the control input using a numerical procedure to be detailed next.

In order to describe this real time numerical procedure, consider the interval of time $t \in [t_{\mathrm{i}}, t_{\mathrm{f}}]$ and let t_{f} be the instant of time when the control input $\mathbf{u}_{\mathrm{T}}$ is applied to system (2.1). Then

$$\mathbf{u}(t) = \begin{cases} \mathbf{0}, & t_{\mathrm{i}} \leq t < t_{\mathrm{f}}, \\ \mathbf{u}_{\mathrm{T}}, & t_{\mathrm{f}} \leq t \leq t_{\mathrm{f}} + \Delta t \end{cases} \tag{2.6}$$

is the control input applied to the system over the period of time $t \in [t_{\mathrm{i}}, t_{\mathrm{f}} + \Delta t]$. If Δt is very small, then the change in the value of the function $\mathcal{V}(\mathbf{x}(t))$ over the interval $t \in [t_{\mathrm{i}}, t_{\mathrm{f}} + \Delta t]$ can be approximated as

$$\Delta\mathcal{V}(\mathbf{x}(t_{\mathrm{f}} + \Delta t)) \approx \int_{t_{\mathrm{i}}}^{t_{\mathrm{f}}} \mathcal{V}'(\mathbf{x}(t)) f(t, \mathbf{x}(t), \mathbf{0}) \mathrm{d}t + \mathcal{V}'(\mathbf{x}(t_{\mathrm{f}})) f(t_{\mathrm{f}}, \mathbf{x}(t_{\mathrm{f}}), \mathbf{u}_{\mathrm{T}}) \Delta t, \tag{2.7}$$

where $\mathbf{x}(t)$ in the first integral is the solution to (2.1) with $\mathbf{u}(t) \equiv \mathbf{0}$, $\mathbf{x}_0 = \mathbf{x}(t_{\mathrm{i}})$, and $t_0 = t_{\mathrm{i}}$. Next, the change in $g(t)$ over the same time interval $t \in [t_{\mathrm{i}}, t_{\mathrm{f}} + \Delta t]$ is given by

$$\Delta g(t_\mathrm{f} + \Delta t) \triangleq g(t_\mathrm{f} + \Delta t) - g(t_\mathrm{i}) < 0. \tag{2.8}$$

The control objective is to maintain

$$\mathcal{V}(\mathbf{x}(t_\mathrm{f} + \Delta t)) < g(t_\mathrm{f} + \Delta t). \tag{2.9}$$

We select the required control parameters $\alpha_1, \ldots, \alpha_m$ such that inequality (2.9) is satisfied as an equality or, practically speaking, until the left-hand side is smaller than the right-hand side by a small margin set a priori by the designer, that is,

$$(1 - \varepsilon)g(t_\mathrm{f} + \Delta t) < \mathcal{V}(\mathbf{x}(t_\mathrm{f} + \Delta t)) < g(t_\mathrm{f} + \Delta t), \tag{2.10}$$

where $\varepsilon > 0$ is a small design parameter whose value can be adjusted according to a specific application. Once the control parameters are selected so that (2.10) is satisfied, the control vector $\mathbf{u}_\mathrm{T}$ is applied to the system (2.1). Then we repeat this process by setting the last value of $t_\mathrm{f} + \Delta t$ as the new t_i and redefining the new t_f as $t_\mathrm{f} = t_\mathrm{i} + t_\mathrm{s}$ and following the above protocol.

Note that, at each instant of time, the numerical algorithm ensures that if $\mathbf{u}_\mathrm{T}$ were to be applied to (2.1), then inequality (2.9) would be satisfied. However, the actual implementation of $\mathbf{u}_\mathrm{T}$ occurs *only* at the instances t_f when the inequality (2.10) is satisfied. Also, note that in between the control implementation instants t_f, the value of $\mathcal{V}(\mathbf{x}(\cdot))$ can increase. However, the trigger condition (2.10) for implementation of the control input $\mathbf{u}_\mathrm{T}$ can be augmented by additional condition restricting the value of $\mathcal{V}(\mathbf{x}(\cdot))$ in between the instants of control application. Finally, since $\lim_{t\to\infty} g(t) = 0$, the above control logic guarantees that $\lim_{t\to\infty} \mathcal{V}(\mathbf{x}(t)) = 0$ which ensures the convergence of the closed-loop subsystem state $\mathbf{x}(t)$ to zero as $t \to \infty$. In many practical applications, it is an acceptable outcome of the control especially since no a priori Lyapunov-like conditions need to be verified.

2.3 Control of Spacecraft Rendezvous and Docking

In this section, the iterative control framework (ICF) is applied to control spacecraft rendezvous and docking using impulsive thrusters. By applying short-duration forces, these thrusters alter spacecraft position, velocity, attitude, and angular rate. The precision of navigation is directly related to the number of thrusters. Each thruster can be used multiple times, and the magnitude of the thruster is adjustable during the control process. Thruster management for such spacecraft presents an ideal case study for verifying and validating the ICF.

A rendezvous maneuver typically involves the passive target vehicle and the active chase vehicle, which executes the necessary movements to reach and align with the target. A notable example is the space vehicle serving as the chaser while rendezvousing with the international space station as the passive target. We start with the relative

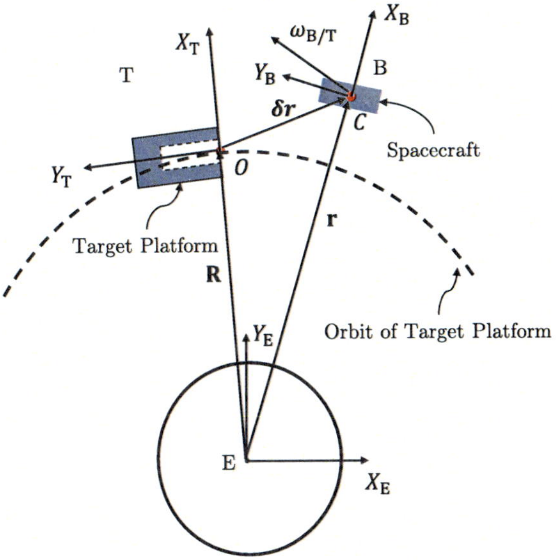

Fig. 2.1 Reference frames of the spacecraft rendezvous and docking

motion equations specialized to the spacecraft maneuvering problems [8]. First, as shown in Fig. 2.1, we introduce the Earth-fixed inertial reference frame E, the moving target frame T attached to the target platform with the coordinate system fixed at the docking port O, and the body-fixed reference frame B attached to the spacecraft with its coordinate system fixed at the mass center C. The $X_E Y_E$ plane of the inertial frame E coincides with the target vehicle's orbital plane, while the Z_E-axis is perpendicular to this plane. The position vectors of the passive target platform and the active spacecraft with respect to the inertial Earth-fixed reference frame are denoted as $\mathbf{R}$ and $\mathbf{r}$, respectively. And $\delta\mathbf{r}$ represents the position vector of the chase spacecraft relative to the target platform. Note that the inequality

$$\frac{\delta r}{R} << 1, \tag{2.11}$$

where $\delta r = ||\delta\mathbf{r}||$ and $R = ||\mathbf{R}||$, is satisfied for the final approach of the chase vehicle to the target vehicle during the rendezvous and docking maneuver [8]. In order to achieve the control objective, that is, to guarantee safe and precise docking of the spacecraft, we resolve the relative position vector $\delta\mathbf{r}$ in the moving target frame T, which indicates observing the moving and rotating spacecraft from the target reference frame. The relative position vector $\delta\mathbf{r}$ of the chase spacecraft relative to the target platform with respect to the target frame T is given by

$$\delta\mathbf{r} = \delta x\mathbf{i} + \delta y\mathbf{j} + \delta z\mathbf{k}, \tag{2.12}$$

where **i**, **j**, and **k** are unit vectors of the coordinate system X_T, Y_T, and Z_T moving with the target. The X_T-axis is aligned with the radial vector **R** directed from the origin of the Earth-fixed frame towards the target vehicle, the Z_T-axis is normal to the orbital plane of the target platform, and the Y_T-axis is directed towards the local horizon of the target platform. In this work, we assume that the orbit of the target platform is circular. Therefore, the following Clohessy-Wiltshire (CW) equations referred to the moving of the target frame T can be used to describe the relative translational motion between the target platform and the chasing spacecraft

$$\begin{aligned} \delta\ddot{x} - 3\Omega^2\delta x - 2\Omega\delta\dot{y} &= \frac{F_x}{m}, \\ \delta\ddot{y} + 2\Omega\delta\dot{x} &= \frac{F_y}{m}, \\ \delta\ddot{z} + \Omega^2\delta z &= \frac{F_z}{m}, \end{aligned} \tag{2.13}$$

where m is the mass of the spacecraft and

$$\Omega = \sqrt{\frac{\mu}{R^3}} \tag{2.14}$$

represents the orbital angular velocity in which μ and R denote the gravitational constant of the Earth and the radius of the circular orbit of the target vehicle, respectively. Projection of the resultant guidance control force onto the moving target frame T is expressed as $\mathbf{F} = [F_x, F_y, F_z]^T$ and can be represented as

$$\mathbf{F} = \mathbf{A}(\Phi)\mathbf{f}_{\text{guid}}, \tag{2.15}$$

where

$$\mathbf{A}(\Phi) \triangleq \begin{bmatrix} \cos\Delta\theta\cos\Delta\psi & -\cos\Delta\varphi\sin\Delta\psi + \sin\Delta\varphi\sin\Delta\theta\cos\Delta\psi & \sin\Delta\varphi\sin\Delta\psi + \cos\Delta\varphi\sin\Delta\theta\cos\Delta\psi \\ \cos\Delta\theta\sin\Delta\psi & \cos\Delta\varphi\cos\Delta\psi + \sin\Delta\varphi\sin\Delta\theta\sin\Delta\psi & -\sin\Delta\varphi\cos\Delta\psi + \cos\Delta\varphi\sin\Delta\theta\sin\Delta\psi \\ -\sin\Delta\theta & \sin\Delta\varphi\cos\Delta\theta & \cos\Delta\varphi\cos\Delta\theta \end{bmatrix}, \tag{2.16}$$

$\Delta\varphi$, $\Delta\theta$, and $\Delta\psi$ are Euler angles of the spacecraft with respect to the target frame T, and

$$\mathbf{f}_{\text{guid}} = \begin{bmatrix} 2\alpha_{g,x} f_{g,x}(t) \\ 2\alpha_{g,y} f_{g,y}(t) \\ 2\alpha_{g,z} f_{g,z}(t) \end{bmatrix} \tag{2.17}$$

is the same resultant control force from the guidance thrusters resolved in the body-fixed reference frame B. In Fig. 2.2, three groups of guidance thrusters, representing thrusters in the X_B direction (denoted in blue), thrusters in the Y_B direction (denoted in yellow), and thrusters in the Z_B direction (denoted in green), are shown. All three groups of guidance thrusters are positioned around the circumference of the spacecraft at the mass center, and each group contains two pairs of thrusters, with

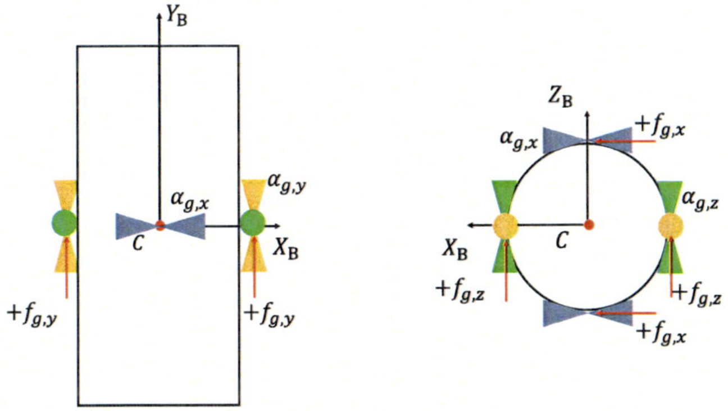

Fig. 2.2 Allocation of guidance thrusters

one pair producing forward thrust and the other generating reverse thrust. Note that to prevent the rotation of the spacecraft due to the firing of guidance thrusters, each pair of thrusters must fire simultaneously with equal magnitudes. Forces from each pair of thrusters $f_{g,x}$, $f_{g,y}$, and $f_{g,z}$ are non-zero during thruster firing and are zero otherwise. These forces can be approximated as constant values $f^*_{g,x} > 0$, $f^*_{g,y} > 0$ and $f^*_{g,z} > 0$ over the duration of firing time Δt. Finally, the control parameters $\alpha_{g,i}$, $i = \{x, y, z\}$, related to the guidance thrusters are specified as:

$$\alpha_{g,i} = \begin{cases} 1, & \text{if } f^*_{g,i} \text{ is applied in the positive direction,} \\ -1 & \text{if } f^*_{g,i} \text{ is applied in the negative direction,} \\ 0 & \text{if } f^*_{g,i} \text{ is not applied.} \end{cases} \tag{2.18}$$

Furthermore, to derive the equations of relative rotational motion, orientation angles of the spacecraft relative to the target frame T are defined by Euler angles $\Delta\varphi$, $\Delta\theta$, and $\Delta\psi$ and are written into the vector of rotational angles $\Phi = [\Delta\varphi, \Delta\theta, \Delta\psi]^{\mathrm{T}}$. Projections onto the body-fixed frame B of the spacecraft angular velocity vector $\omega_{\mathrm{B/T}}$ relative to the target frame T are given by $\omega_{\mathrm{B/T}} = [\Delta\omega_x, \Delta\omega_y, \Delta\omega_z]^{\mathrm{T}}$. Therefore, equation of rotational kinematics is given as

$$\dot{\Phi} = \mathbf{B}(\Phi)\omega_{\mathrm{B/T}}, \tag{2.19}$$

where

$$\mathbf{B}(\Phi) \triangleq \begin{bmatrix} 1 & \sin\Delta\varphi\tan\Delta\theta & \cos\Delta\varphi\tan\Delta\theta \\ 0 & \cos\Delta\varphi & -\sin\Delta\varphi \\ 0 & \sin\Delta\varphi\sec\Delta\theta & \cos\Delta\varphi\sec\Delta\theta \end{bmatrix} \tag{2.20}$$

To avoid singularity in matrix $\mathbf{B}(\Phi)$, we restrict $\Delta\theta \in (-\pi/2, \pi/2)$.

In order to obtain equations of rotational motion for the chase vehicle, introduce the angular velocity $\omega_{\mathrm{B/E}}$ of the spacecraft with respect to the Earth-fixed inertial reference frame E and resolved in B as $\omega_{B/E} = [\omega_x, \omega_y, \omega_z]^T$. Furthermore, angular velocity $\omega_{T/E}$ of the target vehicle relative to E and resolved in T is given by

$$\omega_{T/E} = [0, 0, \Omega]^T. \tag{2.21}$$

By addition theorem,

$$\omega_{B/E} = \omega_{B/T} + \omega_{T/E} \tag{2.22}$$

or, equivalently, in projection onto B

$$\begin{bmatrix} \omega_x \\ \omega_y \\ \omega_z \end{bmatrix} = \begin{bmatrix} \Delta\omega_x \\ \Delta\omega_y \\ \Delta\omega_z \end{bmatrix} + \mathbf{A}^T(\Phi) \begin{bmatrix} 0 \\ 0 \\ \Omega \end{bmatrix}. \tag{2.23}$$

Recall that equations of rotational motion of the space vehicle relative to the inertial reference frame E are given by standard Euler equations

$$\dot{\omega}_{B/E} = -\mathbf{I}^{-1}\tilde{\omega}_{B/E}\mathbf{I}\omega_{B/E} + \mathbf{I}^{-1}\tau_{att}, \tag{2.24}$$

where $\mathbf{I}$ is the inertia matrix of the spacecraft,

$$\tilde{\omega}_{B/E} \triangleq \begin{bmatrix} 0 & -\omega_z & \omega_y \\ \omega_z & 0 & -\omega_x \\ -\omega_y & \omega_x & 0 \end{bmatrix}, \quad \tau_{att} = \begin{bmatrix} 2\alpha_{a,x} l_1 f_{a,x}(t) \\ 2\alpha_{a,y} l_2 f_{a,y}(t) \\ 2\alpha_{a,z} r_a f_{a,z}(t) \end{bmatrix}, \tag{2.25}$$

and τ_{att} denotes the resultant moment of control forces from attitude thrusters resolved in the body-fixed frame B. Using (2.23), we can express the time derivative of $\omega_{B/T}$ in projection onto B as

$$\begin{bmatrix} \Delta\dot{\omega}_x \\ \Delta\dot{\omega}_y \\ \Delta\dot{\omega}_z \end{bmatrix} = -\mathbf{I}^{-1} \begin{bmatrix} 0 & -\omega_z & \omega_y \\ \omega_z & 0 & -\omega_x \\ -\omega_y & \omega_x & 0 \end{bmatrix} \mathbf{I} \begin{bmatrix} \omega_x \\ \omega_y \\ \omega_z \end{bmatrix} - \mathbf{A}^T(\Phi) \begin{bmatrix} 0 \\ 0 \\ \Omega \end{bmatrix} + \mathbf{I}^{-1}\tau_{att}, \tag{2.26}$$

where $\omega_x, \omega_y, \omega_z$ are represented via $\Delta\omega_x, \Delta\omega_y, \Delta\omega_z$ using (2.23).

The spacecraft has three sets of attitude thrusters, each responsible for a different orientation angle, as shown in Fig. 2.3. Each set includes two pairs of thrusters, producing torque around the X_B (depicted in blue), Y_B (depicted in yellow), or Z_B axes (depicted in green). Similarly, in order to prevent extra forces from being generated on the spacecraft during the firing of attitude thrusters, both thrusters within each pair must fire simultaneously with equal magnitudes. Thrusters produce forces $f_{a,x}(t)$, $f_{a,y}(t)$ and $f_{a,z}(t)$ which can be considered as constant values $f^*_{a,x} > 0$, $f^*_{a,y} > 0$ and $f^*_{a,z} > 0$ during the firing time period Δt, while the control parameters $\alpha_{a,i}, i = \{x, y, z\}$ in (2.25) are defined as

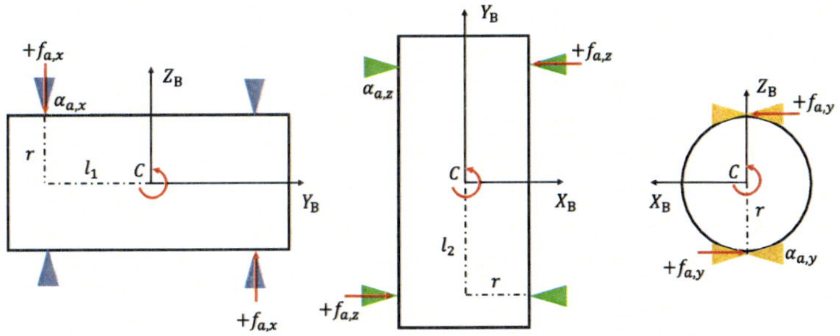

Fig. 2.3 Allocation of attitude thrusters

$$\alpha_{a,i} = \begin{cases} 1, & \text{if } f^*_{a,i} \text{ is applied to generate positive torque,} \\ -1 & \text{if } f^*_{a,i} \text{ is applied to generate negative torque,} \\ 0 & \text{if } f^*_{a,i} \text{ is not applied,} \end{cases} \tag{2.27}$$

and l_1, l_2, and r are the moment arms of the thruster forces $f_{a,x}$, $f_{a,y}$, and $f_{a,z}$ around the corresponding axes of rotation, respectively.

We divide the control of spacecraft rendezvous and docking into two parts: guidance control and attitude control. Guidance control involves controlling the velocity and position of the spacecraft to enable it to approach and synchronize its speed with the target vehicle; while attitude control involves controlling the spacecraft orientation and the rates of rotation to ensure it is correctly aligned with the target vehicle for docking. Both guidance control and attitude control are critical for the success of the spacecraft rendezvous and docking mission.

2.3.1 *Guidance Control*

The objective of guidance control is to control the spacecraft's velocity and position to approach and align with the target platform. Eliminating the relative position and velocity between the target platform and the spacecraft to meet the required docking criteria is crucial for the successful rendezvous and docking of the spacecraft. Let the desired relative position and velocity be represented by $\delta\mathbf{X}_{\text{des}} = [\delta x_{\text{des}}, \delta y_{\text{des}}, \delta z_{\text{des}}]^{\text{T}}$ and $\delta\dot{\mathbf{X}}_{\text{des}} = [\delta\dot{x}_{\text{des}}, \delta\dot{y}_{\text{des}}, \delta\dot{z}_{\text{des}}]^{\text{T}}$, respectively. Typically, for successful docking, both $\delta\mathbf{X}_{\text{des}}$ and $\delta\dot{\mathbf{X}}_{\text{des}}$ must be zero (or close to zero according to specific docking criteria). However, we keep these values arbitrary for more general maneuvers (e.g., formation flying). Meanwhile, the current relative position and velocity of the spacecraft are denoted by $\delta\mathbf{X} = [\delta x, \delta y, \delta z]^{\text{T}}$ and $\delta\dot{\mathbf{X}} = [\delta\dot{x}, \delta\dot{y}, \delta\dot{z}]^{\text{T}}$, respectively. Given that both desired and current vectors are resolved in the target reference frame T, the relative position error vector $\delta\mathbf{X}_{\text{err}}$ and velocity error vector $\delta\dot{\mathbf{X}}_{\text{err}}$ that need to be minimized are expressed as

$$\delta \mathbf{X}_{\text{err}} = \delta \mathbf{X}_{\text{des}} - \delta \mathbf{X} = [\delta x_{\text{des}} - \delta x, \delta y_{\text{des}} - \delta y, \delta z_{\text{des}} - \delta z]^{\text{T}}, \tag{2.28}$$

and

$$\delta \dot{\mathbf{X}}_{\text{err}} = \delta \dot{\mathbf{X}}_{\text{des}} - \delta \dot{\mathbf{X}} = [\delta \dot{x}_{\text{des}} - \delta \dot{x}, \delta \dot{y}_{\text{des}} - \delta \dot{y}, \delta \dot{z}_{\text{des}} - \delta \dot{z}]^{\text{T}}, \tag{2.29}$$

respectively.

Therefore, define a positive definite function of the state of the system (2.13), (2.19) and (2.26) given by

$$\mathcal{V}_{\text{g}}(\delta \mathbf{X}, \delta \dot{\mathbf{X}}, \Phi, \omega_{\text{B/T}}) = \delta \mathbf{X}_{\text{err}}^{\text{T}} \mathbf{D}_{\mathbf{x}} \delta \mathbf{X}_{\text{err}} + \delta \dot{\mathbf{X}}_{\text{err}}^{\text{T}} \mathbf{D}_{\mathbf{v}} \delta \dot{\mathbf{X}}_{\text{err}}, \tag{2.30}$$

where $\mathbf{D}_{\mathbf{x}} = \text{diag}[\alpha_{\delta x}(t), \alpha_{\delta y}(t), \alpha_{\delta z}(t)]$ and $\mathbf{D}_{\mathbf{v}} = \text{diag}[\alpha_{\delta \dot{x}}(t), \alpha_{\delta \dot{y}}(t), \alpha_{\delta \dot{z}}(t)]$ are diagonal weight matrices related to relative position and velocity, respectively. Clearly, driving the value of $\mathcal{V}_{\text{g}}(\delta \mathbf{X}, \delta \dot{\mathbf{X}}, \Phi, \omega_{\text{B/T}})$ to zero will force the spacecraft to approach the target platform with zero position and velocity errors. The diagonal elements in the matrices $\mathbf{D}_{\mathbf{x}}$ and $\mathbf{D}_{\mathbf{v}}$ can be constant or variable values that change based on the current state and are used to assign weight to the corresponding system state. The larger the value of a diagonal entry, the more weight it is assigned to.

2.3.2 Attitude Control

The goal of attitude control is to accurately align the spacecraft's orientation and angular rates with the target platform so it can dock safely and successfully. To achieve this objective, the relative orientation angle and angular velocity between the spacecraft and the target platform must be eliminated. Similar to the guidance control, we set the desired relative orientation angle and angular velocity vectors as $\Phi_{\text{des}} = [\Delta\varphi_{\text{des}}, \Delta\theta_{\text{des}}, \Delta\psi_{\text{des}}]^{\text{T}}$ and $\omega_{\text{des}} = [\Delta\omega_{x,\text{des}}, \Delta\omega_{y,\text{des}}, \Delta\omega_{z,\text{des}}]^{\text{T}}$, respectively. As with guidance control, for the docking maneuver, both Φ_{des} and ω_{des} are zero but, for more general problems, we keep these values arbitrary. Finally, the objective of attitude control is to minimize the error vectors defined as:

$$\Phi_{\text{err}} = \Phi_{\text{des}} - \Phi = [\Delta\varphi_{\text{des}} - \Delta\varphi, \Delta\theta_{\text{des}} - \Delta\theta, \Delta\psi_{\text{des}} - \Delta\psi]^{\text{T}}, \tag{2.31}$$

and

$$\omega_{\text{err}} = \omega_{\text{des}} - \omega_{\text{B/T}} = [\Delta\omega_{x,\text{des}} - \Delta\omega_x, \Delta\omega_{y,\text{des}} - \Delta\omega_y, \Delta\omega_{z,\text{des}} - \Delta\omega_z]^{\text{T}}, \tag{2.32}$$

where $\Phi = [\Delta\varphi, \Delta\theta, \Delta\psi]^{\text{T}}$ and $\omega_{\text{B/T}} = [\Delta\omega_x, \Delta\omega_y, \Delta\omega_z]^{\text{T}}$ denote the current relative orientation angles and angular velocity, respectively.

Similarly, we define a positive definite function of the state of the system (2.19) and (2.26) for the attitude control given by

$$\mathcal{V}_{\text{a}}\left(\Phi, \omega_{\text{B/T}}\right) = \delta \Phi_{\text{err}}^{\text{T}} \mathbf{D}_{\Phi} \delta \Phi_{\text{err}} + \delta \omega_{\text{err}}^{\text{T}} \mathbf{D}_{\omega} \delta \omega_{\text{err}}, \tag{2.33}$$

where diagonal weight matrices $\mathbf{D}_\Phi = \text{diag}[\alpha_{\Delta\varphi}(t), \alpha_{\Delta\theta}(t), \alpha_{\Delta\psi}(t)]$ and $\mathbf{D}_\omega = \text{diag}[\alpha_{\Delta\omega_x}(t), \alpha_{\Delta\omega_y}(t), \alpha_{\Delta\omega_z}(t)]$ are comprised of different weight coefficients related to the corresponding system states. Additionally, the values of the weight coefficients in matrices $\mathbf{D}_\Phi$ and $\mathbf{D}_\omega$ can be constant or variable and adjusted based on the current system state. Adjusting the weight coefficients allows different priorities to be assigned to the respective system states, leading to a control strategy prioritizing specific system states. Once the value of $\mathcal{V}_\text{a}\left(\Phi, \omega_{\text{B/T}}\right)$ reaches zero, the spacecraft's relative attitude is perfectly aligned with the target platform with zero relative angular rates, thus achieving the desired attitude control objective.

2.4 Numerical Simulations

This section presents simulation results for the iterative control framework proposed in this paper. The simulations are conducted using a cylinder-shaped spacecraft with the body frame axes being the principal axes of inertia. All inertial and geometric parameters are given in Table 2.1. The orbital angular velocity Ω is 1.15×10^{-3} rad/s, corresponding to the circular orbit of 300 km altitude above the Earth. The flight process for a complete rendezvous and docking mission typically involves several phases: launch, far-range rendezvous, near-range autonomous control, docking, assembly operation, departure, and re-entry phases [9]. Our simulations focus on the near-range autonomous control phase, specifically the closing sub-phase, corresponding to the relative axial distance between the spacecraft and the target platform, δy in Equation (2.12), ranging from 0 to 130 m. The guidance and attitude control thrusters are activated following the procedure outlined in Sect. 2.2 to ensure the success of the rendezvous and docking mission. Any slight changes in the spacecraft's mass and inertia characteristics resulting from the thruster firings are disregarded.

In the simulation, we utilize the exponential time-varying function $g(t)$ given by

$$g(t) = \mathcal{V}(\mathbf{x}_0)e^{-\lambda(t-t_0)}, \quad t \geq t_0, \tag{2.34}$$

where the value of the design parameter λ is specified as follows:

Table 2.1 Inertial and geometrical parameters of the spacecraft

Parameter	Symbol	Value	Unit
Spacecraft mass	m	2950	kg
Length	L	2.24	m
Diameter	d	2.17	m
Principal moments of inertia	I_{xx}	2101.69	kg m^2
Principal moments of inertia	I_{yy}	1736.41	kg m^2
Principal moments of inertia	I_{zz}	2101.70	kg m^2

Fig. 2.4 Docking corridor

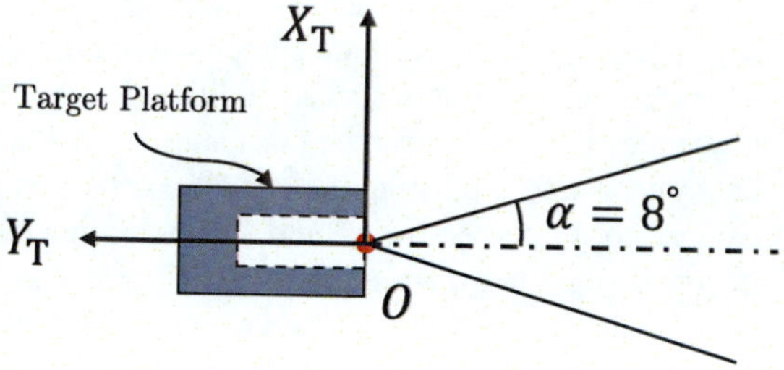

$$\lambda = -\frac{\ln(0.02)}{t_{\text{end}} - t_0}\tau, \tag{2.35}$$

where

$$t_{\text{end}} = \frac{d_0}{v_{\text{m}}}, \tag{2.36}$$

where d_0 represents the relative distance between the target platform and the initial position of the spacecraft, v_{m}, which denotes the average approaching speed of the rendezvous and docking process, is set to be 0.33 m/s, and τ is a design parameter, which is set to be 3 in this paper. All guidance thrusters are allocated around the circumference of the spacecraft at the mass center, as shown in Fig. 2.2. Moreover, the moment arms of the attitude thrusters, labeled as l_1, l_2, and r in Fig. 2.3, are set to be 0.89 m, 1.08 m, and 0.89 m, respectively.

The weighting matrices $\mathbf{D_x} = \text{diag}\left[\alpha_{\delta x}(t), \alpha_{\delta y}(t), \alpha_{\delta z}(t)\right]$ and $\mathbf{D_v} = \text{diag}\left[\alpha_{\delta \dot{x}}(t), \alpha_{\delta \dot{y}}(t), \alpha_{\delta \dot{z}}(t)\right]$ introduced in (2.30) are expressed in the forms

$$\mathbf{D_x} = \begin{bmatrix} 50.62 & 0 & 0 \\ 0 & 1 & 0 \\ 0 & 0 & 50.62 \end{bmatrix}, \quad \mathbf{D_v} = \begin{bmatrix} \frac{w_{\delta \dot{x}}}{0.01+e_{\delta x}} & 0 & 0 \\ 0 & \frac{w_{\delta \dot{y}}}{0.01+e_{\delta y}} & 0 \\ 0 & 0 & \frac{w_{\delta \dot{z}}}{0.01+e_{\delta z}} \end{bmatrix}, \tag{2.37}$$

respectively, where the values of $\alpha_{\delta x}$ and $\alpha_{\delta z}$ are calculated based on the geometrical properties of the docking corridor, which is typically a cone aligned with the target docking axis and has a half-angle of approximately 8° (see Fig. 2.4). The weighting parameters $\alpha_{\delta \dot{x}}(t)$, $\alpha_{\delta \dot{y}}(t)$, and $\alpha_{\delta \dot{z}}(t)$ are inversely related to the error distances along the X_T-, Y_T-, and Z_T-axis, which are defined by $e_{\delta x} = \delta x_{\text{des}} - \delta x$, $e_{\delta y} = \delta y_{\text{des}} - \delta y$, and $e_{\delta z} = \delta z_{\text{des}} - \delta z$, respectively, meaning that smaller errors result in larger weighting values. Moreover, the coefficients $w_{\delta \dot{x}}$, $w_{\delta \dot{y}}$, and $w_{\delta \dot{z}}$ introduced in matrix $\mathbf{D_v}$ are given by

$$w_{\delta \dot{x}} = w_{\delta \dot{z}} = 0.05 \tag{2.38}$$

$$w_{\delta \dot{y}} = \begin{cases} 0.1, & \text{if } e_{\delta x} < 0.1 \wedge e_{\delta z} < 0.1 \wedge e_{\delta y} > 0.5, \\ 3, & \text{otherwise.} \end{cases} \tag{2.39}$$

It is evident (2.37) that a larger value of either coefficient w corresponds to a larger value of the weighting parameter, while the smaller error distance in either direction increases the weight on the relative velocity in this direction. The rationale for selecting weighting parameters in this manner is to minimize the lateral distance error during the early stage of docking, which involves reducing errors along the X_T- and Z_T-axis while slowing down the spacecraft approach speed to the target platform. After stabilizing the lateral error within the required range, greater weighting is given to rapidly decrease errors along the docking axis (i.e., the Y_T-axis direction). Once the Y_T-axis error is reduced to the required value, the spacecraft is decelerated and stabilized.

Furthermore, the weighting matrices $\mathbf{D}_\Phi = \mathrm{diag}[\alpha_{\Delta\varphi}(t), \alpha_{\Delta\theta}(t), \alpha_{\Delta\psi}(t)]$ and $\mathbf{D}_\omega = \mathrm{diag}[\alpha_{\Delta\omega_x}(t), \alpha_{\Delta\omega_y}(t), \alpha_{\Delta\omega_z}(t)]$ introduced in (2.33) are given by

$$\mathbf{D}_\Phi = \begin{bmatrix} 1 & 0 & 0 \\ 0 & 1 & 0 \\ 0 & 0 & 1 \end{bmatrix}, \quad \mathbf{D}_w = \begin{bmatrix} \frac{w_{\Delta\omega_x}}{0.01+e_{\Delta\phi}} & 0 & 0 \\ 0 & \frac{w_{\Delta\omega_y}}{0.01+e_{\Delta\theta}} & 0 \\ 0 & 0 & \frac{w_{\Delta\omega_z}}{0.01+e_{\Delta\psi}} \end{bmatrix}, \tag{2.40}$$

where the values of coefficients $w_{\Delta\omega_x}$, $w_{\Delta\omega_y}$, and $w_{\Delta\omega_z}$ are all set as 0.02. The objective of setting up the parameters in such a way is also to slow down the rotation of the spacecraft when the angular deviation is small.

For all simulations, the design parameter ε shown in inequality (2.10) is set to be 0.1, and we apply the genetic algorithm [10] to find the appropriate thruster magnitude u_T that satisfies this inequality. We set the maximal number of iterations of the genetic algorithm to be 1500 and the bound for each guidance and attitude thruster to be 6×10^5 N, that is, $\forall i \in \{x, y, z\}$, $f^*_{g,i} < 6 \times 10^5 \wedge f^*_{a,i} < 6 \times 10^5$. Furthermore, to validate the ICF concept, we introduce the rendezvous and docking criteria for the Soyuz spacecraft according to [9] as follows (all values below pertain to the terminal state, that is, the end of the docking maneuver)

$$\begin{aligned} &\text{Lateral position: } |e_{\delta x}| < 0.34\text{m} \wedge |e_{\delta z}| < 0.34\text{m}, \\ &\text{Lateral speed: } |e_{\delta \dot{x}}| < 0.1\text{m/s} \wedge |e_{\delta \dot{z}}| < 0.1\text{m/s}, \\ &\text{Angular misalignment: } |e_{\Delta\phi}| < 5^\circ \wedge |e_{\Delta\theta}| < 5^\circ \wedge |e_{\Delta\psi}| < 5^\circ, \\ &\text{Angular rate: } |e_{\Delta\omega_x}| < 1^\circ\text{s}^{-1} \wedge |e_{\Delta\omega_y}| < 1^\circ\text{s}^{-1} \wedge |e_{\Delta\omega_z}| < 1^\circ\text{s}^{-1}. \end{aligned} \tag{2.41}$$

Additionally, in order to prevent the spacecraft from colliding with the target platform, it is required that there is no overshoot in the Y_T direction throughout the entire docking process, that is:

$$\forall t \in [t_0, t_\mathrm{end}], \ (\delta y_\mathrm{des} - \delta y) * \mathrm{sgn}(y_0) < 0, \tag{2.42}$$

where t_0 and t_end represent the simulation's start and the end time, respectively, and δy_0 is the initial position of the spacecraft relative to the target resolved in the target frame T.

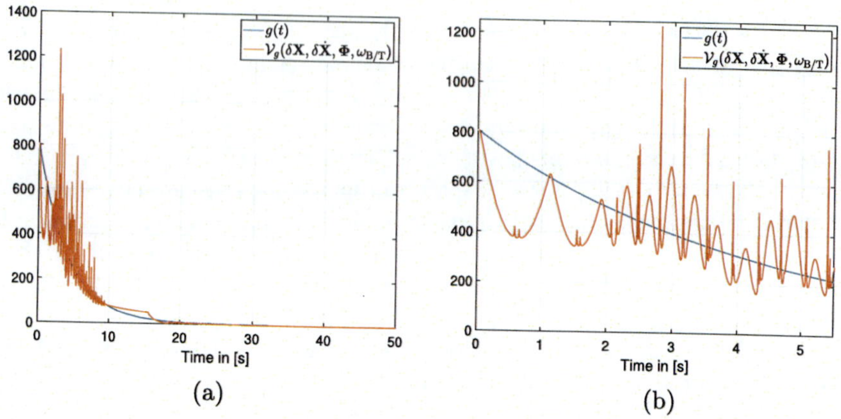

Fig. 2.5 $\mathcal{V}_g(\delta\mathbf{X}, \delta\dot{\mathbf{X}}, \Phi, \omega_{B/T})$ and exponential function $g(t)$ versus time. **a** $\mathcal{V}_g$ and $g(t)$ over entire simulation. **b** $\mathcal{V}_g$ and $g(t)$ for $t \in [0, 5]$

Given the above definitions, we initiate the simulations with the following initial conditions:

$$\begin{aligned} &(\delta x,\ \delta y,\ \delta z) = (2\text{m},\ -20\text{m},\ 2\text{m}),\\ &(\delta\dot{x},\ \delta\dot{y},\ \delta\dot{z}) = (-0.1\text{m/s},\ 0.1\text{m/s},\ -0.1\text{m/s}),\\ &(\Delta\phi,\ \Delta\theta,\ \Delta\psi) = (10^\circ,\ 10^\circ,\ 10^\circ),\\ &(\Delta\omega_x,\ \Delta\omega_y,\ \Delta\omega_z) = (2^\circ\text{s}^{-1},\ 2^\circ\text{s}^{-1},\ 2^\circ\text{s}^{-1}). \end{aligned} \tag{2.43}$$

Figure 2.5a illustrates the positive definite function $\mathcal{V}_g$ for guidance control and the monotonically decreasing function $g(t)$ over the entire simulation duration. We can observe that, in general, the trend of $\mathcal{V}_g$ over time is the same as that of equation $g(t)$, that is, the value decreases with time. Also, this variation tendency follows the procedure outlined in Sect. 2.2, that is, appropriate thrusters are fired to satisfy inequality (2.10). Furthermore, Fig. 2.5b presents the detailed plot of functions $\mathcal{V}_g$ and $g(t)$ in the interval from 0 to 5 s. The spikes in the curve of function $\mathcal{V}_g$ can be attributed to the relative velocity weight coefficients increasing as the distance decreases.

Figure 2.6a shows the relative distances along the X_T-, Y_T-, and Z_T-axis, respectively. It can be seen that in the early stage of the docking process, the algorithm is committed to reducing and stabilizing the errors in the X_T- and Z_T-axis and making the spacecraft approach the target platform at a relatively slow speed. When the error distances in the X_T and Z_T directions stabilize within the required ranges, at about 16 seconds, the algorithm adjusts the weight distribution to accelerate the spacecraft toward the platform along the Y_T-axis. Furthermore, as the spacecraft approaches the platform, more weight is assigned to the Y_T-axis relative velocity to decelerate and stabilize the spacecraft at around 20 s. The final state stability error values in the X_T-, Y_T-, and Z_T-axis directions meet the docking requirements proposed by (2.41).

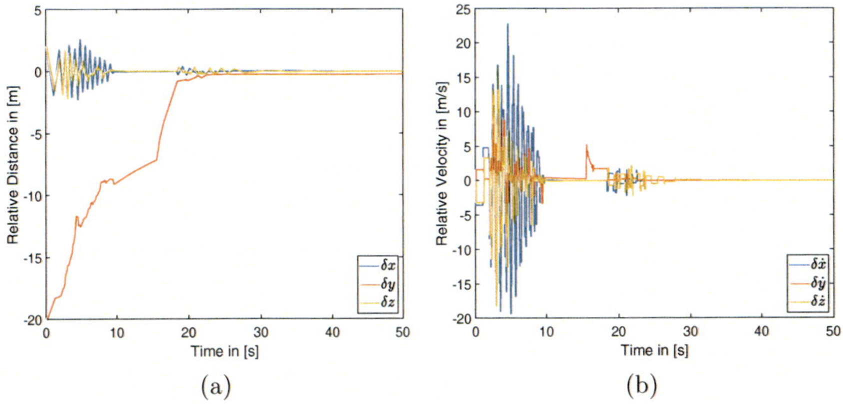

Fig. 2.6 Relative distances and velocities. **a** Relative distances. **b** Relative velocities

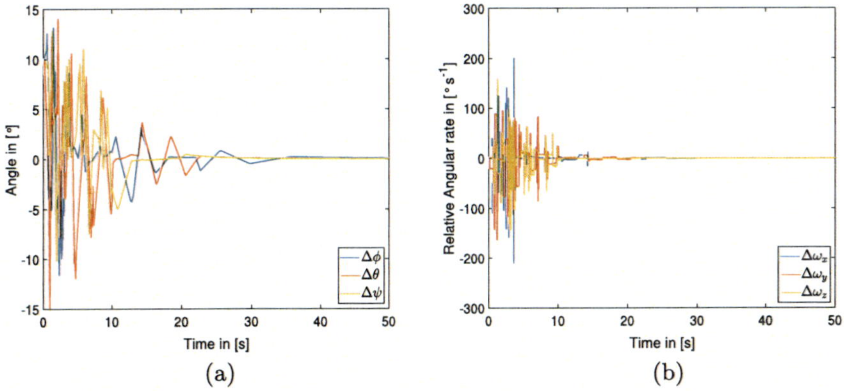

Fig. 2.7 Orientation angles and angular velocities of the spacecraft relative to the target platform. **a** Orientation angles. **b** Augular velocities

Figure 2.6b shows the plots of relative velocities along the X_T-, Y_T-, and Z_T-axis, respectively. Similarly, we observe that all relative velocities $\delta\dot{x}$, $\delta\dot{y}$, and $\delta\dot{z}$, converge to steady states that meet the Soyuz docking requirements.

Figure 2.7 shows the spacecraft orientation angles and angular velocities relative to the target frame T. It can be seen that similar to the relative distances and velocities, the spacecraft orientation angles and angular velocities eventually converge to the steady-state values that satisfy the docking criteria (2.41).

Figure 2.8a, b present the control forces applied on the spacecraft from the guidance and attitude thrusters, respectively. It can be seen that the force exerted on the spacecraft decreases as the docking process progresses. At the end of the docking process, when the errors of the spacecraft relative to the docking platform are already minimal, the forces exerted by the guidance and attitude thrusters on the spacecraft tend to zero.

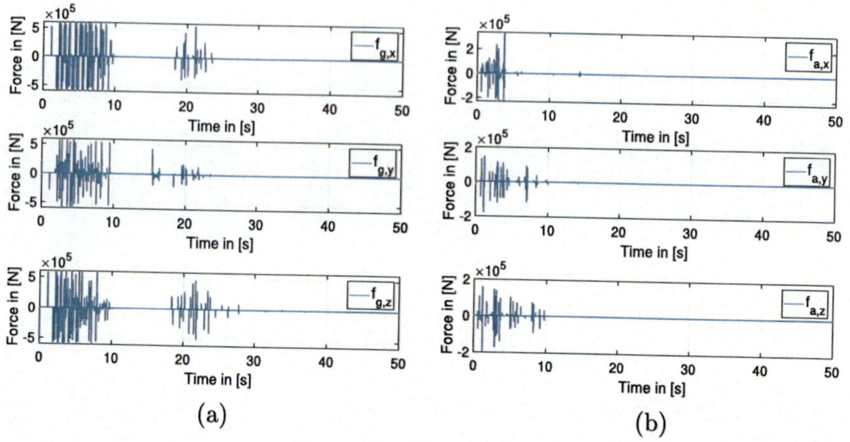

Fig. 2.8 Control forces of guidance and attitude thrusters versus time. **a** Guidance thrusters. **b** Attitude thrusters

2.5 Conclusion

In this paper, we developed the Iterative Control Framework (ICF) for real-time control of nonlinear dynamical systems using numerical algorithms. The ICF determines the value of the control input vector and the timing of its implementation to the dynamical system based on a numerical procedure. The algorithm ensures the non-monotonic decrease of a positive-definite function of the system state, thus guaranteeing convergence of the system state to zero. The main advantage of the approach is that it requires no a priori verification of any Lyapunov-like conditions, thus, making it applicable to real-world systems with complex nonlinear dynamics. It is especially suitable for systems with short-duration impulsive control actions such as spacecrafts controlled by multiple-use guidance and attitude thrusters. We applied the ICF to the problem of rendezvous and docking of a spacecraft to the target vehicle (e.g., space station). Specifically, we presented simulation results of the spacecraft rendezvous and docking control using ICF with the integration of the genetic algorithm. The simulations show that the ICF reaches the terminal docking criteria in all six degrees of freedom for the Soyuz spacecraft.

References

1. Utkin, V.I.: Variable structure systems with sliding modes. IEEE Trans. Auto. Contrl. **22**, 212–222 (1977)
2. Freeman, R., Kokotović, P.: Backstepping design of robust controllers for a class of nonlinear systems. In: Proceedings of the 2nd IFAC World Congress (Bordeaux, France), pp. 431–436 (1992)

3. Krstić, M., Kanellakopoulos, I., Kokotović, P.V.: Nonlinear and Adaptive Control Design. Wiley, New York (1995)
4. Khalil, H.K.: Nonlinear Control. Pearson, Upper Saddle River, NJ (2015)
5. Haddad, W.M., Chellaboina, V., Nersesov, S.G.: Impulsive and Hybrid Dynamical Systems. Stability, Dissipativity, and Control. Princeton University Press, Princeton, NJ (2006)
6. Sanfelice, R.: Hybrid Feedback Control. Princeton University Press, Princeton, NJ (2020)
7. Liu, X., Ashrafiuon, H., Nersesov, S.: Iterative control framework with application to guidance control of rockets with impulsive thrusters. In: Proceedings of IFAC Modeling, Estimation, and Control Conference, Jersey City, NJ, pp. 247–252 (2022)
8. Curtis, H.D.: Orbital Mechanics for Engineering Students, 3rd edn. Butterworth-Heinemann, Boston (2014)
9. Xie, Y., Chen, C., Liu, T., Wang, M.: Guidance, Navigation, and Control for Spacecraft Rendezvous and Docking: Theory and Methods. Springer (2021)
10. Kramer, O.: Genetic Algorithm Essentials, 1st edn. Springer Publishing Company, Incorporated (2017)

Chapter 3
ESA Technology Developments in Vision-Based Navigation

Olivier Dubois-Matra, Massimo Casasco, Manuel Sanchez Gestido, and Irene Huertas Garcia

Abstract The European Space Agency (ESA) has been developing on-board Guidance, Navigation and Control (GNC) technologies to support space activities in Earth orbit and beyond. One of the areas of development has been the use of vision-based systems to improve the performance and the autonomy of the navigation function. This paper will focus on two classes of missions enabled by vision-based navigation (VBN): rendezvous (in Earth or planetary orbit, with collaborative or non-collaborative targets) and precision descent and landing on planetary surfaces. It will present an overview of both recently completed and on-going ESA technology activities raising the Technology Readiness Level (TRL) of VBN systems for these missions.

Keywords GNC · Vision-based · Navigation · Rendezvous · Landing

3.1 Introduction

Hardware and software innovation have made possible the use of autonomous, on-board vision-based system for the navigation of spacecraft. Cameras are relatively lightweight, power sober and cheap sensors which can potentially provide an impor-

O. Dubois-Matra (✉)
Telespazio Belgium, European Space Research and Technology Centre, 2200 AG Noordwijk, The Netherlands
e-mail: Olivier.Dubois-Matra@ext.esa.int

M. Casasco · M. S. Gestido · I. H. Garcia
European Space Agency, European Space Research and Technology Centre, 2200 AG Noordwijk, The Netherlands
e-mail: Massimo.Casasco@esa.int

M. S. Gestido
e-mail: Manuel.Sanchez.Gestido@esa.int

I. H. Garcia
e-mail: Irene.Huertas@esa.int

D. Azimov (ed.), *Proceedings of the IUTAM Symposium on Optimal Guidance and Control for Autonomous Systems 2023*, IUTAM Bookseries 40,
https://doi.org/10.1007/978-3-031-39303-7_3

tant amount of data relevant to the estimation of the navigation state of an orbiting or landing vehicle, beyond what inertial or GNSS sensors can provide, and without the power consumption required by active optical sensors such as LiDAR. The drawback of course is the corresponding complexity at software level of the algorithms and methods required to extract the information, which in turn leads to significant requirements in term of hardware implementation and verification and validation strategy. Nevertheless, the missions enabled by the flexibility and autonomy provided by vision-based navigation (VBN) justify the development of related technology. The European Space Agency (ESA) and the European space industry and academia have been developing VBN for small body missions [1]. This paper will highlight some of the most recent technology developments at ESA in VBN applied to two main type of missions: first, orbital rendezvous with cooperative or uncooperative targets for docking, capture or servicing; second, descent and landing on a planetary surface. Other missions that are not addressed here but were enabled by VBN include the upcoming Jupiter satellites exploration mission JUICE [2].

3.2 Missions and Methods

3.2.1 Rendezvous

VBN for rendezvous in orbit provides relative measurements between the active (or chaser) vehicle and the passive (or target) one. Until the moment that the target becomes visible in the navigation camera, the chaser vehicle's GNC has to rely on its own navigation sensors (typically IMU, star tracker and GNSS) and on estimates of the target's orbital position. The image processing and navigation techniques used from then on will depend on the distance and the size (and thus pixel resolution on the camera) of the target. A long range, pixel-size image of the target will only provide bearing information on its localization, whereas the observability of the target's position and velocity requires several measurements that must be spaced by means of maneuvers. The GUIBEAR activity [3] investigated the applicability of bearings-only navigation in the context of a rendezvous in a Near Rectilinear Orbit (NRO) between a lunar ascender and the Lunar Gateway station. The resulting performance of the navigation algorithm, based on a batch filter, is one order of magnitude better than those obtained without GUIBEAR.

At closer ranges, when the target vehicle covers enough pixels in the camera's detector, it becomes possible to estimate not only the position but also the relative attitude of the fully resolved object. For example, if the target is a known satellite, its 3D model can be first rendered in 2D first, and then matched against the edges extracted from the camera image. VBN systems which have to cover both long and close range approach to the target often rely on two cameras with different apertures for each domain, as well as different image processing techniques. This is the case of the Mars Sample Return (MSR) mission [4], where the Earth Return Orbiter (ERO) will have to rendezvous and capture a cylindrical Orbiting Sample (OS) canister. The GNC system designed for ESA by GMV for the MSR-ERO system study includes a Narrow Angle (NAC) and a Wide Angle camera (WAC). The Long Range Image Processing (LRIP) has to detect, track and distinguish two objects at long distance—the OS and the Mars Ascent Vehicle. At short distances, the shape, size and thus range of the OS can be estimated through the following algorithm: a training set of images of the OS is produced on the ground, for different poses and illumination conditions; a similarity measure is then computed between the camera image and all the reference images with the illumination corresponding to the current Sun direction, the highest scoring template is then selected.

3.2.2 Descent and Landing

Camera navigation for a lander in a powered descent toward a planetary surface falls into two broadly categories: either the terrain observed by the camera has been previously mapped (which is, for example, the case for the Moon, thanks to previous orbiter missions, and up to a resolution of 5m for certain regions) or it has not been mapped (again, in the case of the Moon, this would be true of most of its surface when observed below an altitude of a few kilometers due to the lack of resolution). In the first case, a map of the region flown over can be stored on-board the lander and remarkable elements of the terrain (or features) can be matched between the camera images and the map in order to estimate the position of the lander within a local terrain frame. These features can be terrain geological elements easily recognizable by the human eye, such as craters, or they can be local variations in elevation or albedo which can be identified by image processing. This technique is called *absolute navigation*. In the second case, features cannot be identified on a map; however, it is possible to track them from one image to the next, using their apparent motion through the camera field of view to estimate the lander velocity vector with respect to the ground—this is *relative navigation*. Typically, there is a handover from absolute to relative navigation as a lander descends closer to the surface, with some overlap possible between the two functions.

The Generic Vision-Based Navigation for Descent and Landing (GENEVIS) is a technology activity co-funded between ESA and Airbus Defence and Space [5]. The goal is to develop a camera-based navigation system for Lunar landing, usable from main braking in parking orbit at about 15 km altitude to touchdown. It contains a relative navigation subsystem—THEMIS (Fig. 3.1)—based on a Kanade-Lucas-Tomasi tracker which is robust to rotations, scaling and large translations, and an absolute navigation subsystem—Discover (Fig. 3.2)—which uses landmarks as small patches of features, rendered onboard with current illumination conditions and matched against the camera images. Both image processing output are then pre-processed then fused with measurement from other GNC sensors in a Schmidt

Fig. 3.1 GENEVIS: feature tracking with THEMIS during a field test with an helicopter. In green, feature path tracked through several images (copyright Airbus)

Extended Kalman Filter. GENEVIS has been implemented on a space-qualified processor (LEON4 CPU embedded on a GR740) and successfully tested with synthetic images. The relative navigation was also tested over actual terrain onboard an helicopter.

3.2.3 "Classic" Approach Versus Machine Learning

Recent advances in machine learning specifically the convolutional neural networks (CNN) and their application to image processing for ground-based applications has naturally led to assessing their applicability to VBN for space missions. The Artificial Intelligence for Terrain Relative Navigation in Unknown Environment (ATENA) activity tested, among other approaches, an hybrid relative navigation with a D2-Net CNN as a feature extractor. The scenario was the one of HERA missions to the binary asteroid Didymos; however, it did not perform as well as the others VBN algorithms [6].

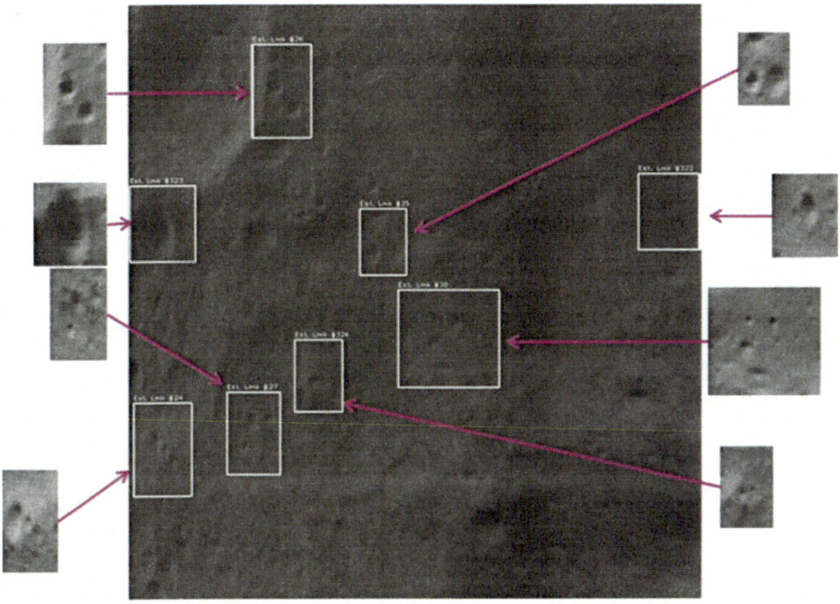

Fig. 3.2 GENEVIS: landmark matching with DISCOVER on a simulated lunar terrain (copyright Airbus)

Although not per se VBN since the main sensor is a LiDAR and not a camera, an ESA funded activity by City, University of London [7] demonstrated pose estimation of a lander above a lunar terrain by feeding 3D LiDAR data to a Deep Recurrent Convolutional Network (DRCNN). The system was successfully tested with synthetic images and also with a drone above a small terrain mockup.

The Ph.D. thesis Monocular Vision-Based Pose Estimation of Uncooperative Spacecraft [8] explored the use of CNN for feature extraction in a rendezvous scenario. Pre-selected features on the spacecraft images are sent to the navigation filter as measurements, as well as their covariance which are computed from the CNN's heatmaps. Tests show a good coupling between the translational and rotational filter states. The thesis also deals with the so-called domain shift, which is the drop in performance of CNN trained in a certain domain is tested in another one. Typically, this applies to CNN trained with synthetic images and tested with real ones. The set of synthetic training images was augmented with a randomization of texture and lighting conditions, resulting in increase robustness of CNN to previously unseen target features. This robustness was checked in robotic test bench tests with a physical mockup performed at GRALS (see Sect. 3.4.3).

3.3 Hardware Development

3.3.1 Cameras

Other types of cameras are being considered beyond the monocular type working in the visible spectrum. Infrared, and specifically thermal infrared, can overcome the issue of sun illumination affecting orbital operations [9]. A multispectral camera, fusing a visible and near-infrared channel with a thermal channel has been preliminary developed [10], and subsequently developed as a breadboard with specifications for a rendezvous with the Lunar Gateway in NRO [11]. Plenoptic, or light field cameras, have also been assessed for space application, although their use would be limited to short distance (below 20 m) and in conjunction with other navigation sensors [12].

3.3.2 Processing—GMVISION

Not only have existing space-qualified processor been assessed for running image processing and navigation algorithms, but also new, task-specific processors have been built around elements such as Field Programmable Gate Arrays (FPGAs), and they are also being designed to improve the runtime performance of computationally intensive tasks. The technology development activity GMVISION, led by ESA in collaboration with the company GMV is the most recent from a series of activities that had the objective of developing and testing a vision-based navigation camera with image processing capabilities, partially implemented in the firmware of an FPGA [13]. The GMVISION unit is built around 2 NanoXplore FPGAs: a BRAVE NG-Medium handling interfaces, and a NG-Large for feature detection and tracking (Fig. 3.3). Typically the complete vision-based navigation algorithm is not entirely implemented in GMVISION: only the functions of the image processing that benefit from the acceleration provided by the FPGA implementation are hosted as firmware; the rest is running on the on-board computer. As an example, the real-time acceleration achieved by having a feature tracking algorithm running on an FPGA rather than a processor like the LEON2 is two order of magnitude. Two pairs of redundant cameras can be plugged in, depending on the application [orbital rendezvous or descent and landing (Fig. 3.4)]. The goal of the activity, to be concluded in March 2023, is to reach a Technology Readiness Level of 6 (engineering model demonstrating the critical functions of the unit in a relevant environment).

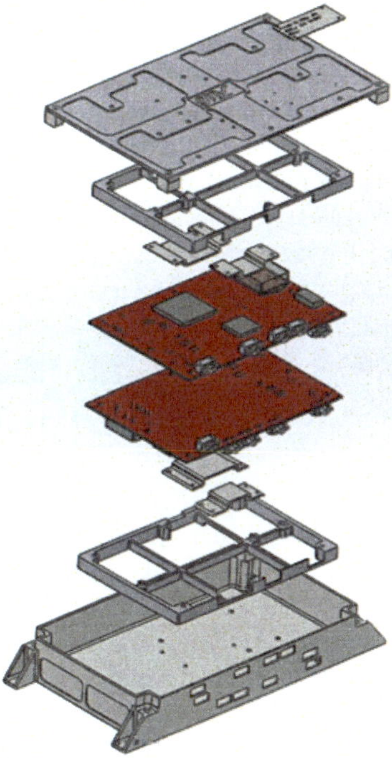

Fig. 3.3 Exploded view of the GMVISION unit showing the two printed circuit boards (in red), the stiffeners and the top and bottom parts of the enclosure (copyright GMV)

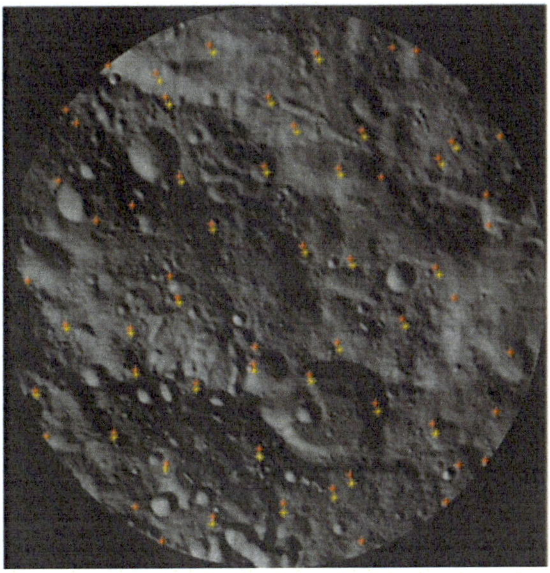

Fig. 3.4 GMVISION: example of relative navigation with feature tracking on the lunar surface (copyright GMV)

3.4 Realistic Images for Training and Validation

The development of VBN systems relies on the possibility to test them at various point of their development cycle. Unlike Earth applications, where realistic training and testing images can be easily obtained, there is a relative scarcity in Europe of usable space images during rendezvous and EDL operations. There are a few sets of orbital images with relevant metadata, but these must be supplemented with synthetic images, or images acquired on optical, robotic or field test campaigns. The following are examples of training and testing image generation means used at ESA.

3.4.1 Actual Space Images Acqusition—LIRIS

The LIRIS (Laser InfraRed Imaging Sensors) experiment was a passenger on the last ATV (Automated Transfer Vehicle) supply mission to the International Space Station [14]. A scanning Lidar and a set of visible and thermal infrared cameras made acquisition of the ISS during a fly-under maneuver and the rendezvous and docking. 72 GB of data, including 18,000 pictures were recorded and are now available in a database which also include all relevant metadata, including the ATV state vector estimated by its GNC function.

3.4.2 Synthetic Images Generation—PANGU

PANGU (Planet and Asteroid Natural Scene Generation Utility [15]) is the main tool at ESA for generating synthetic space images for navigation applications (Fig. 3.5). Developed by the University of Dundee under several ESA contracts, PANGU has two main functions: (a) surface modeling, where it can generate either from scratch or from an existing terrain or whole-body models a realistic and detailed landscape, including craters, rocks, dunes and overall roughness, and (b) rendering, where realistic images of the previously modeled terrain are produced, taking into account surface characteristics, illumination sources, shadows, atmospheric effects and a camera model. Recent versions of PANGU now include the realistic representation of satellites and their surfaces (MLI, mirror, solar arrays) and the generation of thermal infrared images for both natural and artificial objects (Fig. 3.6). PANGU runs on any computer that can accommodate a GPU (Graphics Processing Unit)—based renderer, and is capable of real-time performance.

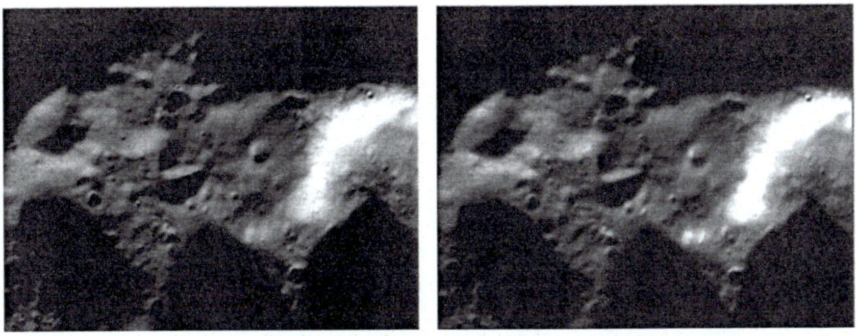

Fig. 3.5 Reproduction with PANGU (left) of an image of the lunar surface acquired by the UVVIS camera of the Clementine orbiter (right) (copyright University of Dundee)

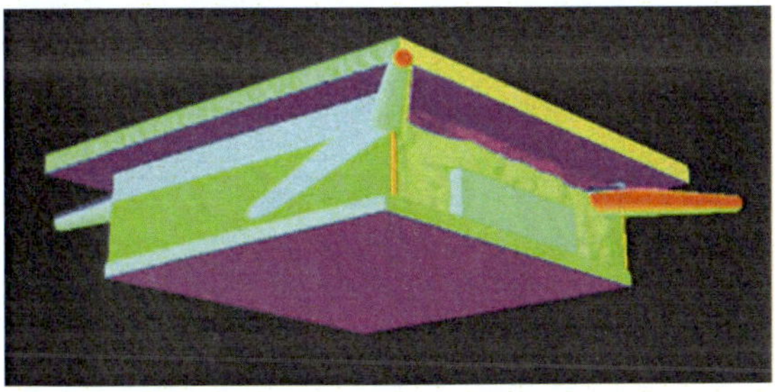

Fig. 3.6 Example of thermal infrared simulation with PANGU (copyright University of Dundee)

3.4.3 Robotic Test Bench—GRALS

The GNC Rendezvous, Approach and Landing Simulator (GRALS) is part of the ESA GNC Test Facilities located at the European Space Research and Technology Centre (ESTEC) in Noordwijk in the Netherlands. Its purpose is to reproduce scaled or full size trajectories for EDL or rendezvous missions so as to be able to study, develop and investigate GNC systems, including software, processors and sensors. GRALS is a robotic test bench, whose main elements are two KUKA robots with a 10-Kg carrying capacity, each mounted on a linear track and thus each providing $6 + 1$ degrees of freedom. They are installed in a 33m long dark room (Fig. 3.7), where a realistic light source such as the sun can be simulated. Ground truth is provided by an array of VICON infrared cameras, illuminating and tracking small markers fixed on the elements of the simulation. A typical scenario replicates the relative kinematics between a sensor (camera, LiDAR or laser ranging equipment, etc.) and a scaled or full-size target (satellite or element of satellite mock-up, terrain model, etc.). An example of a scaled rendezvous scenario with a camera and a satellite is shown in Fig. 3.8.

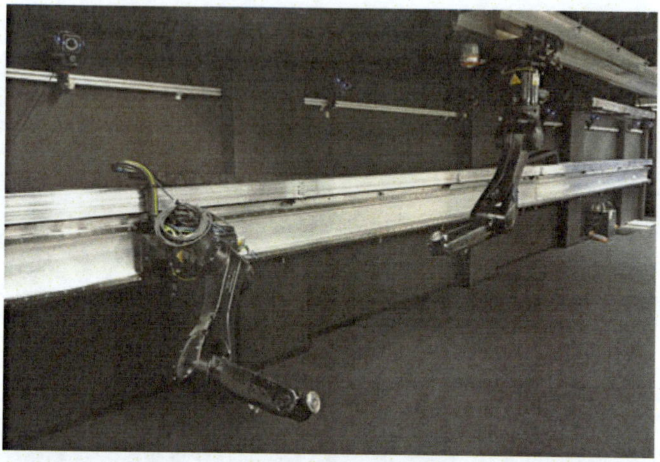

Fig. 3.7 The GRALS testbed, showing the 2 KUKA robots on their respective linear tracks and several of the VICON tracking cameras (copyright ESA)

Fig. 3.8 Rendezvous simulation with GRALS (copyright ESA)

3.5 Conclusion

The VBN technology activities presented here were recently concluded, others are in progress or being approved. Future development are likely to include further miniaturization, more applications of machine learning and additional generation of even more realistic training images, either synthetic or in laboratory. Future flight opportunities for VBN systems include Earth-based and Lunar landers, and in-orbit servicing, in-orbit assembly and in-space transportation missions.

Acknowledgements ESA would like to acknowledge the outstanding work and contributions performed by the following partners in the frame of various technology development activities with the Agency: Airbus Defence and Space, GMV, University of Dundee, City, University of London, Technische Universiteit Delft, Cosine Measurement Systems, and Universidad de La Laguna. The views expressed in this article cannot be considered as the official view of the European Space Agency

References

1. Gil-Fernandez, J., Ortega, G., Casasco, M., Huertas, I., Dubois-Matra, O., Carnelli, I.: ESA GNC technologies for asteroid characterization, sample-return, and deflection missions. In: IEEE Aerospace Conference Proceedings, vol. 2016 (2016). https://doi.org/10.1109/AERO.2016.7500508
2. Gorog, F., Arnolfo, M.C., Belmana, S., Dervaux, S., Gherardi, D.: Juice navigation camera design. **11180**, 1745–1757 (2019). https://doi.org/10.1117/12.2536086, https://www.spiedigitallibrary.org/conference-proceedings-of-spie/11180/111804N/JUICE-navigation-camera-design/10.1117/12.2536086.fullhttps://www.spiedigitallibrary.org/conference-proceedings-of-spie/11180/111804N/JUICE-navigation-camera-design/10.1117/12.2536086.short
3. Serra, P., Lourenço, P., Branco, J., Valero, J.B., Mammarella, M., Peters, T.V., Portela, D.M., Manara, S., Bernardini, D., Bemporad, A., Manzillo, P.F., Witteveen, J., Cropp, A.: Computational guidance and navigation for bearings-only rendezvous—methods and outcomes of GUIBEAR. In: 11th International ESA Conference on Guidance, Navigation and Control Systems (2021). https://doi.org/10.5281/ZENODO.6471464, https://zenodo.org/record/6471464
4. Strippoli, L., Pellacani, A., Pesce, V., Rodrigalvarez, M.A., Gonzalez-Arjona, D., Bajanaru, P., Gogu, D.G., Casasco, M.: MSR rendezvous and capture phase: the GNC subsystem solution and a snapshot over the image processing algorithms and narrow angle camera elegant breadboard. In: 70th International Astronautical Congress (2019). https://www.researchgate.net/publication/338925521_MSR_Rendezvous_and_Capture_phase_the_GNC_subsystem_solution_and_a_snapshot_over_the_Image_Processing_algorithms_and_Narrow_Angle_Camera_Elegant_Breadboard
5. Djafari-Rouhani, D., Duteïs, P., Brochard, R., Tiberio, S., Sanchez-Gestido, M.: Genevis generic vision-based navigation for descent and landing. In: 11th International ESA Conference on Guidance, Navigation and Control System Sensors Data Fusion and Autonomous Navigation (2021)
6. Artificial intelligence for terrain relative navigation in unknown environment (atena). Nebula Public Library. https://nebula.esa.int/content/artificial-intelligence-terrain-relative-navigation-unknown-environment-atena
7. Chekakta, Z., Zenati, A., Aouf, N., Dubois-Matra, O.: Robust deep learning lidar-based pose estimation for autonomous space landers. Acta Astronautica **201** (2022). https://doi.org/10.1016/j.actaastro.2022.08.049
8. Cassinis, L.P.: Monocular vision-based pose estimation of uncooperative spacecraft (2022)
9. Özgün Yılmaz, Aouf, N., Checa, E., Majewski, L., Sanchez-Gestido, M.: Thermal analysis of space debris for infrared-based active debris removal. Proc. Inst. Mech. Eng. Part G J. Aerosp. Eng. **233**, 811–822 (2019). https://doi.org/10.1177/0954410017740917
10. Multispectral sensing for relative navigation. https://indico.esa.int/event/128/attachments/734/865/02_-ME-01-cleanspace-issue1.1.pdf

11. ESA—new multispectral camera. https://www.esa.int/Enabling_Support/Space_Engineering_Technology/Shaping_the_Future/New_MultiSpectral_Camera
12. Delgado, A., Martinez, N., Lüke, J., Rodríguez-Ramos, J., Gestido, M., Bullock, M.: On the use of plenoptic imaging technology for 3d vision based relative navigation in space. In: 10th ESA Conference on Guidance, Navigation and Control Systems (2017)
13. On-board complex image processing based on FPGA acceleration for autonomous navigation in space. https://indico.esa.int/event/225/contributions/4314/attachments/3373/4420/OBDP2019-S06-02_GMV_Bajanaru_On-Board_Complex_Image_Processing_Based_on_FPGA_Acceleration_for_Autonomous_Navigation_in_Space.pdf
14. Mongrard, O., Ankersen, F., Casiez, P., Cavrois, B., Donnard, A., Vergnol, A., Southivong, U.: Liris flight database and its use toward noncooperative rendezvous. Progress in Flight Dynamics, Guidance, Navigation, and Control **10**, 21–34 (2018)
15. Martin, I., Dunstan, M., Gestido, M.S.: Planetary surface image generation for testing future space missions with Pangu. In: 2nd RPI Space Imaging Workshop, Saratoga Springs, NY (2019)

Chapter 4
Extending Explicit Guidance Methods to Higher Dimensions, Additional Conditions, and Higher Order Integration

Evan Kawamura and Dilmurat Azimov

Abstract Guidance functions play critical roles in autonomy to steer vehicles and aircraft to the intended target or destination. Explicit guidance (E Guidance) solves the two-point boundary value problem with initial and final conditions for position and velocity. The original formulation of E Guidance involves translational acceleration commands with a direct relationship to time, and it is possible to modify E Guidance for rotational acceleration. Other extensions for E Guidance include higher dimensions, additional conditions, and higher-order integration of the linearly independent E Guidance functions. The most promising extension involves higher-order integration of the E Guidance functions, but it may be physically impractical by initially moving away from the target. This paper provides a brief overview of some methods that extend E Guidance to higher dimensions, utilize additional conditions, or perform higher-order integration, and if they satisfy the two-point boundary value problem.

Keywords Explicit guidance · Polynomials · Unmanned aerial vehicles · Boundary conditions · Higher-order integration · Additional conditions

4.1 Introduction

Explicit Guidance (E Guidance) started before the famous Apollo missions and became the fundamentals for Apollo lunar guidance. The primary goal for E Guidance involves determining a thrust acceleration vector to solve the rendezvous two-point boundary problem for rocket-propelled spacecraft at a specified terminal time based

E. Kawamura (✉)
Intelligent Systems Division, NASA Ames Research Center, Moffett Field, Mountain View, CA, USA
e-mail: evan.t.kawamura@nasa.gov

D. Azimov
Department of Mechanical Engineering, University of Hawai'i at Manoa, 2540 Dole Street, Honolulu, HI, USA
e-mail: azimov@hawaii.edu

D. Azimov (ed.), *Proceedings of the IUTAM Symposium on Optimal Guidance and Control for Autonomous Systems 2023*, IUTAM Bookseries 40,
https://doi.org/10.1007/978-3-031-39303-7_4

on the spacecraft's current position and velocity. Even though the original E Guidance derivations were for spacecraft, they can be applied to all types of vehicles [1]. This paper will discuss methods for extending E Guidance: intermediate positions and velocities, desired acceleration and jerk, and higher-order integration of the E Guidance functions [2].

Here is an overview of the E Guidance extensions. The intermediate positions and velocities approach cuts T_{go}, in half and uses desired position and velocity as the third and fourth sets of conditions, which extends **E** from $\mathbb{R}^{2\times 2}$ to $\mathbb{R}^{4\times 4}$. Including desired acceleration and jerk resembles Cherry's formulation with final attitude guidance with the velocity and acceleration vectors aligned [1]. Still, the acceleration, velocity, and position profiles do not satisfy the desired conditions. The most promising E Guidance extension involves higher order integration of the E Guidance functions, but it has some impractical movements initially away from the target. The initial movement away from the target may be useful for incoming obstacle avoidance only if there are no obstacles behind or below the vehicle. Numerous combinations with polynomials, exponentials, and trigonometric functions for $p_i(t)$, $p_j(t)$ do not satisfy the boundary conditions, but some combinations partially fulfill the boundary conditions. Other combinations do not reveal real numbers, so they yield impractical solutions.

This paper describes the proposed E Guidance extensions in Ref. [2] with more details and some modifications, and it is organized as follows. Section 4.2 discusses the original equations, and Sect. 4.3 describes methods to extend E Guidance. Section 4.4 provides simulation results for the proposed extensions, and finally, Sect. 4.5 ends with concluding remarks.

4.2 Original Formulation for E Guidance

The translational commanded thrust acceleration, $a_{Tx}(t)$, in the x-direction is [1]:

$$a_{Tx}(t) = c_1 p_1(t) + c_2 p_2(t) - g_x(t)\,, \tag{4.1}$$

such that $p_1(t)$ and $p_2(t)$ are linearly independent functions and $c_1, c_2 \in \mathbb{R}$.

$$p_1(t) = \tau^m\,,\ \ p_2(t) = \tau^n \tag{4.2}$$

with $m, n \in \mathbb{Z}$ (integers) and $\tau = (T - t)$ such that T is the terminal time and t is the current time. The time-to-go to reach the desired position and velocity is defined by the difference between the terminal and current time: $T_{go} = T - t$. The general E Guidance procedure starts with setting the initial position and velocity boundary conditions at an initial time, t_0, and the desired position and velocity at some terminal time, T. Second, choose linearly independent functions, $p_i(t)$, $p_j(t)$, for $i = 1, 3, 5$ and $j = 2, 4, 6$ for three dimensional E Guidance. E Guidance in one dimension requires only $p_1(t)$, $p_2(t)$. Third, integrate $p_i(t)$, $p_j(t)$ to determine $\mathbf{F} = [f_{ij}]$ and $\mathbf{E} = [e_{ij}]$ as the inverse of the **F** matrix:

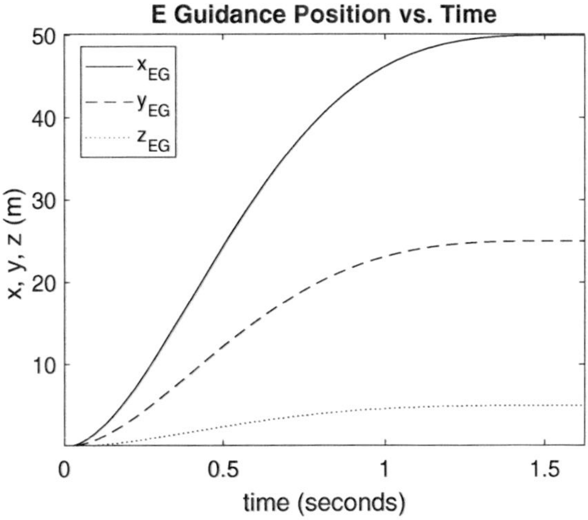

Fig. 4.1 Guided trajectory example

$$
\begin{aligned}
&e_{11} = f_{22}/\triangle \ , \ e_{12} = -f_{12}/\triangle, e_{21} = -f_{21}/\triangle \ , \ e_{22} = f_{11}/\triangle \ , \\
&f_{11} = \int_{t_0}^{T} p_1(t)\mathrm{d}t \ , \ f_{12} = \int_{t_0}^{T} p_2(t)\mathrm{d}t \ , \ f_{21} = \int_{t_0}^{T} \left[\int_{t_0}^{t} p_1(s)\mathrm{d}s \right] \mathrm{d}t \ , \\
&f_{22} = \int_{t_0}^{T} \left[\int_{t_0}^{t} p_2(s)\mathrm{d}s \right] \mathrm{d}t \ , \ \triangle = det(\mathbf{F}) = f_{11}f_{22} - f_{12}f_{21}.
\end{aligned} \tag{4.3}
$$

Fourth, determine the coefficients, c_1, c_2 (x-direction), from the **E** matrix components and the boundary conditions:

$$
\begin{bmatrix} c_1 \\ c_2 \end{bmatrix} = \begin{bmatrix} e_{11} & e_{12} \\ e_{21} & e_{22} \end{bmatrix} \begin{bmatrix} \dot{x}_D - \dot{x}_0 \\ x_D - x_0 - \dot{x}_0 T_{go} \end{bmatrix} , \tag{4.4}
$$

in which "0" denotes initial, and "D" denotes desired. Finally, use the computed coefficients to compute the thrust acceleration profile as shown in Eq. (4.1). Integrating yields the guided velocity profile, and integrating again generates the guided position profile. Figure 4.1 shows an example of translational E Guidance.

4.3 Description of E Guidance Extensions

It is possible to extend E Guidance to rotational acceleration commands [2–4]:

$$\begin{aligned}\alpha_\phi(t) &= c_1\tau^2 + c_2\tau^3 - \frac{I_{yy} - I_{zz}}{I_{xx}}\omega_y\omega_z\,,\\ \alpha_\theta(t) &= c_3\tau^2 + c_4\tau^3 - \frac{I_{zz} - I_{xx}}{I_{yy}}\omega_x\omega_z\,,\\ \alpha_\psi(t) &= c_5\tau^2 + c_6\tau^3 - \frac{I_{xx} - I_{yy}}{I_{zz}}\omega_x\omega_y\end{aligned} \tag{4.5}$$

in which $\alpha_\phi, \alpha_\theta, \alpha_\psi$ are the angular acceleration commands and $\omega_x, \omega_y, \omega_z$ are the body angular velocities. See Ref. [2–4] for more details on rotational E Guidance. The next subsections discuss four translational E Guidance extensions.

4.3.1 Desired Intermediate Positions and Velocities

The first method considers intermediate conditions halfway during the maneuver such that $T_{go,int} = T_{go}/2$. The **E** matrix utilizes the original formulation defined by Cherry as: $\mathbf{E} = \mathbf{F}^{-1}$ [1]. Including intermediate conditions yields $\mathbf{E}, \mathbf{F} \in \mathbb{R}^{4\times4}$ with the upper left submatrix containing the intermediate conditions and the lower right submatrix containing the endpoint (terminal) conditions:

$$\mathbf{F} = \begin{bmatrix} \mathbf{F1}_{2\times2} & 0_{2\times2} \\ 0_{2\times2} & \mathbf{F2}_{2\times2} \end{bmatrix} \tag{4.6}$$

with submatrices **F1**, **F2** defined as:

$$\begin{aligned}\mathbf{F1} &= \begin{bmatrix} \int_{t_0}^{T/2} p_1(t)\mathrm{d}t & \int_{t_0}^{T/2} p_2(t)\mathrm{d}t \\ \int_{t_0}^{T/2}\left(\int_{t_0}^{t} p_1(s)\mathrm{d}s\right)\mathrm{d}t & \int_{t_0}^{T/2}\left(\int_{t_0}^{t} p_2(s)\mathrm{d}s\right)\mathrm{d}t \end{bmatrix}\\ \mathbf{F2} &= \begin{bmatrix} \int_{t_0}^{T} p_1(t)\mathrm{d}t & \int_{t_0}^{T} p_2(t)\mathrm{d}t \\ \int_{t_0}^{T}\left(\int_{t_0}^{t} p_1(s)\mathrm{d}s\right)\mathrm{d}t & \int_{t_0}^{T}\left(\int_{t_0}^{t} p_2(s)\mathrm{d}s\right)\mathrm{d}t \end{bmatrix}\end{aligned} \tag{4.7}$$

The vector with the intermediate (I) and final (F) conditions for the x-direction is defined by the first two rows for the intermediate conditions and the last two rows for the final conditions:

$$x_{I,F}(t) = \begin{bmatrix} \dot{x}(T/2) - \dot{x}(t_0) \\ x(T/2) - x(0) - \dot{x}(t_0)T_{go}/2 \\ \dot{x}(T) - \dot{x}(T/2) \\ x(T) - x(T/2) - \dot{x}(T/2)T_{go}/2 \end{bmatrix}. \tag{4.8}$$

with $x(0), x(T/2), x(T), \dot{x}(0), \dot{x}(T/2), \dot{x}(T)$ defined as the initial position, intermediate position, final position, initial velocity, intermediate velocity, and final velocity, respectively. Similar forms exist for the y, z directions but are not shown for brevity.

Overall, the method presented here in this paper differs from the method in Ref. [2] by:

1. integrating two linearly independent polynomials instead of four linearly independent polynomials
2. using the intermediate conditions for the computations of the final conditions, i.e., last two rows of Eq. (4.8)

The default and simplest set of linearly independent polynomials is:

$$p_1(t) = 1 \;, \; p_2(t) = (T - t). \tag{4.9}$$

Testing this method with the polynomials in Eq. (4.9) shows that the desired final and intermediate position conditions and intermediate velocity conditions are not satisfied. However, the final velocity conditions are satisfied with zero acceleration at the end of the maneuver. Preliminary tests also demonstrate extremely large final position values such that this proposed method is not physically feasible or practical. Consequently, Sect. 4.4 will not show these simulation results for brevity.

4.3.2 Final Desired Jerk Vector

This method augments the 2×2**E** matrix to 4×4 with final desired acceleration and jerk values as the third and fourth set of conditions for computing the c_i. The first two conditions are the original conditions in the column vector on the right-hand-side of Eq. (4.4). The jerk vector is simply the time derivative of the acceleration vector, which is straightforward to compute because acceleration is a function of only time and constants:

$$\begin{aligned} a_x &= c_1 p_1(t) + c_2 p_2(t) + c_3 p_1(t) + c_4 p_4(t) \\ j_x &= \frac{da_x}{dt} = c_1 p_1'(t) + c_2 p_2'(t) + c_3 p_1'(t) + c_4 p_4'(t) \end{aligned}. \tag{4.10}$$

This method algebraically solves for the coefficients without utilizing the **E** or **F** matrices, i.e., four coefficients (unknowns) and four equations: two from the column vector on the right-hand-side of Eq. (4.4) and two from Eq. (4.10). This approach follows Cherry's method for final desired attitude guidance by aligning the final velocity and acceleration vectors [1]. Overall, preliminary tests for this method yield drastic divergence and does not satisfy the boundary conditions. Therefore, Sect. 4.4 will not show these simulation results.

4.3.3 Four Integrations of the F Matrix

This method differs from the previous method by including the **E** and **F** matrices by utilizing four integrations for four polynomials:

$$\mathbf{F} = \begin{bmatrix} f_{11} & f_{12} & f_{13} & f_{14} \\ f_{21} & f_{22} & f_{23} & f_{24} \\ f_{31} & f_{32} & f_{33} & f_{34} \\ f_{41} & f_{42} & f_{43} & f_{44} \end{bmatrix}$$
$$= \begin{bmatrix} \int p_1(t)\mathrm{d}t & \ldots & \int p_4(t)\mathrm{d}t \\ \int\int p_1(s)\mathrm{d}s\mathrm{d}t & \ldots & \int\int p_4(s)\mathrm{d}s\mathrm{d}t \\ \int\int\int p_1(u)\mathrm{d}u\mathrm{d}s\mathrm{d}t & \ldots & \int\int\int p_4(u)\mathrm{d}u\mathrm{d}s\mathrm{d}t \\ \int\int\int\int p_1(r)\mathrm{d}r\mathrm{d}u\mathrm{d}s\mathrm{d}t & \ldots & \int\int\int\int p_4(r)\mathrm{d}r\mathrm{d}u\mathrm{d}s\mathrm{d}t \end{bmatrix}. \quad (4.11)$$

The third condition of **E** is final desired acceleration, and the fourth condition is the final desired jerk. Here is the equation for computing the coefficients for the x-direction:

$$\mathbf{c}_x = \begin{bmatrix} e_{11} & e_{12} & e_{13} & e_{14} \\ e_{21} & e_{22} & e_{23} & e_{24} \\ e_{31} & e_{32} & e_{33} & e_{34} \\ e_{41} & e_{42} & e_{43} & e_{44} \end{bmatrix} \begin{bmatrix} \dot{x}_D - \dot{x}_0 \\ x_D - x_0 - \dot{x}_0 T_{go} \\ a_{x_D} + g_x \\ j_{x_D} \end{bmatrix}. \quad (4.12)$$

The final desired position, velocity, and jerk conditions are not satisfied, but the final desired acceleration is. Thus, Sect. 4.4 does not show these simulation results.

4.3.4 Three Integrations of the F Matrix

This method resembles the previous method but integrates three polynomials such that the rows of the **F** matrix have one, two, or three integrations for $p_1(t)$, $p_2(t)$, $p_3(t)$:

$$\mathbf{F} = \begin{bmatrix} f_{11} & f_{12} & f_{13} \\ f_{21} & f_{22} & f_{23} \\ f_{31} & f_{32} & f_{33} \end{bmatrix} = \begin{bmatrix} \int p_1(t)\mathrm{d}t & \int p_2(t)\mathrm{d}t & \int p_3(t)\mathrm{d}t \\ \int\int p_1(s)\mathrm{d}s\mathrm{d}t & \int\int p_2(s)\mathrm{d}s\mathrm{d}t & \int\int p_3(s)\mathrm{d}s\mathrm{d}t \\ \int\int\int p_1(u)\mathrm{d}u\mathrm{d}s\mathrm{d}t & \int\int\int p_2(u)\mathrm{d}u\mathrm{d}s\mathrm{d}t & \int\int\int p_3(u)\mathrm{d}u\mathrm{d}s\mathrm{d}t \end{bmatrix} \quad (4.13)$$

The third condition involves the final desired acceleration such that the coefficients for the x-direction are:

$$\mathbf{c}_x = \begin{bmatrix} e_{11} & e_{12} & e_{13} \\ e_{21} & e_{22} & e_{23} \\ e_{31} & e_{32} & e_{33} \end{bmatrix} \begin{bmatrix} \dot{x}_D - \dot{x}_0 \\ x_D - x_0 - \dot{x}_0 T_{go} \\ a_{x_D} + g_x \end{bmatrix}. \quad (4.14)$$

The other coefficient vectors, $\mathbf{c}_y$ and $\mathbf{c}_z$, have similar forms and are not shown for brevity. This method demonstrates promising results since it satisfies the position and velocity boundary conditions. However, the final desired acceleration is not satisfied. Overall, since this method has the best results of the four methods, Sect. 4.4 shows the simulation results for this method.

4.4 Simulation Results

This section shows simulation results for the proposed extension with three polynomials for three UAV maneuvers: a 360° roll maneuver, a vertical one-dimensional takeoff maneuver, and a waypoint maneuver.

4.4.1 Roll Maneuver

The coefficients, c_1, c_2, c_3, resemble the original 2×2 expression for the $\mathbf{E}$ matrix but with desired angular acceleration as the third element in the vector multiplied by the $\mathbf{E}$ matrix:

$$
\begin{aligned}
c_\phi &= \begin{bmatrix} e_{11} & e_{12} & e_{13} \\ e_{21} & e_{22} & e_{23} \\ e_{31} & e_{32} & e_{33} \end{bmatrix} \begin{bmatrix} \dot{\phi}_D - \dot{\phi}_0 \\ \phi_D - \phi_0 - \dot{\phi} T_{go} \\ \ddot{\phi}_D \end{bmatrix}, \\
c_\theta &= \begin{bmatrix} e_{11} & e_{12} & e_{13} \\ e_{21} & e_{22} & e_{23} \\ e_{31} & e_{32} & e_{33} \end{bmatrix} \begin{bmatrix} \dot{\theta}_D - \dot{\theta}_0 \\ \theta_D - \theta_0 - \dot{\theta} T_{go} \\ \ddot{\theta}_D \end{bmatrix}, \\
c_\psi &= \begin{bmatrix} e_{11} & e_{12} & e_{13} \\ e_{21} & e_{22} & e_{23} \\ e_{31} & e_{32} & e_{33} \end{bmatrix} \begin{bmatrix} \dot{\psi}_D - \dot{\psi}_0 \\ \psi - \psi - \dot{\psi} T_{go} \\ \ddot{\psi}_D \end{bmatrix}.
\end{aligned} \tag{4.15}
$$

For three polynomials, the boundary conditions for a 360° roll maneuver are:

$$
\begin{aligned}
\phi_D = -2\pi\,,\ \dot{\phi}_D = 0\,,\ \theta_D = 0\,, \\
\dot{\theta}_D = 0\,,\ \psi_D = 0\,, \dot{\psi}_D = 0\,, \\
\ddot{\phi} = 0\,,\ \ddot{\theta} = 0\,,\ \ddot{\psi} = 0.
\end{aligned} \tag{4.16}
$$

Figure 4.2 shows the Euler angles and time derivatives for a 360° roll maneuver for a quadcopter UAV with polynomials: $p_1(t) = 1$, $p_2(t) = (T - t) + 1$, $p_3(t) = (T - t)^2 + (T - t) + 1$. The pitch angle diverges and has a final pitch angle of 0.595 rad or 34.1°, which is not even close to the final desired pitch angle of 0°.

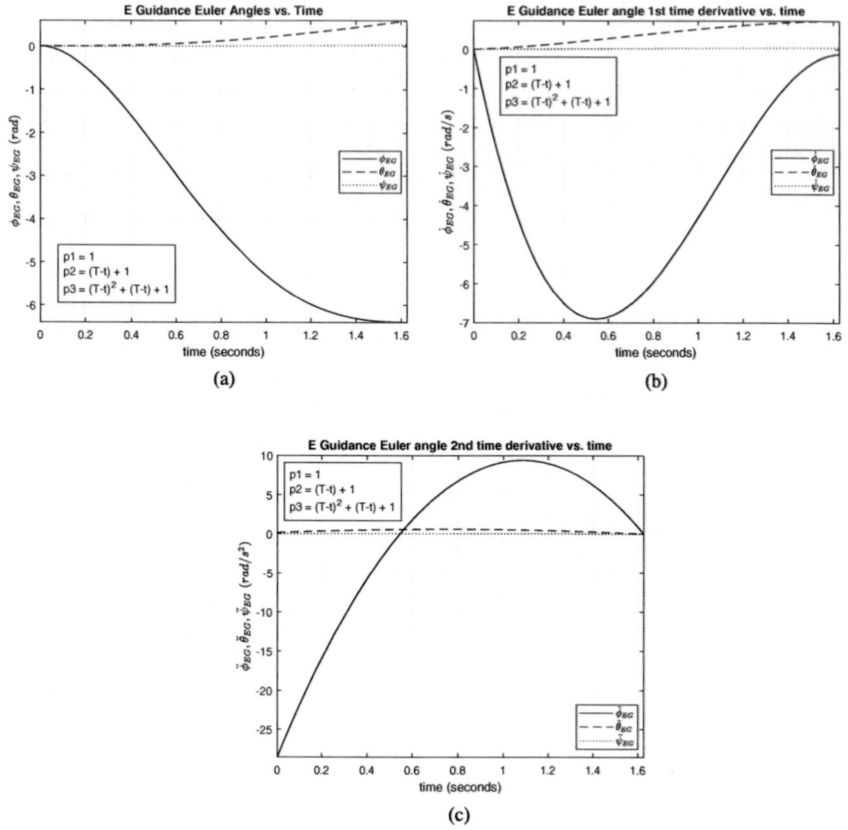

Fig. 4.2 Guided roll maneuver: $p_1(t) = 1$, $p_2(t) = (T - t) + 1$, $p_3(t) = (T - t)^2 + (T - t) + 1$. **a** Guided 360 ° roll maneuver: euler angles. **b** Guided 360° roll maneuver: 1st time derivative. **c** Guided 360 ° roll maneuver: 2nd time derivative

4.4.2 *Takeoff Maneuver*

This subsection describes a one-dimensional vertical ascent (takeoff maneuver) to stop at a desired altitude of 15 m. The polynomial selection is:

$$p_1(t) = 1 \ , \ p_2(t) = (T - t) + 1 \ , \ p_3(t) = (T - t)^2 + (T - t) + 1. \tag{4.17}$$

Figure 4.3 shows the guided takeoff maneuver with the altitude, vertical velocity, and vertical acceleration. The guided takeoff is very smooth, and the largest speed is approximately 0.85 m/s. Decreasing the time-to-go will increase the magnitude of the velocity profile because the vehicle has less time to reach the desired altitude. The acceleration is nearly constant at around 10 m/s^2 due to gravity. The final altitude is 15 m from the initial height, and the final vertical velocity is 0 m/s, which satisfies the final desired boundary conditions.

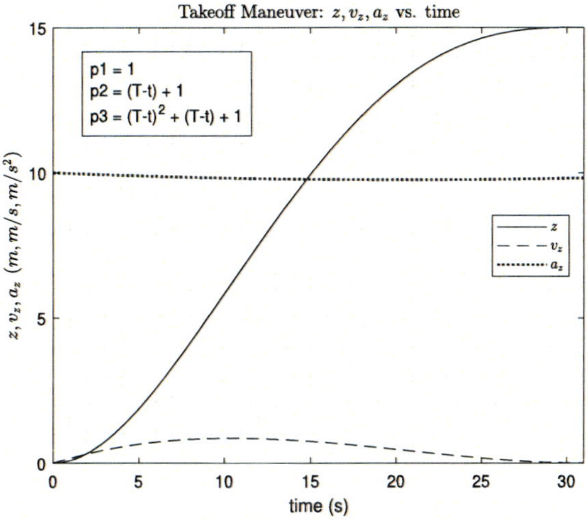

Fig. 4.3 Guided takeoff maneuver

4.4.3 Waypoint Maneuver

This subsection describes a three-dimensional waypoint maneuver for a vehicle to stop at a waypoint located 50 m away in the x-direction, 40 m away in the y-direction, and increases its altitude (z-component) by 15 m. The choice for $p_1(t)$, $p_2(t)$, $p_3(t)$ is the same as the polynomial selection in Eq. (4.17). Figure 4.4 shows the guided waypoint trajectory profile. Figure 4.4a shows the trajectory, which ends at the intended waypoint. It has an initial backward motion away from the target for approximately the first 4 s before moving toward the target. This initial reverse may not be practical or physically possible if there are obstacles behind the vehicle. Figure 4.4b shows the velocity profile, which shows a max reverse speed of around −10 m/s at approximately 1.7 s, a max forward speed of around 20 m/s at approximately 7.5 s, and satisfaction of the initial and final velocity conditions, i.e., 0 m/s. Figure 4.4c shows the acceleration profile, and the final accelerations are nonzero.

Appendix A of Ref. [2] has numerous possible combinations of exponential, sinusoidal, and polynomial functions for $p_1(t)$, $p_2(t0 < p_3(t)$. The trigonometric and exponential functions perform poorly by diverging, oscillating, or not satisfying the boundary conditions.

4.5 Conclusion

In summary, there were four methods for extending E Guidance. The first method utilized intermediate and final conditions with a sparse **E** matrix and four functions: $p_1(t)$, $p_2(t)$, $p_3(t)$, $p_4(t)$. The second method utilized final desired jerk and followed Cherry's formulation for final desired attitude without including **E**. The third method

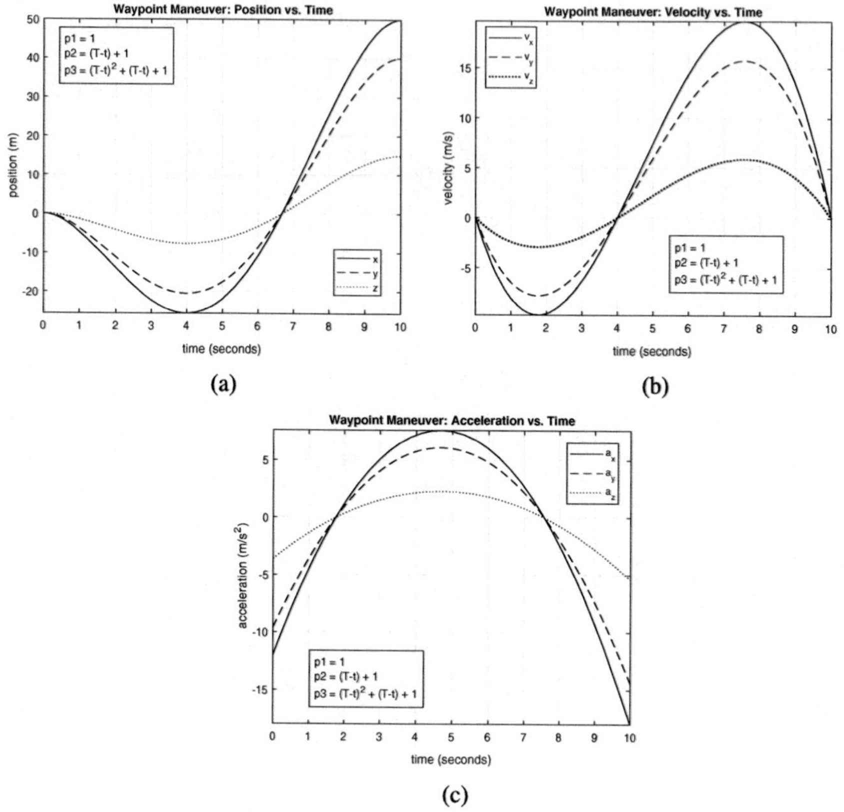

Fig. 4.4 Guided waypoint maneuver: $p_1(t) = 1$, $p_2(t) = (T - t) + 1$, $p_3(t) = (T - t)^2 + (T - t) + 1$. **a** Guided waypoint: position versus time. **b** Guided waypoint: velocity versus time. **c** Guided waypoint: acceleration versus time

resembled the second method but included four integrations in **F**. The fourth method used three integrations for the three functions, $p_1(t)$, $p_2(t)$, $p_3(t)$, and included final desired acceleration.

The fourth method is the most promising since it satisfies the position and velocity boundary conditions but not the final desired acceleration. However, the fourth proposed method has an initial reverse, which might not be practical due to vehicle motion constraints, actuator dynamics, or obstacles behind the vehicle. For instance, fixed-wing aircraft cannot reverse in mid-flight, but multicopter UAVs can reverse in mid-flight. Conversely, Cherry's original formulation with just two functions lacks the initial backward motion and is more practical. A potential application for this extended E Guidance method involves avoiding incoming obstacles while the initial plan was to move forward. In this case, applying E Guidance discretely with regular navigation and obstacle detection updates ensures the dynamic obstacle is out of the vehicle's intended path before proceeding forward.

Overall, extending E Guidance to three or four functions does not yield better performance than Cherry's original formulation with $p_1(t)$, $p_2(t)$. Increasing the exponents to higher powers typically generates higher accuracy like a Taylor series expansion, but that is not the case for the proposed E Guidance methods. It is possible that there are other combinations for the functions that satisfy all the boundary conditions, so the authors might not have been lucky in their selection of the functions. Even though three of the four methods do not satisfy the boundary conditions, they are presented here to prevent future researchers from reinventing inadequate methods, which will save time researching innovative methods for extending E Guidance.

References

1. Cherry, G.: A general, explicit, optimizing guidance law for rocket-propelled spaceflight. In: Astrodynamics Guidance and Control Conference, p. 638 (1964)
2. Kawamura, E.: Integrated targeting, guidance, navigation, and control for unmanned aerial vehicles. Ph.D. dissertation, University of Hawai'i at Manoa (2020)
3. Kawamura, E., Azimov, D.: Integrated targeting, guidance, navigation, and control for unmanned aerial vehicles. In: Volume 168 of the Advances in the Astronautical Sciences Series, pp. 4259–4277. Univelt (2019)
4. Kawamura, E., Azimov, D.: Extremal control and modified explicit guidance for autonomous unmanned aerial vehicles. J. Autonomous Veh. Syst. **2**(1) (2022)

Part II
Robotic Technologies

Chapter 5
Identifying Consumer Drones via Encrypted Traffic

David Liang and Yingfei Dong

Abstract To address the threats of consumer drones, effective drone countermeasures are required. As with such a process, identifying the specific type of invading drone is the critical initial step. While existing drone identification systems show some advantages, they also have several limitations, such as cost and deployability. In this paper, a unique method is proposed to identify drones based on features in the encrypted traffic between a drone and its controller. These features are constructed based on an in-depth understanding of drone communication protocols, which aids in classifying different types of popular ArduPilot drones. We collected plaintext and encrypted traffic from multiple drones, analyzed their communication patterns, and extracted features to identify specific drones. The experimental results show that the proposed framework can precisely identify target drones while outperforming existing methods. While this paper focuses on drone communications, the proposed method can also be applied to classify many other auto-control devices that exhibit similar communication behaviors.

Keywords Drone identification · MAVLink · Machine learning · WiFi traffic classification

5.1 Introduction

In recent years, consumer drones have been utilized in many applications as they have been made more affordable and easier to operate. However, they have also been abused in various attacks. Effective drone management schemes must be in place to address these issues, e.g., the type of drone must be accurately identified before

D. Liang (✉) · Y. Dong
University of Hawai'i at Manoa, Honolulu, HI, USA
e-mail: david947@hawaii.edu

Y. Dong
e-mail: yingfei@hawaii.edu

D. Azimov (ed.), *Proceedings of the IUTAM Symposium on Optimal Guidance and Control for Autonomous Systems 2023*, IUTAM Bookseries 40,
https://doi.org/10.1007/978-3-031-39303-7_5

applying corresponding countermeasures. Therefore, in this work, we focus on the drone identification issue.

Many consumer drones use WiFi for communication and control due to its strong encryption, reasonable range, and interoperability with popular personal devices. Furthermore, many vendors manufacture their drones by taking advantage of the most popular open-source drone control system (*ArduPilot* [1]) that provides reliable control and operations and reduces costs. Meanwhile, most drones use the Micro Air Vehicle Link (*MAVLink* [2]) protocol for communication due to its lightweight and versatile design. It allows different vendors to define customized message sets and semantics on top of the basic set of MAVLink messages. This project focuses on three popular open-sourced drones: *SkyViper Journey*, *Intel Aero*, and *3DR Solo*, each with a flight controller stack built with ArduPilot and MAVLink.

In the initial research, deep packet inspection (DPI) and protocol reverse engineering (PRE) techniques were deployed to analyze plaintext traffic traces of consumer drones, and concrete patterns were discovered in the frame size, inter-arrival timing, and message frequency distributions. These results showed that drones from different vendors are distinguishable based on patterns in their control traffic, despite each using the same ArduPilot and MAVLink stack for control and messaging. In addition, we found that the AES-CCMP encryption used in the popular WPA2-Personal standard increases the length of a WiFi frame by a constant value, which allowed us to associate the length of an encrypted frame with its corresponding plaintext. We incorporated these findings into our approach. In this paper, we analyzed the characteristics of a drone communication protocol and investigated methods to identify key patterns in the plaintext traffic that provide insight into the classification of encrypted traffic between a drone and its controller. Then, a framework is proposed to identify a specific type of drone among a set of known drones based on its encrypted WiFi traffic.

The main idea behind this approach is to match the communication patterns in encrypted traffic to patterns in plaintext traffic. First, we developed a procedure to capture known drone traffic and identify significant patterns in plaintext traffic traces. Next, we used these key patterns as features and established a drone class profiling scheme. Lastly, we proposed two classification methods using these profiling schemes to classify an encrypted traffic trace.

This paper is organized as follows. In Sect. 5.2, the relevant technical background is presented. Then, a survey of related works is conducted. In Sect. 5.3, the system design is introduced, which includes the phases of data collection, data preprocessing, pattern identification, drone traffic profiling, and classification. In Sect. 5.4, our framework is implemented, evaluated against real traces, and compared to a related work reimplementation. Finally, some conclusions are drawn in Sect. 5.5.

5.2 Background and Related Work

In this section, we discuss the technical concepts related to our research. First, the WPA2-Personal encryption standard is defined, and we describe how its properties contributed to extracting drone-control information from encrypted traffic. Second, the relevant characteristics of the MAVLink protocol are described. Lastly, a brief survey of related drone classification methods is provided for comparison.

5.2.1 WPA2-Personal Encryption in 802.11 Standards

The standard method for protecting the communications between a drone and its control station is by encrypting its traffic with the WiFi Protected Access 2 (WPA2) security standard. WPA2 uses AES-CCMP, a protocol that combines the Advanced Encryption Standard (AES) with Counter Mode CBC-MAC (CCM).

Before each transmission, the AES-CCMP algorithm [3] processes the plaintext data frame to generate an encrypted frame to send. The plaintext frame comprises both the MAC header (frame information) and the plaintext UDP datagram (MAVLink message). The process begins by generating an eight-byte CCMP encryption header and placing it between the MAC header and plaintext datagram. Next, the message integrity code (MIC) is computed based on the MAC header, CCMP header, and plaintext datagram. Then, the MIC is appended to the plaintext datagram, and both are encrypted using AES. Finally, the MAC header, CCMP header, and the now encrypted ciphertext datagram are combined to form the encrypted frame for transmission.

In this process, the size of the UDP datagram remains unchanged before and after the AES encryption step. Therefore, *we can infer the size of the drone communication message, i.e., a MAVLink message, based on the size of the encrypted datagram.* Furthermore, each datagram header has a fixed size, and the combined size of extra headers is a constant of 44 bytes. From this, we deduced that a MAVLink message with a size of $L0$ bytes corresponds to an encrypted UDP datagram with a size of $N = L0 + 44$ bytes. This characteristic has been confirmed on each drone. More details are available in [4].

5.2.2 Micro Air Vehicle Link (MAVLink) Protocol

MAVLink [2] is an application-layer communication protocol designed for bidirectional communication between a drone and its Ground Control Stations (GCSs). MAVLink is open-source, highly configurable, and a popular option for consumer drone vendors. Messages containing telemetry or command/control information are serialized in a binary format comparable to a network packet structure. This also ensures lightweight delivery with minimal overhead. Although MAVLink is a versatile protocol that supports many media, our study is focused on MAVLink over WiFi using the User Datagram Protocol (UDP).

MAVLink is currently on its second version (v2) but is still compatible with its first version (v1), to support legacy devices. MAVLink v1 uses an 8-byte serialization header format with a 1-byte Message ID field, which only allows Message ID values between 1 and 256; it quickly became insufficient over time as more message types were created. Meanwhile, MAVLink v2 was released later with the primary intention of amending the flaws of the previous version. The second version uses a 14-byte serialization format with a 3-byte Message ID field, which supports over 16 million possible message types [2]. Both versions of MAVLink were discovered in the captured traffic, which is dependent on vendor specifications.

The version number of a MAVLink message can be determined by the initial byte in its message header, known as the *start-of-text (STX)* field. A MAVLink v1 frame has an STX value of **0xFE**, and a MAVLink v2 frame has an STX value of **0xFD**. The Message ID field indicates the defined message type. The definitions and serialization structures for these message types are defined by MAVLink contributors, and can be found in the official documentation formatted as *XML* files. Certain vendors developed their customized message sets to incorporate unique features for their specific drones. As is the case with the 3DR Solo drone [5].

5.2.3 Existing Methods for Encrypted WiFi Traffic Classification

We surveyed both WiFi-based drone detection studies and generalized WiFi traffic classification methods and drew comparisons of these existing techniques against our work. In [6], a machine learning framework was proposed to identify a type of drone among a known set of drones solely by comparing the encrypted WiFi communication traffic. Each traffic trace is partitioned into segments with a uniform number of frames. This method selected features by computing statistics from the frame sizes and inter-arrival timings of each segment. Such statistics include mean, median, skewness, kurtosis, min/max, etc. Then, these statistical results are concatenated to form feature vectors. Similarly, in [7], a drone detection framework was proposed that aimed to distinguish drone transmissions from non-drone transmissions within an encrypted traffic trace. This method utilized statistical features to classify the directional traffic flows found in each trace. Each flow is partitioned into multiple segments via a timing window function, and each feature vector comprises statistics computed from each segment.

In addition to traditional machine learning techniques, representational machine learning, i.e., deep learning, approaches were also surveyed. For instance, Rezaei and Lui [8] proposed a deep learning methodology that uses a generalized WiFi classification template, which could also be applied to classify encrypted traffic across various applications. This method involves extracting the size and timing from each packet in a trace to be used as features for training various deep learning models. Deep learning methods do not rely on human-extracted features, which is convenient for users but leads to difficulty in logistically and holistically justifying the reasoning and impact behind the technique.

Existing techniques completely depend on statistical features without considering concrete patterns in drone communication. These statistics are either extracted from each frame individually or by grouping sequential frames into segments. Unlike existing methods, *our idea applies the knowledge of drone-specific characteristics to obtain more meaningful patterns in drone communication to construct high-quality and focused features based on these patterns, which ultimately improves the effectiveness of classification.*

5.3 System Overview

This work is focused on identifying key patterns in plaintext traffic and leveraging them to provide insight into the classification of encrypted traffic. This section introduces the proposed three-phase framework, which comprises the traffic capture and preprocessing, pattern identification and profile creation, and classification phases. Each phase is described in detail and provides insight to justify the approach.

5.3.1 Traffic Capture and Preprocessing

Data Collection

We conducted this study on drones from three vendors—*SkyViper* (VIPER), *Intel Aero* (INTEL), and *3DR Solo* (SOLO). Although all of these drones use ArduPilot and MAVLink, each drone has a specific control message set for command and telemetry messages, which is unique to each vendor and model. The collection procedure must be uniform despite these differences. For the VIPER and INTEL, the drone acts as the Access Point (AP), while its controller or ground control station (GCS) connects as a client. The SOLO drone works in an opposite manner, where the controller acts as the AP, and the drone or the GCS connects as a client. This distinction does not impact traffic collection, as the controller AP directly relays all communication traffic between the ground control station and the SOLO drone.

Each drone was flown in the AUTO mode [9] and followed a pre-defined waypoint mission. The open-source flight control program *Mission Planner* was used to build and execute these waypoint missions. The *Mission Planner* application also served as the GCS, which allows for arming/disarming the drone and starting/stopping the execution of missions. Using the interface, we monitored the current path of the drone against its pre-planned flight path. In considering the capability of each drone,

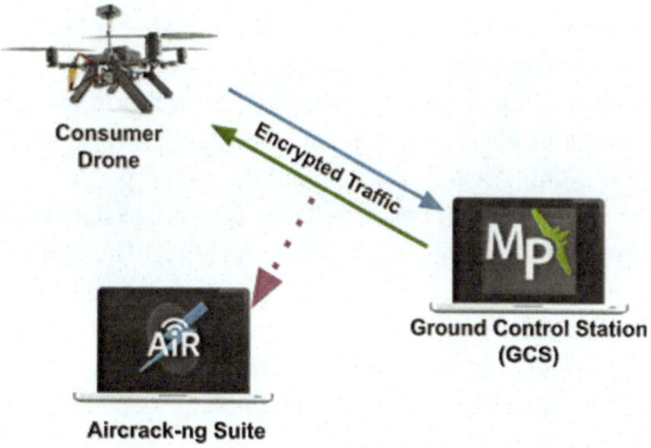

Fig. 5.1 Traffic capture diagram

which involves factors such as stability, battery life, and motor speeds, we designed a single mission plan to be flown by each drone (Fig. 5.1).

The *airodump-ng* tool from the *Aircrack-ng* suite [10] was used to capture traffic between a drone and its GCS. The traffic capture procedure is as follows. First, the WiFi channel and BSSIDs of the drone and controller are obtained and verified to filter the capture. While each BSSID is constant, the WiFi channel resets at each AP startup. Next, the *airodump-ng* capture command is activated. At this point, the GCS laptop connects to the drone AP. Then, the drone is armed, and the waypoint mission begins execution. Once the mission has been completed, as indicated by the GCS, the airodump-ng capture is terminated. To collect another trial, the capture command is restarted, and as stated previously, the GCS reconnects to the AP.

Because we start our analysis with plaintext traffic, we need to obtain the plaintext traffic first, in addition to capturing encrypted traffic. To decrypt WPA2-encrypted data, the initial WPA authentication between a client and its AP, i.e., the four-way handshake, must be captured. During this process, the drone and controller generate keys for encrypting data across the medium [11]. The four-way handshake is captured during the collection procedure by starting the *airodump-ng* capture before the GCS connects to the drone AP. Alternatively, the *aireplay-ng* tool can be used to initiate a deauthentication attack, which will break the current WiFi session and force a reconnection between the client and the AP such that we can capture the handshake messages.

Figure 5.2 summarizes the data collection procedure. The drone flight begins once the capture command is activated, and the four-way handshake is captured. As the drone executes the waypoint mission, the output of the capture command and the amount of data frames collected is monitored by *airodump-ng*. Once 30 trials from each drone type were collected, we proceeded to the data preprocessing step.

Data Preprocessing

The captured traffic is in *CAP* format, the binary format supported by *libcap* to record captured packets. We leveraged *tshark*, the CLI utility of the *Wireshark* open-source packet analyzer tool, to decrypt, filter, and export trial traces as *CSV* files for further

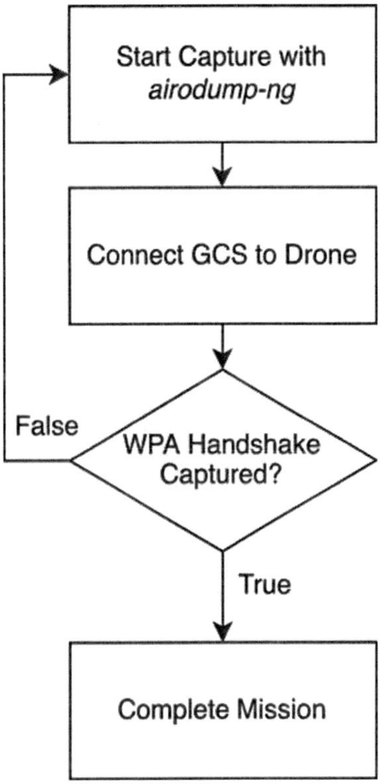

Fig. 5.2 Data collection flow chart

processing in Python. We configured the exported frame columns to display the frame number, inter-arrival timing, data payload in bytes, payload length in bytes, and the source and destination MAC addresses. Most importantly, two *CSV* files were exported for each traffic trace, a plaintext (decrypted) and a ciphertext (encrypted) version. Having both decrypted and encrypted versions of a trial allow us to compare the characteristics between these two.

Both encrypted and decrypted filter rules were applied to remove irrelevant frames, and we ensured that these filter rules were compatible across each drone type. The decrypted trace filter includes IEEE 802.11 data frames and removes retransmitted frames. By examining the first byte of the data payload, i.e., the start-of-text byte, the filter isolates frames containing MAVLink version 1 or 2. Lastly, it specifies the source and destination MAC addresses to be of the drone and ground control station, respectively. In the encrypted trace filter, most rules are identical. However, as it is unable to isolate MAVLink frames, it instead isolates frames carrying a data payload. After processing, we maintained the original *CAP* file, the decrypted *CSV*, and the encrypted *CSV* for each traffic trace. This filtering process is summarized in Fig. 5.3.

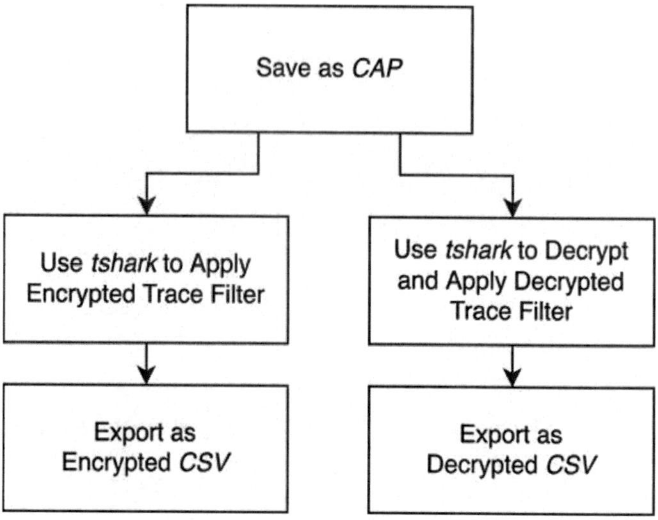

Fig. 5.3 Data preprocessing flow chart

5.3.2 Pattern Identification and Profiling

Analysis of Control Message Patterns

A consumer drone functions as a common control system, i.e., it operates on a feedback loop in which the drone awaits commands from a controller device (e.g., GCS) and responds with status and telemetry information. Specifically, consumer drones periodically relay information to inform the controlling device of its velocity, position, battery health, and other status information. In the preliminary analysis, we compared the amount of data sent from controller-to-drone (C2D) and drone-to-controller (D2C) and discovered that ~98% of the collected traffic is D2C. As such, this study focuses on the MAVLink messages sent as D2C traffic.

We also discovered that each drone vendor decides the type of information prioritized and sent in the control loop; therefore, they configured the control loop of different drone types with distinct message patterns. In the following sections, we discuss how the distinct patterns in these unique message types can be used as features to generate profiles of each drone type.

Proof of Stationarity

We aimed to create features based on the statistical properties of each trace, such as the distribution of MAVLink message sizes. However, we must first verify that the flow of each message type in each trace is a stationary process. Specifically, a stationary process refers to any time series possessing a mean, variance, and autocorrelation that remain constant through time. i.e., it must be free of seasonality and have no trend. The issue of a non-stationary time series is that it is dependent on factors beyond that of controlled message patterns. To mathematically verify the data is stationary, we leveraged a standard statistical test called a *unit root test*. A unit root is a statistical characteristic of a time series that, if possessed, denotes it as non-stationary.

In this study, we leveraged the *Augmented Dickey-Fuller* (*ADF*) test [12], a commonly used unit root test. The ADF test attempts to prove the null hypothesis H_0, i.e., that a process possesses a unit root, which implies non-stationarity. The ADF test forms an autoregressive model that outputs a *p*-value statistic representative of the data. We determined the result of this test by comparing this *p*-value to a specific threshold value referred to as the *critical value*. If the *p*-value is lower than this threshold value, it rejects the null hypothesis, and the system is stationary. Otherwise, if the *p*-value is higher than the threshold value, the null hypothesis has passed, and the system is non-stationary. In this research, we implemented the ADF test in Python using the *adfuller()* function from the *statsmodels* library. The results of these tests showed that the subset of messages of interest to us within each trial in the dataset were stationary, which allowed us to proceed with the analysis [4].

Unique and Non-Varying Message Size Patterns

Previously, we established that the length of the encrypted payload is precisely 44 bytes larger than that of the plaintext payload. This section builds upon this fact to propose the concept of *Unique and Non-Varying (UNV)* message types. These UNV patterns are used as feature characteristics for drone profiling and classification.

The first step is to find all MAVLink message types with *unique* lengths, i.e., no other message type within a given drone class has the same length. However, there is another discrepancy with MAVLink messages in that a single message type may possess varying lengths at different instances, e.g., status text messages. As such, a second qualifier is that unique message types must also be *non-varying*, i.e., of a fixed size. Since the MAVLink message definitions do not explicitly define the size of messages, we experimentally determined which message types are non-varying and separated them, which added a discrepancy of message types that are either inherently fixed length versus only shown to be fixed length. The plaintext data was processed to find such unique and non-varying message types.

Specifically, a message type X of size L is a UNV message type if: (1) no other message type has the data length L, and (2) the message type X is shown to have a fixed length. The main idea behind UNV is that, due to the nature of the WPA2 encryption standard, they are easily distinguishable in plaintext and ciphertext. The patterns of these UNV message types were leveraged to build drone class profiles for the classification of consumer drone traffic, as discussed in the next section. Figure 5.4 depicts the frequency of each UNV message type within a subset of VIPER traffic traces. As shown, the frequency of each UNV message sustains relative uniformity across each trial. We argue that this relative uniformity will hold for each additional trial collected of the corresponding drone class.

Profile Creation

The concept of UNV message types is first used to generate profiles for each drone class. To determine the class of a testing trace, we also extract the UNV profile of the

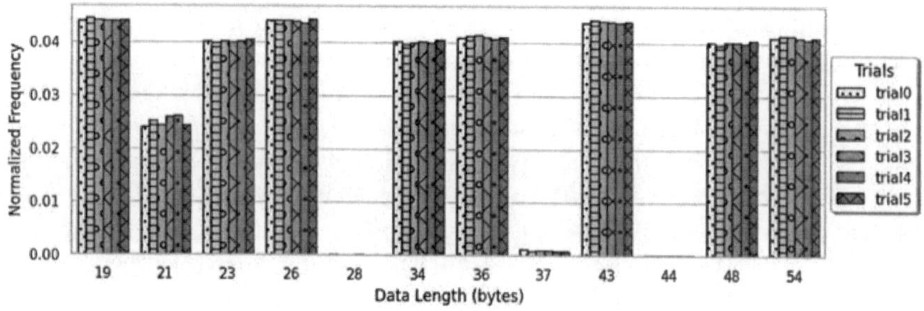

Fig. 5.4 Normalized frequencies of UNV message types for VIPER

trace as its target profile. Then, the target profile is compared against a set of known drone class profiles for identifying the drone. The initial step is to generate a *UNV message list* for each trial in the dataset, which is a list of normalized frequencies for each UNV message size. To create a drone class profile, the UNV lists of the same drone class are merged using an outer join to create an $M \times N$ design matrix, where M rows correspond to the UNV lists from each trial in the training set, and N columns correspond to the UNV message sizes. A similar procedure is used to create a target profile for a target trace. However, since these traces are encrypted, the UNV list must be generated from frame sizes instead of message sizes. This results in a $1 \times N$ vector, where the N column corresponds to the UNV frame sizes. Note that the frame sizes of the target profile are subtracted by 44 bytes when compared to the drone class profile. The main idea of profiling is to create a comparable entity that may represent each drone class or trace. So, we can compare a drone class profile to another drone class profile or a target profile. We built upon the profiling concept to create two different classification methods, as discussed in the following (Fig. 5.5).

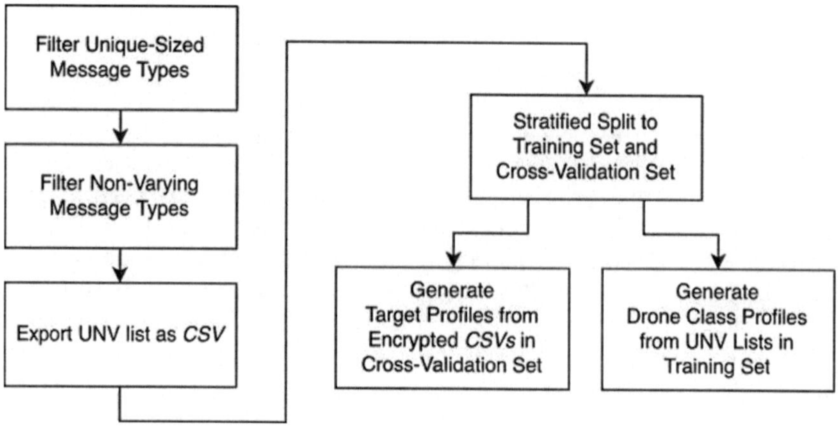

Fig. 5.5 Traffic profiling flow chart

5.3.3 *Drone Classification*

Two methods are proposed to classify encrypted drone traffic, each using the drone class profiles and target profiles built from UNV message patterns. The first method is scoring-based and compares a target profile to each drone class profile by generating similarity scores. It determines the classification based on these similarity scores. The second method is feature-based and uses the drone class profiles to construct feature vectors. Then, it classifies a target trace using standard machine learning algorithms. In each method, the initial step is to split the trials of the dataset into a training set and a cross-validation set. Most importantly, the trials in the training set are combined to create profiles for each drone class. Meanwhile, individual target profiles are created from each trial in the cross-validation set (Fig. 5.6).

Method 1: Scoring-Based Classification Scheme

The first classification method utilizes similarity scoring functions to classify a target trace. A similarity score is computed between a target profile and each drone class

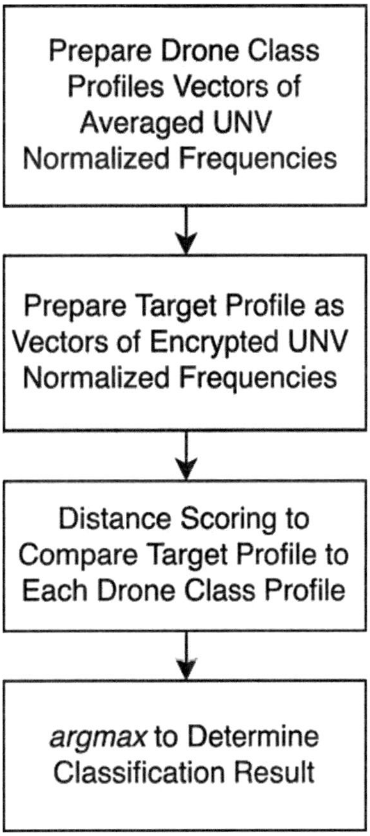

Fig. 5.6 Scoring-based classification flow chart

profile. These profiles are each a real-valued vector, and the comparison between a target profile and each drone class profile is done based on the normalized frequency of each UNV message type. First, the drone class profiles are modified by averaging the normalized frequencies of each trial, producing an N-dimensional vector of mean normalized frequencies for N UNV message types. Next, similarity-scoring functions are used to compare the normalized frequencies of each target profile to the mean normalized frequencies of each drone class profile. The classification is decided based on the *argmax* of the maximum score, i.e., the drone class profile that produced the maximum score with the target profile (Fig. 5.7).

Method 2: Feature-Based Classification Scheme

The second proposed method leverages machine learning (ML) algorithms for classification. First, a *feature template* is created by concatenating the vector of UNV message sizes from each drone class profile and removing duplicates. Each element of the feature template is a distinct UNV message size. The procedure to create a feature vector from a trial in the drone class profile is as follows: for each UNV message size L in the feature template, we obtained the normalized frequency of this size within that trial. In other words, we took the trial's normalized frequency of each size in the feature template to generate the feature vector from that trial. This procedure was repeated to generate a feature vector from each target profile. The feature vectors were used to train machine learning classifiers, which were then evaluated against each target profile. This methodology is further explained in Sect. 5.4.2.

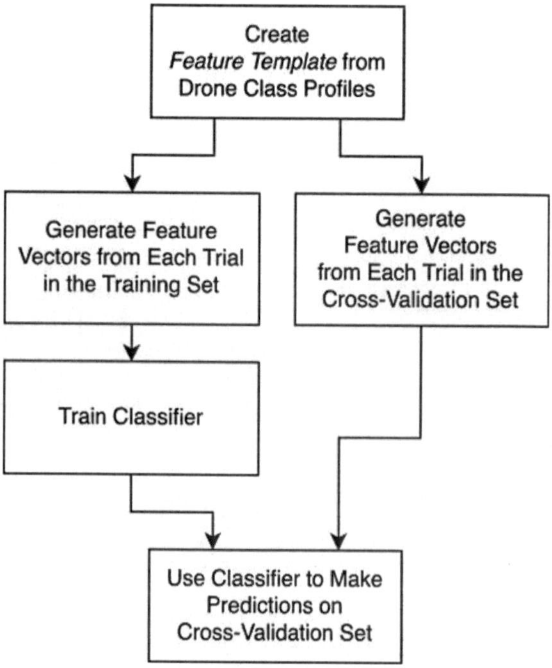

Fig. 5.7 Feature-based classification flow chart

5.4 Performance Evaluation

The previous section defined a procedure for creating profiles from UNV message types and established the foundation of two methods for drone classification. In this section, we walk through the practical implementation of our model, from the preliminary data preprocessing to the evaluation of each classification method. Then, we evaluate the performance of these methods by comparing the results to that obtained from the implementation of a closely related work.

5.4.1 Setup and Preprocessing

A Python script and *tshark* were each used to produce a filtered encrypted *CSV*, a filtered decrypted *CSV*, and a *CSV* of unique and non-varying messages (UNV) from each *CAP* trace in the dataset. In setting up the classifiers, the file paths to these trials are organized into a Python dictionary called the *Dataset Reference*. Each trial has an assigned drone class label and file paths to the respective decrypted *CSV*, encrypted *CSV*, and UNV list *CSV*. This *Dataset Reference* streamlined the later process of sampling and resampling data to create multiple iterations of training and cross-validation datasets. Throughout this section, we use the shorthand term "*component*" to refer to the attributes of each entry within the *Dataset Reference*. For example, the *encrypted component* of a trial refers to the encrypted trace linked to that trial, and the *UNV component* of a trial refers to the UNV message list of that trial.

With the dataset organized and indexable, it can then be sampled and split using stratified K-fold cross-validation. The training dataset is used to generate profiles for each drone class. How to use the drone class profiles and cross-validation dataset depends on the two proposed classification methods. In the following, we present the specific implementation of each classification scheme and analyze their results.

5.4.2 Evaluation of Method 1: Scoring-Based Classification Scheme

This method uses quantitative scoring functions to compare normalized frequencies, or occurrences, of each message type in a trace. Here we argue that the frequency of a UNV message type in a target profile must be similar to the average frequency of the same UNV message type in the matching drone class profile. Each target and drone class profile is structured as a real-valued vector, so we leveraged scoring metrics that are ideal for measuring the distance between two real-valued vectors. This phase compared the performance of the *Euclidean* $(1 - \sqrt{\sum_{i=1}^{n} (x_i - y_i)^2})$, *Manhattan* $(1 - \sum_{i=1}^{n} |(x_i - y_i)|)$, and *Cosine* $(1 - \frac{x \cdot y}{\|x\| * \|y\|})$ distances.

First, the Euclidean distance is a common metric that measures the length of the line segment directly connecting two equal-sized vectors projected in a multi-dimensional space. This is defined as the square root of the sum of squared differences of elements between two equal-sized vectors [13]. Secondly, the Manhattan distance, or City-Block distance, is defined as the sum of the absolute differences between two equal-sized vectors. While the Euclidean distance measures the shortest path between vectors, the Manhattan distance measures a grid-like path and may be preferable when used on high dimensional data [13]. Next, the Cosine distance measures the cosine of the angle between two equal-sized vectors projected in a multi-dimensional space and is defined as the dot product of the two vectors divided by the cross product of each vector's magnitude, which is derived from the formula for the Euclidean dot product [13]. Unlike the Euclidean and Manhattan distances, this metric disregards the vector magnitude, considering only the angular distance, which is beneficial when the size of the drone class profile largely differs from that of the target profile.

The setup begins with stratified eight-fold sampling for cross-validation. This number of folds was determined heuristically. Stratified sampling ensured a balanced representation of each UAV subset within the cross-validation set. For each fold, new drone class profiles are generated from the training set, and a new target profile is generated from each trial in the cross-validation set.

To create a profile for a drone class K, the UNV component of each trial i_K within the training set is merged using an outer join on UNV message types, preserving the message-type norm-frequency values of each trial. Then, the average frequencies for each UNV message type are computed, resulting in a list of message-type norm-frequency pairs. Next, each target profile is obtained by generating the list of UNV message types from the encrypted component of each trial in the cross-validation set.

With the drone class profiles and the target profile established, we continued to the classification step. Each scoring function was used to compute the score of the target profile against each drone class profile, and the result of the classification corresponds to the drone class profile that returns the highest score. Here, we frequencies of two size vectors of UNV message types are compared. This step was repeated for each trial in the cross-validation set. The F1-Score results are shown in Table 5.1. Confusion matrices are provided as an accompanying visual in [4]. We observed that the Cosine distance function performed the best, with a 98.66% macro average F1-Score. This is as expected since, unlike the Euclidean and Manhattan distances, the Cosine distance disregards magnitude and is beneficial when the magnitude of the drone class profile differs from the target profile. In the next section, we discuss a robust method using machine learning algorithms.

Table 5.1 F1-Score results of scoring-based classification

Scoring function	F1-Score (%)			Macro avg. F1-Score (%)
	INTEL	SOLO	VIPER	
Euclidean	88.89	88.89	97.67	91.55
Manhattan	100.00	96.55	95.45	97.18
Cosine	100.00	98.31	97.67	98.66

5.4.3 Evaluation of Method 2: Feature-Based Classification Scheme

In this second method, traditional machine learning techniques were used to classify frequency vectors of UNV message types. As with the first method, the drone class profiles were created by merging the UNV component of each trial into the training set. However, in this method, the UNV message sizes of each drone class profile were then concatenated into a single vector to form the *feature template* of length n, denoted as T. Duplicate message sizes were removed. Each element j of T, where $j \in \{1 \ldots n\}$, is a distinct size of a UNV message type. Next, the *feature function* $V(\bullet)$ is defined, which returns the normalized frequency of some input message size. To generate a feature vector $F^{(i)}$ from a trial i in the training dataset, the UNV component of that trial is used. For each element j in the feature template T, we computed $V(T_j^{(i)})$, i.e., we found the normalized frequency of each UNV message type size in the feature template that also appears in trial i. This process is summarized in Eq. 5.1. Note that if $T_j^{(i)}$ was not found in trial i, then $V(T_j^{(i)}) = 0$.

$$F^{(i)} = \left[V\left(T_1^{(i)}\right), V\left(T_2^{(i)}\right), \ldots, V\left(T_n^{(i)}\right)\right] \tag{5.1}$$

The following example is used to further illustrate the generation of the feature vector. First, the drone class profiles are initialized, each composed of UNV message types: $Y_{VIPER} = \{19, 26, 43\}$, $Y_{SOLO} = \{36, 54, 48\}$, $Y_{INTEL} = \{21, 23, 34\}$. The target profile is given as, $X_{target} = \{19, 26, 43, 23, 34\}$. The next step is to initialize the feature template T, which was done by concatenating each drone class profile and removing duplicate elements: $T = \{19, 26, 43, 36, 54, 48, 21, 23, 34\}$. Next, Eq. 5.1 was used to fit the target profile onto the feature template and compute the normalized frequencies of each feature template element inherent to the target trial, as shown below:

$$\begin{aligned} F^{(\mathrm{target})} &= \{V(19), V(26), V(43), V(36), V(54), V(48), V(21), V(23), V(34)\} \\ &= \{0.35, 0.35, 0.19, 0.00, 0.00, 0.00, 0.00, 0.02, 0.03\}. \end{aligned}$$

Classification leveraged the *Multinomial Logistic Regression* and *Support Vector Machine (SVM)* algorithms. As this research focuses on the feature engineering methodology, these were configured using the classifier functions provided in the *scikit-learn* Python library [14].

First, the Logistic Regression algorithm is a frequently used classification method that fits a logistic function, i.e., the *sigmoid function*, onto each trial in the dataset and maps these probabilities for binary classification [15]. Predictions are made based on where the probability score of a target trial lies on the sigmoid function—if it exceeds the classification threshold value. This method extends to the multinomial case by leveraging the one-vs-rest (OVR) modeling scheme. Here, for k classes, a series of k binary classifiers are created for each class C_k versus not in C_k. A binary classifier is fitted per class independently, and the greatest output score determines the result. To fit the logistic function to the dataset, *solver functions* are leveraged, which optimize the parameters, i.e., *weights*, to obtain the least error in the predicted output. The performances of three standard solver function methods are evaluated: *Library for Large Linear Classification (liblinear)* [16], *Limited-memory Broyden–Fletcher–Goldfarb–Shanno (LBFGS)* [17], *and Stochastic Average Gradient Descent Accelerated (SAGA)* [18]. Each solver function is paired with the *L2*-penalty function.

Another algorithm used is the Support Vector Machine (SVM). For n features, SVM aims to identify an optimal hyperplane in n-dimensional space that best partitions and separates trials by class. SVM is inherently a binary classifier. For a nonlinear feature space, a common issue is that the data is linearly inseparable by a hyperplane. In such a case, the naive approach would be to skew the feature space, converting an n-dimensional feature space to an m-dimensional feature space, where $m \gg n$. However, this would be computationally costly. In such a case, a kernel function is required, which takes the inner product of the feature space, effectively transforming the data to a higher dimensional space without the overhead (otherwise, known as *kernel substitution.*) The performances of three distinct kernel functions were analyzed with each SVM classifier configuration: the linear function, radial basis function (RBF), and sigmoid function.

The setup process for this method is identical to that in the previous section. The dataset was randomized and split into eight folds, with stratification. Then, drone class profiles were created from the training set by grouping and merging the UNV component of each UAV subset. From here, the process deviates from the first method. Next, the *feature template* was created by concatenating each drone class profile and removing duplicate UNV message sizes. Equation 5.1 was used to generate feature vectors from each trial in the training set, then from the cross-validation set. Finally, the classifier configurations are trained and F1-Scores are computed for each drone class.

Table 5.2 shows the F1-Scores results for each configuration. These results show that this feature methodology did not perform as well in classifying the INTEL data. However, the F1-Score for SOLO is consistently high across each configuration, which may allude to the SOLO drone class profile containing more unique sizes of UNV message types that distinguish it from the INTEL and VIPER traces.

Table 5.2 F1-Score results of feature-based classification

Algorithm	Configuration	One-versus-Rest F1-Score (%)			Macro avg. F1-Score (%)
		INTEL	SOLO	VIPER	
Logistic regression	LIBLINEAR	74.14	100.00	100.00	91.14
	LBFGS	74.14	100.00	100.00	91.14
	SAGA	74.14	100.00	100.00	91.14
Support vector machine	Linear	71.28	100.00	98.21	89.83
	RBF	100.00	100.00	76.55	92.18
	Sigmoid	67.56	100.00	76.55	81.37

5.4.4 Comparison to Related Work

The classification model presented in [6] was reimplemented to compare its efficiency to our method when tested on the same dataset. The process began with preparing the dataset to match their criteria. Twelve statistical functions were computed on both the set of frame sizes and inter-arrival timings from each encrypted trial. These statistical results were concatenated into feature vectors. The original model used a custom OVR logistic regression classifier that penalized the computation time of each statistic. As the focus of this comparison is on the feature engineering technique, the same set of classifier configurations in Method 2 was used instead, along with the eight-fold cross-validation process.

The F1-Score results for the related work implementation are displayed in Table 5.3. The proposed Method 2 obtained equal or higher F1-Score for SOLO and VIPER classification across all configurations. Meanwhile, the related work implementation outperformed Method 2 in INTEL classification when Logistic Regression was used with LIBLINEAR or LBFGS, which may allude to an overlap of UNV message types across drone classes. This is also true when SVM is used with the Linear or RBF kernel. The mean AUC results are provided in [4] as an alternate statistic, and those scores correlate well with these F1-Score results. In comparing the macro average F1-Score of each configuration, our Method 2 outperformed the related work implementation in 5 of 6 classifier configurations.

5.5 Conclusions and Future Work

In this paper, we discovered the characteristics of AES-CCMP encryption that allowed us to infer the size of an encrypted MAVLink frame based on the size of a WiFi WPA2 frame. From here, we introduced the concept of Unique and Non-Varying (UNV) message types and used these UNV patterns to build distinct profiles

Table 5.3 F1-Score results of ML algorithms in related work implementation

Algorithm	Configuration	One-versus-Rest F1-Score (%)			Macro avg. F1-Score (%)
		INTEL	SOLO	VIPER	
Logistic regression	LIBLINEAR	74.31	95.80	83.44	84.52
	LBFGS	81.67	97.38	85.32	88.12
	SAGA	59.70	72.91	55.89	62.83
Support vector machine	Linear	85.45	95.83	92.88	91.39
	RBF	76.92	93.81	65.03	78.59
	Sigmoid	52.44	64.84	49.98	55.75

for each drone type. We proposed two profile-based classification methods that leveraged scoring functions (Method 1) and ML algorithms (Method 2) to classify drones via encrypted traffic. The experimental results showed that Method 1 obtained macro average F1-Scores of 91.55–98.66%, while Method 2 obtained macro average F1-Scores of 81.37–92.18%. Furthermore, when compared to a related work that leverages statistical features strictly from encrypted size and timing, Method 2 obtained a greater macro average F1-Score in 5 of 6 classifier configurations.

The methods discussed in this paper possess certain limitations, such as the data preprocessing overhead to extract UNV patterns and construct profiles for each drone class. A possible countermeasure to disrupt UNV patterns would be by enforcing uniformity in message sizes, which can be done by padding existing messages or modifying the message definition set. Given these limitations, this approach holds the advantage of having an insightful feature selection process. That is, it contrasts related works that either build features from encrypted traffic in which the data may be misrepresented or use deep learning algorithms that require greater computation power with a classification paradigm that is neither clear nor insightful.

The following aspects can be further developed in future work. To improve the classification process, a voting ensemble can be implemented, in which the results of multiple classifiers are combined, and prediction is based on the majority vote. This method can work alongside the proposed feature selection process to decrease variance and improve cross-validation accuracy. To improve the speed of classification, the number of frames in each sample trace can be variated to determine the minimum amount required for reliable UNV-based classification. This may also establish the foundation for a real-time classification framework. Next, we also can build upon UNV patterns by discovering other distinct patterns and leveraging them to build drone class profiles. An example is to observe any repeated sequences of UNV message types, i.e., k-adjacent frames. Lastly, while this work focused on the classification of consumer drones, the proposed methods can be generalized to other applications, particularly any device that communicates over WiFi and sends control message frames, e.g., autonomous systems and IoT devices.

References

1. ArduPilot Documentation (2016). https://ardupilot.org/ardupilot
2. Koubaa, A., Allouch, A., Alailan, M., Javed, Y., Belghith, A., Khalgui, M.: Micro air vehicle link (MAVLink) in a nutshell: a survey. IEEE Access **7**, 87658–87680 (2019)
3. How CCMP is used in RSN (2008). http://etutorials.org/networking
4. Liang, D.: Identifying consumer drones via encrypted traffic. M.S. thesis, University of Hawaii at Manoa (2022).
5. Ardupilot-Solo (2013). https://github.com/3drobotics/ardupilot-solo
6. Alipour-Fanid, A., Dabaghchian, M., Wang, N., Wang, P., Zhao, L., Zeng, K.: Machine learning-based delay-aware UAV detection and operation mode identification over encrypted WiFi traffic. IEEE Trans. Info. Forensic Security **15**, 2346–2360 (2020)
7. Bisio, I., Garibotto, C., Lavagetto, F., Levorato, M., Sciarrone, A.: Statistical analysis of wireless traffic: an adversarial approach to drone surveillance. In: 2019 IEEE Global Communications Conference (GLOBECOM), pp. 1–6 (2019)
8. Rezaei, S., Liu, X.: Deep learning for encrypted traffic classification: an overview. IEEE Commun. Mag. **57**(5), 76–81 (2019)
9. DroneKit-Python API Guide (2015). https://dronekit-python.readthedocs.io/en/latest/guide
10. Aircrack-ng (2009). http://aircrack-ng.org
11. Singh, R. R., Moreira, J., Chothia, T., Ryan, M. D.: Modelling of 802.11 4-way handshake attacks and analysis of security properties. In: Markantonakis, K., Petrocchi, M. (eds) Security and Trust Management. STM 2020. Lecture Notes in Computer Science, p. 12386 (2020)
12. Dickey, D., Fuller, W.: Distribution of the estimators for autoregressive time series with a unit root. J. Am. Stat. Assoc. **74**, 366–427 (1979).
13. Vijaymeena, M.K., Kavitha, K.: A survey on similarity measures in text mining. Mach. Learn. Appl. Int. J. **3**(1), 22 (2016)
14. Pedregosa, F.: Scikit-learn: machine learning in Python. JMLR **12**, 2825–2830 (2011)
15. Hosmer Jr., D. W., Lemeshow, S., Sturdivant: R. X.: Applied Logistic Regression. John Wiley & Sons, Inc., Hoboken, New Jersey (2013).
16. Fan, R.-E., Chang, K.-W., Hsieh, C.-J., Wang, X.-R., Lin, C.-J.: LIBLINEAR: a library for large linear classification. J. Mach. Learn. Res. **9**, 1871–1874 (2008)
17. Morales, J. L., Nocedal, J.: L-BFGS-B: Remark on Algorithm 778: L-BFGS-B, FORTRAN routines for large scale bound constrained optimization. ACM Trans. Math. Softw. **38**(1), 1–4 (2011)
18. Defazio, A., Bach, F., Lacoste-Julien, S.: SAGA: a fast incremental gradient method with support for non-strongly convex composite objectives. Adv. Neural Inform. Process. Syst. **27**, 1646–1654 (2014).

Chapter 6
Development of a Simulator for Operator Proficiency Training for Seafloor Exploration by Remotely Operated Vehicle

Ryusei Kamewari, Yusuke Fujishima, Kuniaki Kawabata, Kenta Suzuki, Norimitsu Sakagami, Fumiaki Takemura, and Satoru Takahashi

Abstract This paper describes a new ocean simulator developed to train operators to operate ROV-based seafloor exploration. This simulator has the function to simulate time-varying ocean currents in addition to calculating hydrodynamic effects, which we consider important for the simulation of underwater ROV operations. It can also provide more realistic operational training by applying arbitrary ROV models and

---

R. Kamewari (✉)
Graduate School of Science for Creative Emergence, Kagawa University, 2217-20, Hayashi-Cho, Takamatsu 761-0396, Kagawa, Japan
e-mail: s22g457@kagawa-u.ac.jp

Y. Fujishima · S. Takahashi
Area of Mechanical Systems Engineering, Faculty of Engineering and Design, Kagawa University, 2217-20, Hayashi-Cho, Takamatsu 761-0396, Kagawa, Japan
e-mail: s19t543@kagawa-u.ac.jp

S. Takahashi
e-mail: takahashi.satoru@kagawa-u.ac.jp

K. Kawabata · K. Suzuki
Sector of Fukushima Research and Development, Japan Atomic Energy Agency, 1-22 Nakamaru, Yamadaoka, Naraha, Futaba 979-0513, Fukushima, Japan
e-mail: kawabata.kuniaki@jaea.go.jp

K. Suzuki
e-mail: suzuki.kenta38@jaea.go.jp

N. Sakagami
Mechanical Engineering and Robotics Course, Faculty of Advanced Science and Technology, Ryukoku University, 1-5 Yokotani, Seta, Oe-Cho, Otsu 520-2194, Shiga, Japan
e-mail: sakagami@rins.ryukoku.ac.jp

F. Takemura
Department of Mechanical Systems Engineering, National Institute of Technology, Okinawa College, 905, Henoko, Nago 905-2192, Okinawa, Japan
e-mail: takemura@okinawa-ct.ac.jp

D. Azimov (ed.), *Proceedings of the IUTAM Symposium on Optimal Guidance and Control for Autonomous Systems 2023*, IUTAM Bookseries 40,
https://doi.org/10.1007/978-3-031-39303-7_6

seafloor data as the experimental operating environment. In this paper, we apply 3D seafloor data created by our research group based on seafloor surface images and an operation model created based on a real ROV. Then, an operation training experiment was conducted in a simulated ocean current using this simulator. As a result, we confirmed an increase in the operator's operation accuracy and demonstrated the effectiveness of the operation training on the simulator.

Keywords Underwater robot · Remotely operated vehicle · Ocean simulator · Ocean currents · Operator proficiency training

6.1 Introduction

It is a well-known fact that mineral and energy resources and ancient ruins are abundantly deposited on the seafloor. In addition, coral reefs and corals, which are home to one-fourth to one-third of all marine life, are important marine organisms from the perspective of seafloor resource conservation. Currently, there is a worldwide need for efficient marine investigations of these resources and habitats [1, 2]. Generally, marine investigations are conducted either by divers diving to observe or by pulling divers by boat or other means. However, diving investigations by divers have problems such as limited diving time and physical burden. Therefore, in recent years, Remotely Operated Vehicle (ROV) and Autonomous Underwater Vehicle (AUV) underwater robots are used for wide-area oceanographic investigations to solve the human problem [3]. The Sakagami Laboratory at Tokai University is developing an underwater robot that can replace divers for dexterous tasks [4]. In order to operate these underwater robots, it is necessary to devote time to prior training on how to operate them and to understand their structure. This is because the operation of an underwater robot must be based on camera images attached to its body. In addition, it is necessary to consider the effects of fluid resistance and ocean currents underwater. However, there are few facilities where operation training can be conducted, and conducting such training in the actual ocean involves the risk of unexpected accidents.

Our purpose is to construct an ocean simulator that effectively utilizes actual seafloor observation data and allows users to experience the operation of an underwater robot. Simulation of ROV operations underwater requires dynamics calculations, hydrodynamic calculations, and sensor simulations such as cameras. Therefore, as simulator software, we used Choreonoid [5], an integrated software for robot development, and its plug-in FluidDynamicsPlugin [6–8] to meet these requirements. However, this is not enough to express the ocean currents, which also have the most impact on ROV operations. Therefore, we will add a new function to express ocean currents. This will provide a more realistic operating experience. Our group has developed a towed ROV shown in Fig. 6.1 with a multi-camera unit consisting of five GoPros. We then created a 3D map of the seafloor from the images taken during the experiment on Sesoko Island, Okinawa (see Fig. 6.2) [9]. The underwater robot

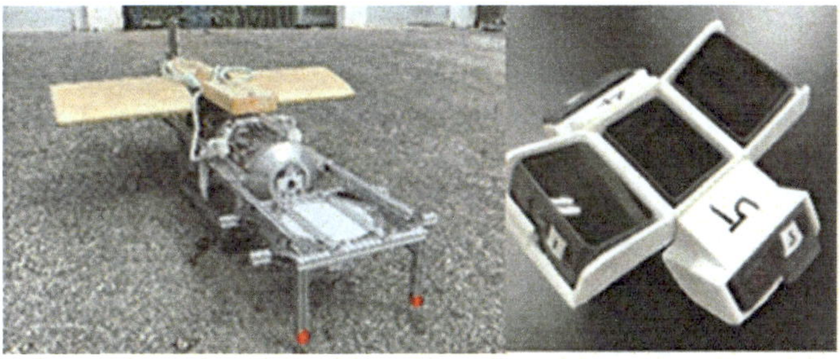

Fig. 6.1 Towed ROV with a multi-camera unit

Fig. 6.2 Experiment on Sesoko Island, Okinawa

model of the simulator is used the ROV model based on DiveUnit 300 of FullDepth Corporation. In our experiments, we combine these elements and conduct operation experiments in a more realistic environment.

6.2 Development of a Simulator

6.2.1 Components of the Simulator

This section describes the components of the simulator developed. Three major elements are considered important to simulate ROV operations in the real ocean (see Fig. 6.3). In addition to the physical calculations necessary for a simulator that

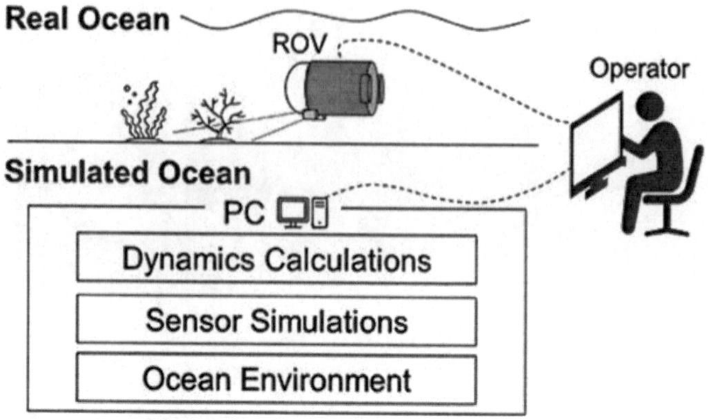

Fig. 6.3 Simulator components

imitates the real world, sensor simulation is also important because ROV operations are performed by viewing images from a camera mounted on the ROV. These can be represented by the software and its plug-ins described in Sect. 6.2.2. However, this is not enough to represent the ocean environment. The ocean environment here refers to ocean currents. Therefore, we have introduced a new function to generate simulated ocean currents. This function for expressing ocean currents is described in Sect. 6.2.3.

6.2.2 The Base Software of the Simulator

This research proposes a simulator that can reproduce complex marine environment on a simulator and can be applied to training for the operation of underwater robots for seafloor investigation. We treat Choreonoid which is an integrated software for robots, as the base of the simulator software. The introduction of "FluidDynamicsPlugin" makes it possible to reproduce the underwater behavior of ROV. FluidDynamicsPlugin is an extension of Choreonoid developed by the Japan Atomic Energy Agency (JAEA) to simulate the decommissioning of the Fukushima Daiichi Nuclear Power Station using remotely operated robot. This Plugin adds to Choreonoid the ability to create fluid space that simulate the buoyancy, fluid resistance, the behavior of ROV and unmanned aerial vehicles. Figure 6.4 shows the fluid effect acting on the robot model in the fluid space, and Fig. 6.5 shows the fluid space created on the simulator. Fluid space calculates the following fluid effects on the ROV.

- Buoyancy force [N] F_b
- Fluid resistance (due to pressure) [N] F_d
- Fluid resistance (due to fluid friction) [N] F_v
- Force applied by steady flow [N] F_s

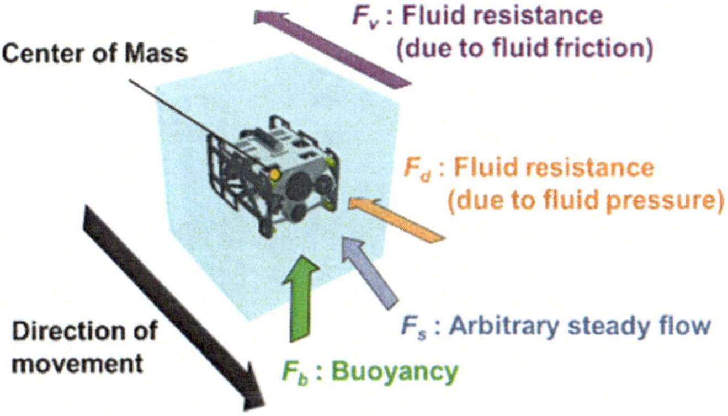

Fig. 6.4 Various parameters in fluid space

Fig. 6.5 Model in fluid space

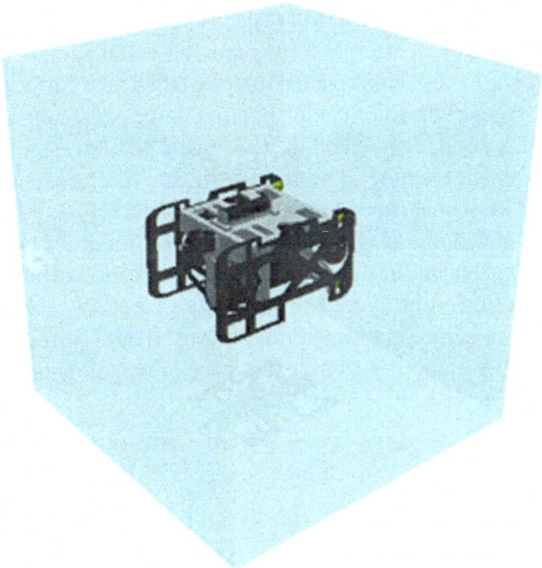

6.2.3 *Function to Express Ocean Current*

In oceanographic investigations using underwater robots, it is important to consider the marine environment at the time of the investigation. If the sea is rough, it is naturally difficult to proceed with the investigation, and the investigation may have to be suspended due to the marine environment in consideration of accidents. In addition, no matter what the marine environment is like, the effects of ocean currents on robot operations cannot be ignored. Therefore, as an example of a complex marine environment, the simulator reproduces underwater currents caused by ocean currents.

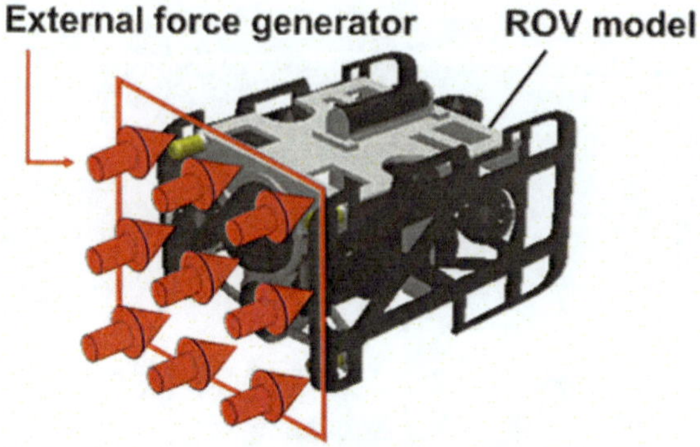

Fig. 6.6 An example of an external force generator

In order to reproduce underwater currents caused by ocean currents, an "External force generator" with a thruster function is installed around the ROV model. An example of an external force generator installation is shown in Fig. 6.6.

The external force generator is installed nine per surface and are installed on all six surfaces of the ROV model. External force generators can exert a force perpendicular to the surface on which they are installed, and each generator represents an ocean current by exerting an appropriate force on each generator. Examples of ocean current and ROV behavior simulated using an external force generator are shown in Fig. 6.7. Simulated ocean current is generated for the ROV model placed at the initial point on the left side of the figure in the left direction of the figure. Then, as shown on the right side of the figure, the behavior of the ROV being swept by the simulated ocean current can be confirmed. In this way, a new current expression function is added, which is important to the experience of operating the ROV underwater.

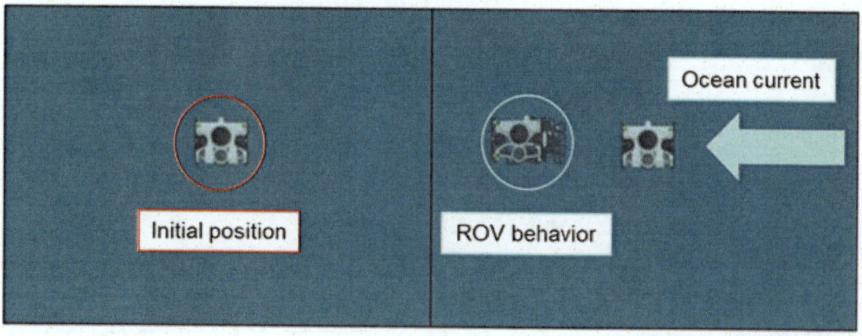

Fig. 6.7 ROV behavior

Fig. 6.8 Appearance model

6.3 Application Data

This chapter describes the underwater robot model and 3D seafloor data used in the construction of the simulator. The screen configuration of the simulator is also described.

6.3.1 Application of ROV Model

We first describe the ROV model applied in this experiment. In recent years, ROV is used for wide-area oceanographic investigations for a variety of purposes and applications [10, 11]. Therefore, in this research, the ROV model is designed based on DiveUnit 300 developed and marketed by FullDepth Corporation. The two models are an appearance model and a physical model. The appearance model is, as the name implies, the model that is projected on the simulator. The physical model is the model used for dynamics calculations in the simulator. The constructed ROV model is shown in Figs. 6.8 and 6.9. The ROV model is equipped with a downward-facing camera and Range Sensor in addition to the front-facing camera and is configured to simultaneously acquire images and distance to the seafloor. Attitude angle control using three vertical thrusters is configured in the ROV model. This reduces the effects of fluid resistance and buoyancy, enabling stable maneuvering and data acquisition.

The constructed ROV model is operated using the controller. The ROV model is operated with the left and right Joystick, as shown in Fig. 6.10.

6.3.2 Application of 3D Seafloor Data

We describe the applied seafloor data. Actual 3D seafloor data is applied to the simulator in order to bring the operation training performed on the simulator close to

Fig. 6.9 Physical model

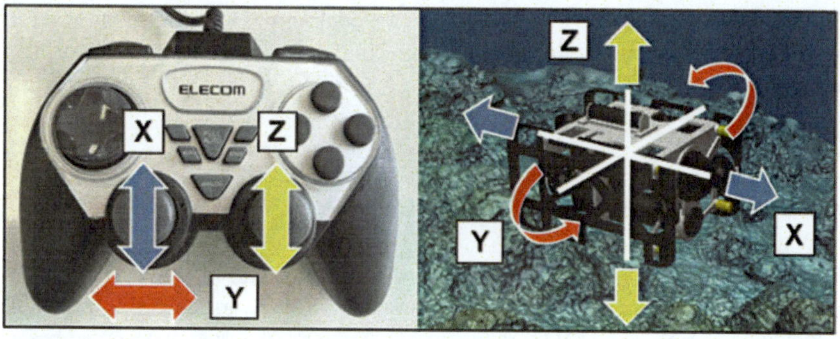

Fig. 6.10 Operation of ROV model

the actual marine environment. Our research group conducted ocean experiments on Sesoko Island, Okinawa in March and September 2019. At that time, the 3D seafloor data created is applied to the bottom of the simulator to represent the seafloor surface. The simulator screen with the 3D seafloor data applied is shown in Fig. 6.11.

6.3.3 Screen Configuration During Operation Experiment

An example of the screen configuration of the operation experiment is shown in Fig. 6.12. The operator operates the ROV model by looking at the camera image equipped on the front of the ROV model in the center of the screen. In addition to the camera image on the ROV, it can also display a bird's-eye view of the entire simulation. This enables the operator to confirm the movement of the ROV model during the operation experiment. In this experiment, we will not use bird's-eye views

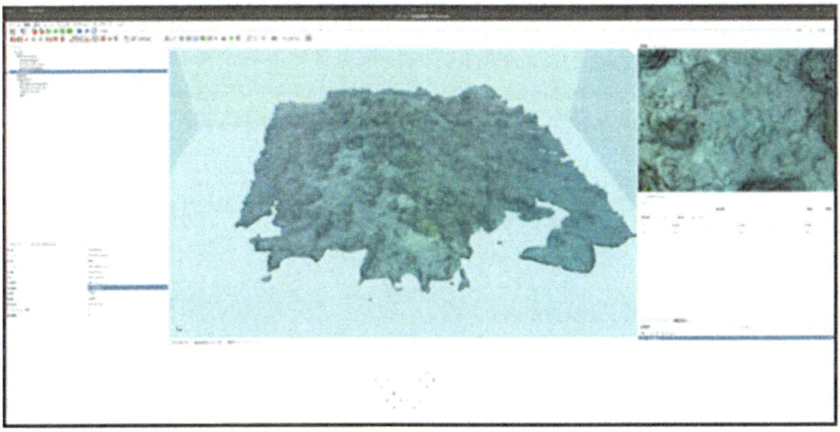

Fig. 6.11. 3D seafloor data in Choreonoid

because we will operate in the same environment as in the reality. In this way, screen configurations can be freely designed to suit the purpose of use.

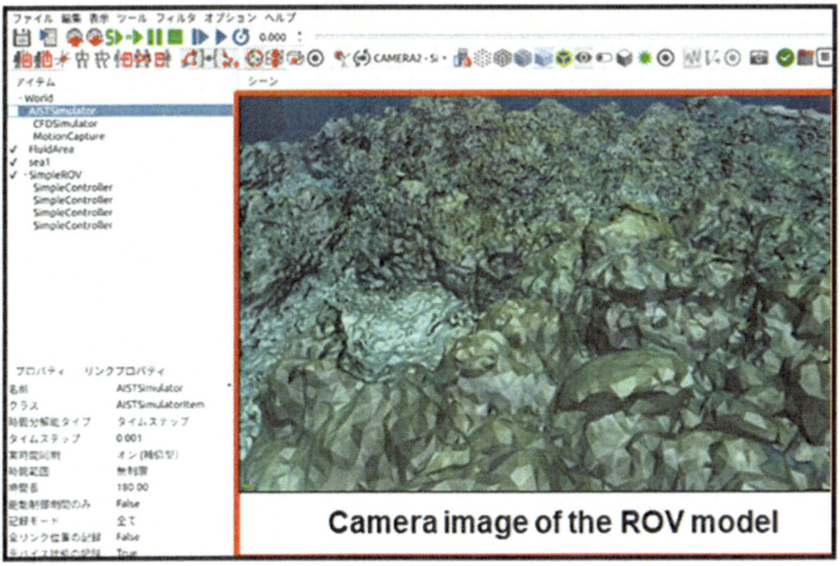

Fig. 6.12 Screen configuration

6.4 Experiments

This chapter describes the experiments using this simulator. Two experiments were conducted as examples of ROV operation experience using the ocean simulator.

6.4.1 *Experiment 1*

6.4.1.1 Condition of Experiment 1

In the experiment 1, the operator operates the ROV model in a simulated ocean current and aims to make the ROV model go as straight as possible. From the start of the simulation, a 0.1 N force is applied to the ROV model from an external force generator. The direction of force application is changed every five seconds in the positive and negative directions of the simulator's Y-axis. The operator's operations are recorded a total of three times, and the operation trajectories are confirmed and compared. The operating trajectory displayed on the simulator is shown in Fig. 6.13. The purpose of this experiment is to verify whether repeated simple operations in simulated ocean currents change the accuracy of the operator's operation of the ROV model.

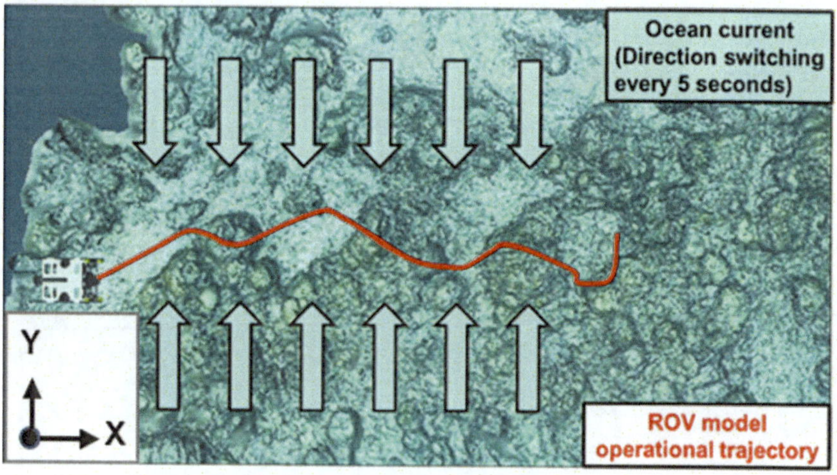

Fig. 6.13 Operational trajectory of experiment 1 (Trajectory marked with red line)

6.4.1.2 Result of Experiment 1

In Experiment 1, each of the two examiners (A, B) performed a total of three operations in which the ROV was moved straight ahead in a simulated ocean current. The graphs summarizing the trajectories for each examiner when the initial position of the ROV model is set to (X, Y) = (0.0, 0.0) are shown in Figs. 6.14 and 6.15. The average values of displacement in the Y-axis direction, the direction affected by the simulated ocean currents, are shown in Table 6.1. The first-time trajectory shows that both examiners A and B are greatly affected by the ocean currents. However, the second and third operational trajectories show that they are gradually approaching a straight line.

The results show that the operation proficiency of the ROV in simulated ocean currents can be improved by repeated operation of the ROV. Also, throughout the whole process, it can be seen that examiner A corrects its trajectory by many operations. In contrast, it can be seen that examiner B is trying to correct the trajectory with

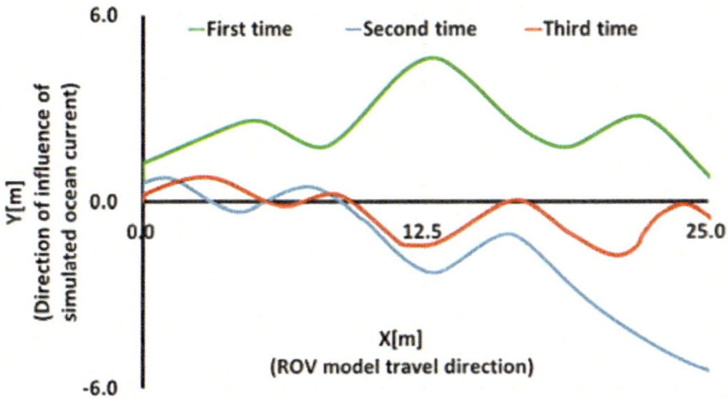

Fig. 6.14 Operational trajectory of ROV model (examiner A)

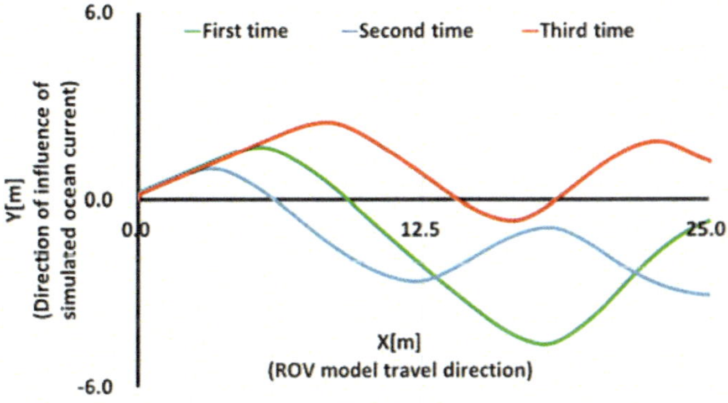

Fig. 6.15 Operational trajectory of ROV model (examiner B)

Table 6.1 Average value of displacement in Y-axis direction

	Examiner A	Examiner B
First time	2.2686	1.9727
Second time	1.8762	1.5023
Third time	0.6482	1.1756

fewer operations and is trying to approach a straight-line movement. The average values of displacement in the Y-axis direction shown in Table 6.1 also indicate that both examiners A and B were able to approach a straight trajectory with each iteration. Furthermore, examiner A, whose trajectory was greatly displaced in the first and second times, was able to operate on a trajectory with less displacement than examiner B in the third time. From these results, it can be analyzed that the fine operation of the ROV is suitable for traveling in the ocean current affected area. In this way, operation training using the simulator allows the operator's habits and effective operation methods to be analyzed, since the operation trajectory can be verified.

6.4.2 *Experiment 2*

6.4.2.1 Condition of Experiment 2

In the second experiment, the operator operates the ROV model to drive across the seafloor, passing near checkpoints set up on the seafloor. The target course of travel and the coordinates of the objects placed on the seafloor as installed checkpoints are shown in Fig. 6.16. As in experiment 1, from the start of the simulation, a 0.1 N force is applied to the ROV model from an external force generator. The direction of force application is always in the negative direction of the simulator's Y-axis.

The purpose of this experiment is to train the ROV to run near the seafloor and approach the research object, assuming an actual oceanographic investigation. The following sections describe the results of the respective experiments.

6.4.2.2 Result of Experiment 2

In Experiment 2, one examiner operated with the goal of running through the vicinity of an object placed on the seafloor. The simulator operation screen for passing near the object is shown in Fig. 6.17. In the experiment, a simulated ocean current in one direction is constantly generated. As opposed to Experiment 1, the ROV model must be operated close to the complex seafloor terrain. Therefore, there were many cases during the operation experiment where the ROV model collided with the seafloor surface along the way. Therefore, in this experiment, we obtained only the operational trajectory of the ROV model that ran through without colliding with the seafloor or any objects. And then, we compared the operation trajectories before and after

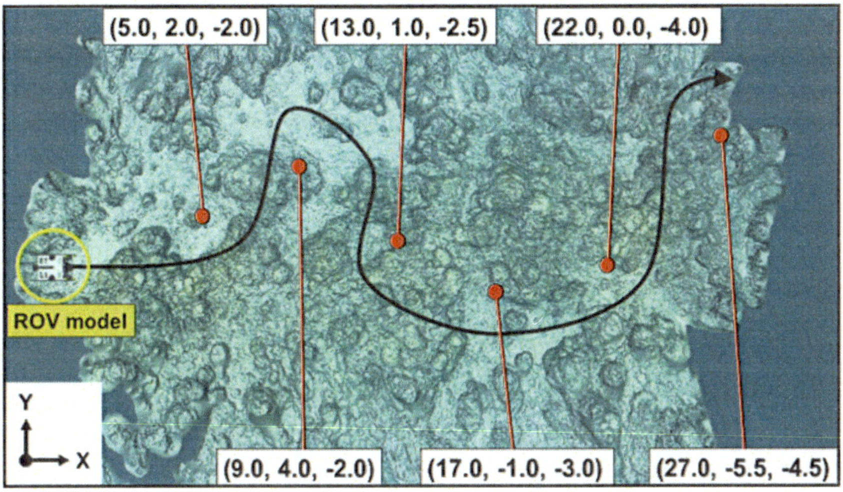

Fig. 6.16 The target course of travel and the coordinates of the objects of experiment 2

repeated operation practice. The operational trajectories are shown in Figs. 6.18 and 6.19. The operation trajectory can also be confirmed from any direction as shown in Fig. 6.20. Before the practice operation, the ROV model was too close to the seafloor surface and was being pushed away by the ocean currents. After the operation practice, the operator was able to grasp the proper distance from the seafloor and the ROV model was able to run smoothly.

The time taken for the run was also reduced from 77 s before the practice to 65 s after the operation practice. Thus, the improvement in proficiency

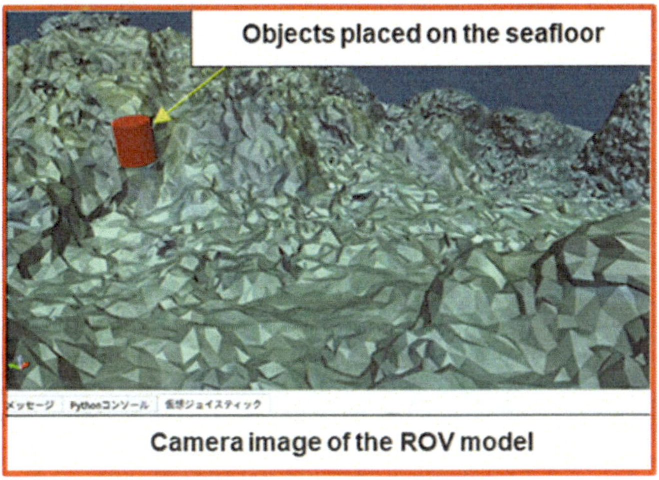

Fig. 6.17 Simulator operation screen when passing near the objects

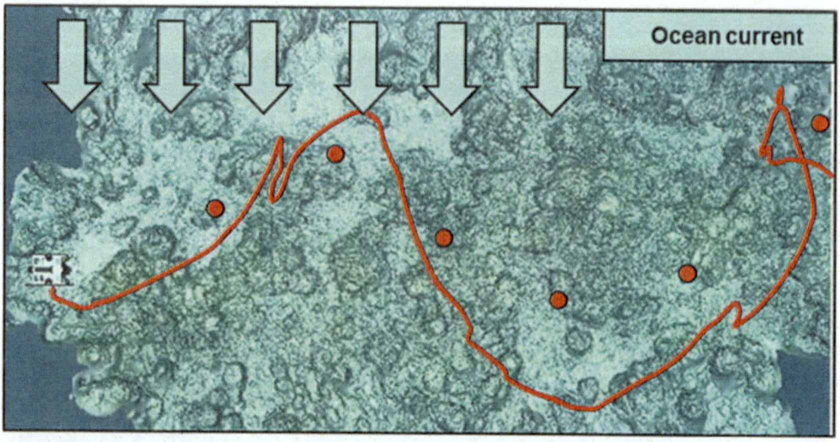

Fig. 6.18 Operational trajectory before the practice (trajectory marked with red line)

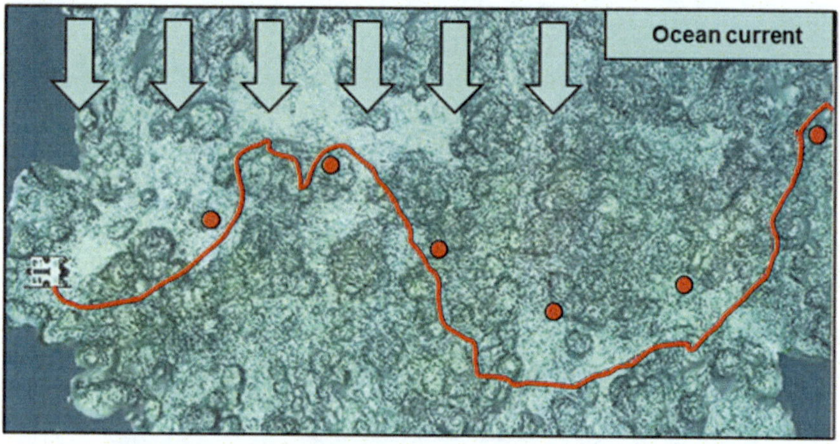

Fig. 6.19 Operational trajectory after the practice (trajectory marked with red line)

was confirmed by practicing operations on this simulator, which reproduces the underwater environment.

The simulator we constructed this time applies a 3D map created based on actual seafloor data. In the future, as new seafloor data is applied, more variations in operational training will be possible.

As for the expression of ocean currents, only one direction of ocean current was generated in this experiment. Therefore, in the future, we are considering controlling the direction and force of the simulated ocean currents according to the position of the ROV traveling along the seafloor.

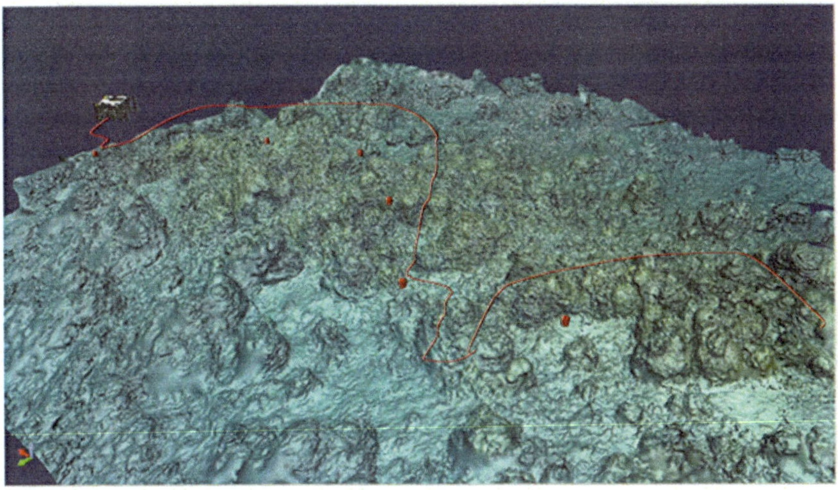

Fig. 6.20. 3D confirmation of operational trajectory

6.5 Conclusion

This paper described the development of an ocean simulator based on Choreonoid for training of underwater robots for seafloor exploration. Choreonoid is a dynamics simulator with a physics engine. It can also simulate sensors such as cameras and distance sensors, and is highly extensible. At this time, the Choreonoid extension "FluidDynamicsPlugin" was introduced to reproduce the behavior of objects underwater. The ROV model was designed based on the ROV used in actual oceanographic research. For the seafloor, seafloor data acquired from the Sesoko Island oseanographic research experiment was applied to the ocean simulator. Thus, we reproduced more realistic oceanographic environment on the simulator by using the ROV model and the actual seafloor data.

In oceanographic investigations using underwater robots, it is important to consider the marine environment at the time of the investigation. Therefore, as an example of a complex marine environment, underwater currents caused by ocean currents were reproduced in the simulator by installing an external force generator on the ROV model. This made it possible to simulate currents in all directions and forces.

The screen of the developed ocean simulator is configured to display a robot camera image to reproduce the operation with an actual underwater robot. In addition, an bird's-eye view of the entire simulator can be viewed, so that the operational behavior can be checked later.

Finally, two experiments were conducted as examples of ROV operation experience using the developed ocean simulator. In the first experiment, the operator operated the ROV model in a simulated ocean current, aiming to make the ROV model move as straight as possible. Therefore, the two examiners each recorded three operations. Both examiners gradually approached a straight trajectory. This result confirmed the improvement of proficiency by manipulating the ROV model on the simulator. The second experiment simulated an actual seafloor investigation and involved an operation that traveled near an object placed at an arbitrary location. Here, we compared the operation trajectories before and after the ROV model operation practice. As a result, before the operation practice, the model made unnecessary movements to adjust the distance to the seafloor, but after the operation practice, the model was able to run smoothly.

In this simulator, it is possible to view and operate the camera image installed in the same position as the actual ROV, and after the operation is completed, the movement of the ROV model can be checked from a bird's eye view. This enables the operator to know the actual behavior of the ROV due to the operator's own operation, and can contribute to improving the operator's proficiency.

In this experiment, the currents expressed by the external force generator were simplified. The physical model of the ROV was also simplified. In the future, we plan to use the external force generator to express ocean currents based on actual ocean currents in order to provide a more realistic operating experience. We also plan to construct a physical model that is close to the actual ROV. We also plan to develop a simulator for more realistic proficiency training by adding collision detection with the seafloor and objects.

References

1. Nakajima, S., Miki, M.: Ocean surveys using AUVs and ROVs. Impact Assess. **18**(2), 10–13 (2020)
2. Mizuno, K., Terayama, K., Hagino, S., Tabeta, S., Sakamoto, S., Ogawa, T., Sugimoto, K., Fukami, H.: An efficient coral survey method based on a large-scale 3-D structure model obtained by Speedy Sea Scanner and U-Net segmentation. Sci. Rep. **10**(1), 12416 (2020)
3. Macreadie, P., McLean, D., Thomson, P., Partridge, J., Jones, D., Gates, A., Benfield, M., Collin, S., Booth, D., Smith, L. L., et al.: Eyes in the sea: unlocking the mysteries of the ocean using industrial, remotely operated vehicles (ROVs). Sci. Total Environ. **634**, 1077–1091 (2018).
4. Sakagami Lab. Homepage. http://www.scc.u-tokai.ac.jp/~sakagami/index_e.html. Accessed 26 Feb 2023
5. Nakaoka, S.: Choreonoid: integrated extensible GUI environment for robots (in Japanese). J. Robot. Soc. Japan **31**(3), 12–17 (2013)
6. Suzuki, K., Kawabata, K.: Development of a robot simulator for decommissioning tasks utilizing remotely operated robots. J. Robot. Mechatron. **32**(6), 1292–1300 (2020)
7. Suzuki, K.: Hairo-world-plugin. https://github.com/k38-suzuki/hairo-world-plugin. Accessed 26 Feb 2023
8. Suzuki, K.: HAIROWorldPlugin Operation Manual (in English). https://k38-suzuki.github.io/hairo-world-plugin-doc/en/index.html. Accessed 26 Feb 2023
9. Sakagami, N., Hirayama, K., Taba, R., Kobashigawa, S., Arashiro, S., Takemura, F., Takahashi, S.: Development and field experiments of a human-portable towed ROV for high-speed and wide area data acquisition. J. Artif. Life Robot. **26**(1), 1–9 (2021)

10. Clague, D., Paduan, J., Caress, D., Chadwick Jr., W., Morgane, S., Dreyer, B., Portner, R.: High-resolution AUV mapping and targeted ROV observations of three historical lava flows at axial seamount. Oceanography **30**(4), 82–99 (2017)
11. Yamaki, A., Sugimura, M., Ito, S.: Ctenophore Lyrocteis imperatoris rediscovered from the type locality, the Sagami Bay, after 79 years since the original description. Nat. Hist. Rep. Kanagawa **42**, 101–108 (2021)

Chapter 7
Optimization of the Four Coil Configuration for Single Magnet Levitation from Below

Peter Berkelman and Steven Kang

Abstract Magnetic levitation systems have the capability to provide six degree-of-freedom rigid-body motion control with high precision in translation, rotation, force, and torque in all directions with only a single moving part and noncontact actuation and sensing. This means of motion control eliminates deleterious effects on performance from mechanical contact such as friction, hysteresis, backlash, and vibration. These motion control capabilities have applications in precision fabrication and manipulation, medical interventions, and haptic human-machine interaction. In this work we investigate a specific system configuration which uses four cylindrical coils to levitate a magnet. The notable features of this configuration are that only four coils and two Hall effect sensors are needed to levitate a single magnet from underneath, and simple linear proportional-derivative (PD) controllers are sufficient to stabilize both translation and rotation in both planar directions in a small motion region. Translation in the vertical direction is stable and rotation about the vertical axis is neutrally stable and uncontrolled, allowing a disk magnet to rotate freely. Optimized parameters are presented for sample values of levitation height, mass, and maximum current, and disturbance response experimental results will be shown for the implemented levitation system with optimized dimensional values.

Keywords Magnetic levitation · Mechatronics · Dynamic control

7.1 Introduction

Magnetic levitation systems can provide many benefits to precision motion control systems due to the lack of direct mechanical contacts which produce vibration, friction, and wear. It is a technical challenge to realize the potential capabilities of

P. Berkelman (✉) · S. Kang
University of Hawai'i at Manoa, 96822 Honolulu, HI, USA
e-mail: peterb@hawaii.edu

S. Kang
e-mail: kang2@hawaii.edu

D. Azimov (ed.), *Proceedings of the IUTAM Symposium on Optimal Guidance and Control for Autonomous Systems 2023*, IUTAM Bookseries 40,
https://doi.org/10.1007/978-3-031-39303-7_7

magnetic levitation due to the complex nonlinearities of magnetic fields, cross coupling between the multiple controlled degrees of freedom, and the inherent passive instability of magnetic levitation systems as established by Earnshaw in 1842 [1]. Furthermore, as the mass and dimensions of a levitated object and levitation system are reduced to table-top scale and below, the speed of the rigid-body motion dynamics to be controlled requires high speed real-time operation of sensing, sensing, actuation, and calculation subsystems. Our previous magnetic levitation work has been focused towards realizing precise levitation at table-top scales over large ranges of motion in translation and rotation, using arrays of cylindrical coils for actuation, and optical motion tracking for feedback control [2].

A sample four coil levitation configuration in operation is shown in Fig. 7.1. In this levitation configuration, an outer ring of permanent magnets provides passive lifting forces to support the levitated magnet in the center. The polarities of the fixed permanent magnet ring and the levitated magnet are in the same direction and it is actually the induced magnetization of the iron cores of the central coils in the opposite direction that provides the lifting force, as shown in Fig. 7.2. The four coils are connected together in two pairs, where the windings of each coil in each pair have opposite polarities, so that a single current through a coil pair generates antagonistic forces to push and pull the levitated magnet in one planar direction, and rotate the magnet about one planar axis. Two Hall effect sensors at the center of the levitation system detect off-center components of the magnetic field from the levitated magnet in the two planar directions, so that the feedback signal from each Hall effect sensor determines the current through its corresponding pair of electromagnetic coils according to its digital PD controller. This type of levitation configuration was patented in [3], and our initial testing, analysis and implementation of this magnetic levitation configuration with digital PD controllers is described in [4].

This levitation configuration is typically only stable in a small region, and the system must be carefully balanced using potentiometers or feedforward terms in the digital controllers. Steady-state coil currents are limited in practice to a fraction of an Ampere to avoid burnout from resistive heating. We have implemented methods using *Radia* analysis [5] of electromagnetic fields, forces, and torques to find the optimal coil and magnet positions and the levitated magnet dimensions to maximize the stable range of levitation for a given levitated mass, height, and maximum coil current with the described four coil levitation configuration.

7.2 Background

A comprehensive survey of magnetic levitation technologies as applied to transportation, precision motion control, human-machine interaction, and instrumentation systems is given in [6]. Electromagnetic actuation and motion control systems and methods in robotics are described in [7]. Tabletop-scale levitation systems include levitated planar motors [8] and Lorentz levitation devices [9]. Levitated planar motor

Fig. 7.1 Implemented four coil magnetic levitation system in operation

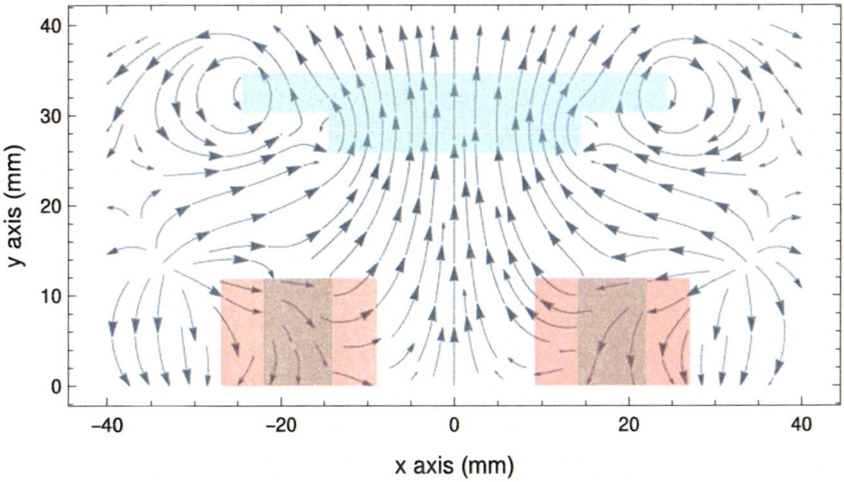

Fig. 7.2 Magnetic field modeling for four coil levitation, with levitated magnet in green, coils in red, and iron cores in gray

systems use rectangular flat coils, magnets arranged in Halbach arrays [10] to maximize magnetic fields on the underside of the levitated platform, and can provide nanometer levels of motion control precision through the use of laser interferometry position sensing. Lorentz force levitation devices use flat oval racetrack coils and horseshoe magnet assemblies arranged in all different orientations to provide significant ranges in all rigid-body motion degrees of freedom. These devices have been developed as fine motion compliant robot wrists [11] and as haptic interaction devices [12].

Medical applications of electromagnetic motion control may be less demanding than levitation systems in that constant support and stability of a magnetic capsule is not needed to direct its motion through bodily tissues. Position sensing is a more difficult issue for these systems, as the accuracy and rapid response of optical sensing is typically not available. Optical position sensing for motion control is available however for the *Octomag* system which controls the motion of minature magnetic capsules through intraocular fluid [13]. The *Stereotaxis* system was developed to use large external electromagnets to control the motion of magnetic capsules through the vascular or gastrointestinal vessels in the human body [14].

In regards to magnet levitation by cylindrical coils from underneath, the most significant early development was undertaken by NASA for noncontact suspension of aerodynamic models in wind tunnels [15]. The advantages of magnetic levitation for wind tunnel suspensions are that there is no support structure to disturb the air flow surrounding the model, and aerodynamic forces and torques can be estimated with high accuracy by monitoring the coil currents needed to keep the model in place. Air gap distances for wind tunnel suspension were further extended to 60 cm in [16].

We have developed a range of systems which use planar arrays of cylindrical coils to levitate platforms containing one to six cylindrical magnets. Our coil array levitation methods are described in detail in [17]. To calculate the total forces and torques generated on the magnets at any position and orientation by any combination of coil currents in real time during operation, a precomputed numerical model of the forces and torques between a single magnet and coil with a 1.0 A current is used. The single magnet and coil model is stored as a multidimensional lookup table with force and torque data to the nearest mm in relative position and the nearest 5 or 10 °C in relative orientation between a single magnet and coil.

At each update of the control system during operation, the relative position and orientation between each magnet and coil is calculated, the corresponding force and torque contribution from a 1.0 A current in the coil is obtained from the lookup table by linear interpolation, and the resulting transformation between the total set of coil currents and the total force and torque vector on the levitated platform is compiled in a transformation matrix A so that

$$F = AI, \tag{7.1}$$

where F is the six-element vector of 3D forces and torques $[f_x f_y f_z \tau_x \tau_y \tau_z]^T$ in the case of a levitated platform with multiple cylindrical magnets, F is the five-element vector $[f_x\, f_y\, f_z\, \tau_x\, \tau_y]^T$ in the case of single cylindrical magnet levitation where the rotation about the cylindrical axis is uncontrolled, and I is the vector of coil currents.

The forces and torques on the magnets from coils are nonlinear with respect to position and orientation, but linear with respect to coil currents at a given rigid-body position and orientation. Since these equations are an underdetermined linear system, the pseudoinverse of the A matrix can be used to calculate the coil currents to produce the desired set forces and torques on both magnets with the minimum power requirements, as the pseudoinverse produces the sum of least squares solution, and the electrical power produced by the coils is proportional to the sum of squared coil currents.

It is necessary to use a numerical lookup table force and torque model obtained from electromagnetic analysis software or direct measurements rather than the commonly used magnetic point dipole model because the distances between the magnets and coils may be similar to their dimensions, so that the point dipole approximation is not sufficiently accurate. Five coils has been shown to be sufficient to levitate a single cylindrical coil with its rotational axis uncontrolled. Position and orientation feedback for control is provided by optical motion tracking systems using three or more LED markers on the levitated rigid body.

A similar system using square rather than cylindrical coils is described in [18]. Levitation of a six-magnet platform with an unlimited range of controlled rigid body rotation was demonstrated in [19]. Multiple cylindrical magnets have been levitated and independently controlled using a single coil array [20].

The four coil configuration for single magnet levitation from below was first proposed and described in [21]. Four coil levitation systems are commonly available as low cost demonstrations. Their most compelling features are that all six degrees of freedom of rigid body motion are stable with only four coils, connected together in two pairs so that only two actuator amplifiers are needed, and two Hall effect sensors. Optimal feedback control parameters for a four coil levitation system were studied in [22].

With the four coil configuration, the linear transformation of Eq. (7.1) is no longer applicable. In this configuration, the six rigid-body motion degrees of freedom are not controlled individually, and the relationships between currents, forces, and torques are not linear due to induced magnetization of iron cores in the coils.

7.3 Modeling

A software model of four coil magnetic levitation configuration shown in Fig. 7.3, with four iron core coils surrounded by eight permanent magnets, levitating two disk magnets stuck together. Forces and torques produced by the permanent magnets and coils are calculated using the *Radia* software package in conjunction with *Mathematica* from Wolfram Research. In this configuration the fixed and levitated permanent magnets are magnetized in the same direction and the iron cores of the coils have induced magnetization in the opposite direction, which produces a repulsive lifting

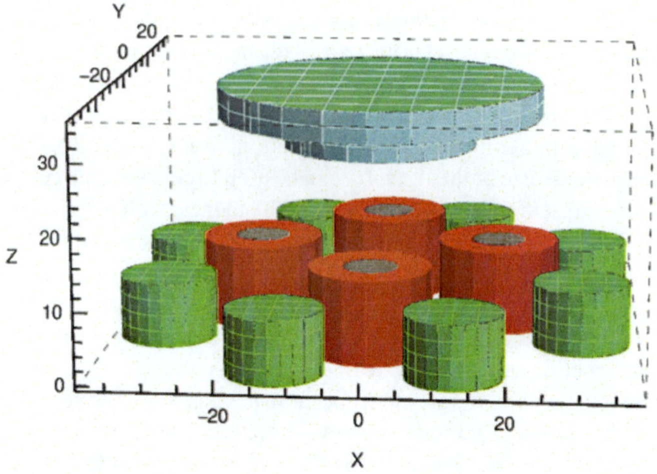

Fig. 7.3 Software model for force and torque calculations

force on the levitated magnets and an unstable equilibrium point on the central vertical axis of the device. The position of the levitated magnet is stabilized by active control using real time measurements from two Hall effect sensors positioned in the center of the four coils to detect components of the magnetic field in the two horizontal directions.

The parameters of the initial configuration are given in Table 7.1.

7.3.1 *Active Force and Torque*

Figure 7.4 shows the horizontal and vertical forces produced on the levitated magnets over a range of heights with a 1.0 A current through one coil pair. At a nominal levitation height of 18 mm, this current would produce approximately 1.5 N of horizontal force. The torque produced on the magnet pair about their common center of mass is within ± 10 Nmm throughout this vertical range, which is relatively low in comparison to the stabilizing torque produced by gravity acting upon tilted levitated magnets.

7.3.2 *Passive Force and Torque*

Because the equilibrium point of the levitated magnets is unstable, small horizontal perturbations without active control will produce forces in the same direction. This effect is shown in the modeled results of Fig. 7.5, where passive horizontal magnetic

Table 7.1 Magnet and coil parameters

Coils:	
Height	12 mm
Inner diameter	8 mm
Outer diameter	18 mm
Distance from center	18 mm
Resistance	4.25 Ω
Inductance	3.25 mH
Fixed magnets:	
Magnetization	N40
Diameter	14 mm
Height	9 mm
Distance from center	33 mm
Levitated magnets:	
Magnetization	N30
Thickness	4.5 mm
Top magnet diameter	59 mm
Bottom magnet diameter	29 mm

forces are produced on the magnets in the same direction when they are displaced horizontally. At levitation height of 18 mm, a horizontal deflection of 5 mm produces a destabilizing force of 1.5 N. According to the results of Sect. 7.3.1 and Fig. 7.4, this force may be neutralized by a coil current of 1.0 A. Therefore, a maximum coil current of 1.0 A should be sufficient to stabilize the horizontal motion of the magnets within a radius of 5 mm of the central axis at a levitation height of 18 mm.

Passive torques on the levitated magnets at a height of 18 mm with horizontal disturbances up to 5 mm are also within ± 10 Nmm, so that it is likely that gravitational torques are sufficient to stabilize the orientation of the levitated magnets within the horizontal disturbance radius up to 5 mm.

7.3.3 Variable Coil and Magnet Positions

We have investigated the effects of increasing the distances of the coils and fixed magnets to determine whether the 5 mm radius of stability can be increased. The passive and active force analysis of Figs. 7.4 and 7.5 were repeated for coil circle radii of 21, 24, and 27 mm and corresponding fixed magnet circle radii of 36, 39, and 42 mm. The active and passive force results from these increased radial distances are shown in Figs. 7.6 and 7.7 as horizontal and vertical forces “1”, “2”, and “3” corresponding to the 21, 24, and 27 mm coil circles.

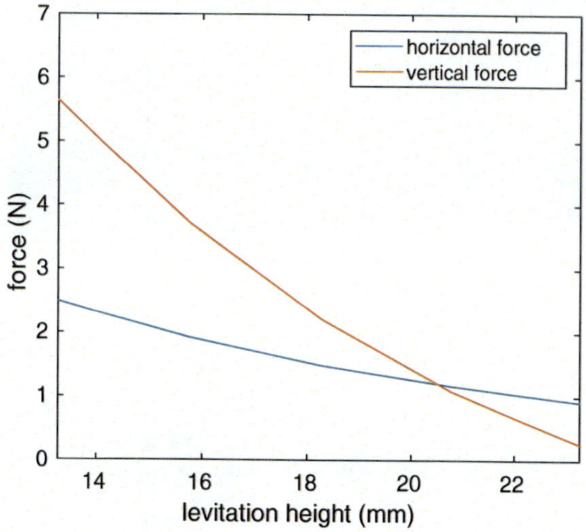

Fig. 7.4 Total forces on magnets with 1.0 A coil current

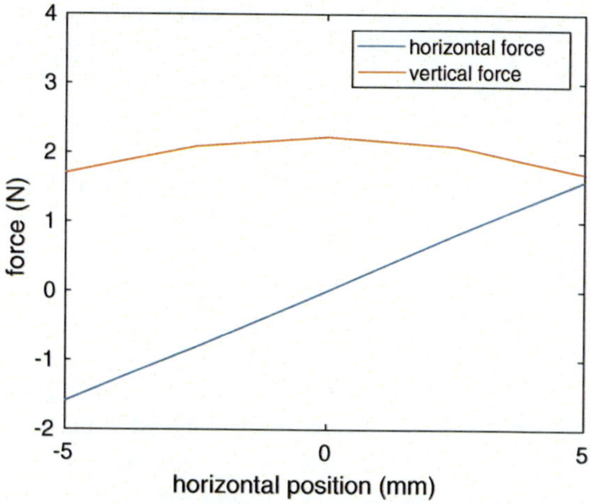

Fig. 7.5 Passive forces on magnets at 18 mm height with no coil currents

The vertical forces produced on the magnet pair by the wider coil arrangements are significantly increased, to a maximum of 3.8 N for the 27 mm coil circle as seen in Fig. 7.7. The passive horizontal forces from the 27 mm coil circle configuration decrease to 1.1 N at a 5 mm horizontal magnet displacement, as expected. The active forces produced with the 27 mm coil circle from a 1.0 coil current also decrease proportionally. As a result the overall stability of the controlled system does not change significantly at the wider coil and fixed magnet configurations.

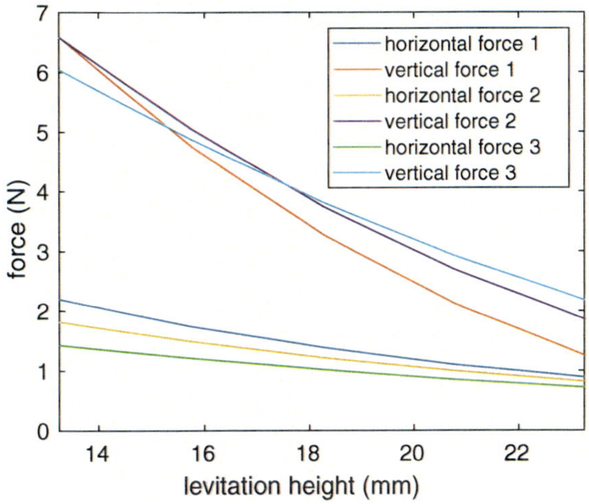

Fig. 7.6 Total forces on magnets with 1.0 A coil current at larger circle radii

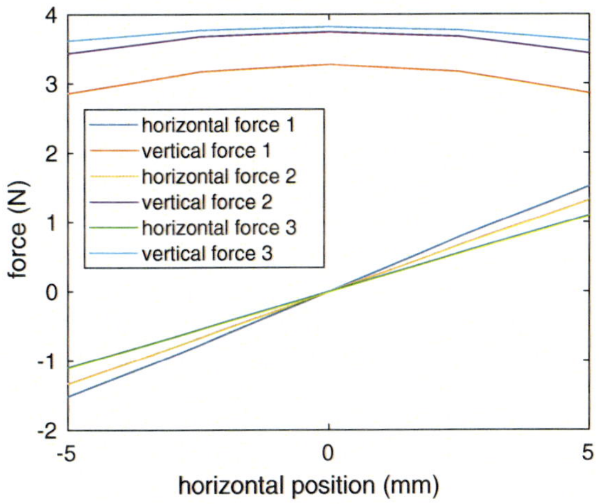

Fig. 7.7 Passive forces on magnets at 18 mm height with no coil currents at larger circle radii

7.4 Results and Discussion

A four coil levitation system was fabricated using the initial configuration parameters as listed in Table 7.1. Digital proportional-derivative controllers were implemented on a PC in C/C++ with a low-latency Linux OS to execute as a real time high priority process. The PCI-1716 Analog Input Card and the PCI-1724U Analog Output Card from Advantech Co. Ltd. were used for analog input from the two Hall effect sensors and analog output to the two current amplifiers respectively. SS495a Hall effect sensors from Honeywell and 4212Z DC motor amplifiers from Copley Controls were used. The controller update rate was 1000 Hz Hz. The magnetic fields produced by the coil currents were measured beforehand to allow the coil fields to be subtracted

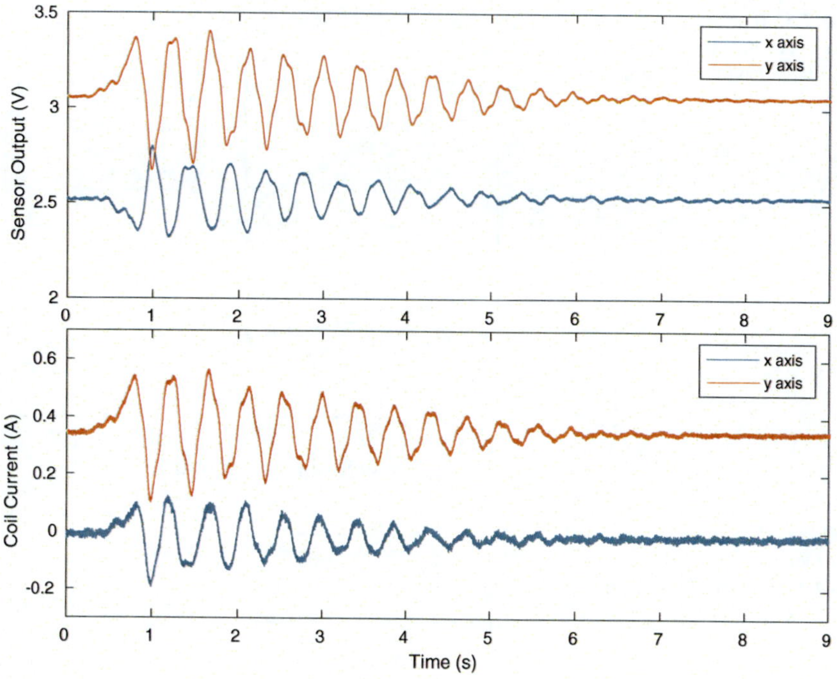

Fig. 7.8 Control results

from the Hall effect sensor measurements during operation, so that accurate measurements corresponding directly to the horizontal magnet positions could be obtained for feedback control.

The position and orientation response of the levitated magnets to disturbances in both position and orientation is shown in Figs. 7.8 and 7.9. Figure 7.8 shows the sensor and actuation signals during the disturbance and settling of the system, and Fig. 7.9 shows the 3D position and orientation of the levitated magnets. The rigid-body motion of the levitated magnets was recorded using a Polaris Vicra optical motion tracking system from Northern Digital Inc. and four reflective markers fixed to the top surface of the magnets.

The motion control disturbance was produced by tapping the edge of the levitated magnet pair by hand. Position oscillations of approximately 2 Hz and angular oscillations of approximately 5 Hz with a settling time of approximately 7 s were observed. The maximum position oscillation amplitudes were not significantly greater than 1 mm and the maximum angular oscillation amplitudes were approaching 20 °. The sensor signal oscillations are synchronized closely with the position oscillations while the angular oscillations had little effect on the sensor signals.

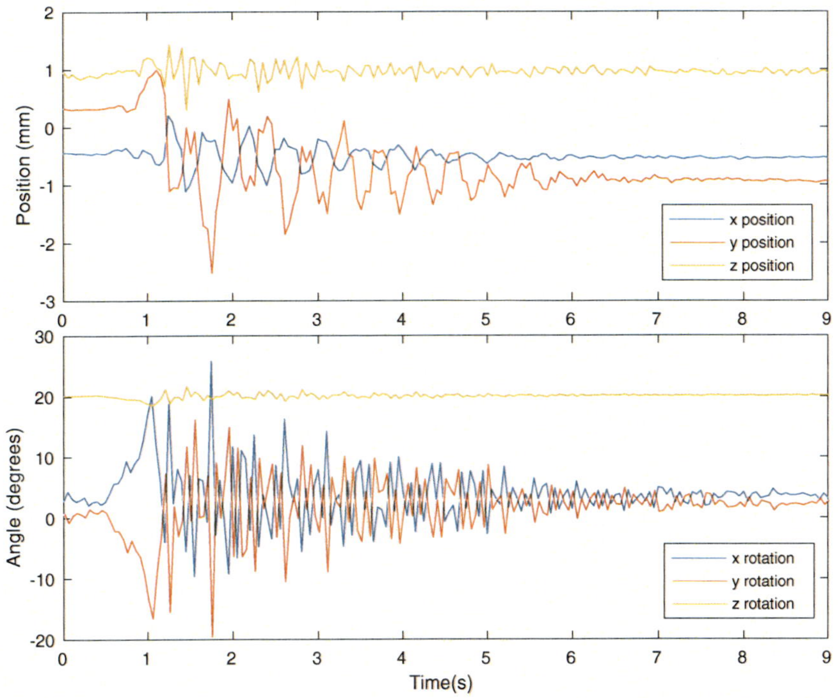

Fig. 7.9 Motion results

The active damping provided by the control system could not be increased further due to the noise level of the sensor signals, the limited response bandwidths of the current amplifiers, and the impedance of the levitation coils. Increasing the proportional control gains of the controller also did not improve the stability, disturbance rejection, or settling time of the response.

7.5 Conclusion

In this four coil magnetic levitation configuration, the experimental results of Sect. 7.4 indicate that the Hall effect sensor signals mainly correspond to the planar position of the levitated magnets. Similarly, the modeling results of Sect. 7.3 indicate that the coil currents produce only minor effects on the rotational dynamics of the levitated magnets. Therefore the controller principally stabilizes the planar motion of the levitated magnet pair. The rotational dynamics of the levitated magnets are relatively unaffected by the controllers, oscillate at a different frequency from the translational motion, and are only lightly damped. The rotational dynamics of the levitated magnets in this coil and magnet configuration appear to be passively stable, with a stabilizing return torque produced by gravity and pendulum dynamics.

Increasing the radial distance of the fixed coils and magnets from the central axis of the device from the initial values as shown in Fig. 7.3 and Table 7.1 produces the nonintuitive result that passive vertical levitation forces are increased. Although the destabilizing passive horizontal forces produced by horizontal displacements of the magnets are decreased with the wider coil configurations, the active control forces produced by coil currents are decreased by a similar proportion, with the result that the overall stability of the controlled magnetic levitation system is not significantly changed by the wider coil configurations.

References

1. Earnshaw, S.: On the nature of the molecular forces which regulate the constitution of the luminiferous ether. Trans. Cambridge Philos. Soc. **7**, 97–112 (1842)
2. Berkelman, P., Dzadovsky, M.: Magnetic levitation over large translation and rotation ranges in all directions. IEEE/ASME Trans. Mechatron. **18**(1), 44–52 (2013)
3. Whitehead, L.: Levitation system with permanent magnets and coils. US Patent 5 168 183 (1992)
4. Berkelman, P., Ohashi, N.: Analysis and testing of a four coil magnetic levitation configuration. In: IECON 2022—Annual Conference of the IEEE Industrial Electronics Society, pp. 1–5 (2022)
5. Chubar, O., Elleaume, P., Chavanne, J.: A three-dimensional magnetostatics computer code for insertion devices. J. Synchrotron Radiat. **5**, 481–484 (1998)
6. Berkelman, P.: Magnetic levitation technologies. In: Chugg, A. (ed.) New Waves in Electromagnetic Technology. The Institution of Engineering and Technology, ch. 1, in press (2023)
7. Abbott, J.J., Diller, E., Petruska, A.J.: Magnetic methods in robotics. Ann. Rev. Control Robot. Auton. Syst. **3**(1), 57–90 (2020). [Online]. Available https://doi.org/10.1146/annurev-control-081219-082713
8. Kim, W.-J., Trumper, D.: High-precision magnetic levitation stage for photolithography. Precis. Eng. **22**, 66–77 (1998)
9. Hollis, R.L., Salcudean, S.E.: Lorentz levitation technology: a new approach to fine motion robotics, teleoperation, haptic interfaces, and vibration isolation. In: Proceedings of 6th International Symposium on Robotics Research, Hidden Valley, PA, October 1993
10. Halbach, K.: Design of permanent multipole magnets with oriented rare earth cobalt material. Nucl. Instrum. Methods **169**(1), 1–10 (1980)
11. Oh, S.-R., Hollis, R.L., Salcudean, S.E.: Precision assembly with a magnetically levitated wrist. IEEE International Conference on Robotics and Automation, Atlanta, May 1993, pp. 127–134
12. Berkelman, P.J., Hollis, R.L.: Lorentz magnetic levitation for haptic interaction: device design, function, and integration with simulated environments. Int. J. Robot. Res. **9**(7), 644–667 (2000)
13. Kummer, M.P., Abbott, J.J., Kratochvil, B.E., Borer, R., Sengul, A., Nelson, B.J.: OctoMag: an electromagnetic system for 5-DOF wireless micromanipulation. IEEE Trans. Robot. **26**(6), 1006–1017 (2010)
14. Meeker, D., Maslen, E., Ritter, R., Creighton, F.: Optimal realization of arbitrary forces in a magnetic stereotaxis system. IEEE Trans. Magn. **32**(2), 320–328 (1996)
15. Britcher, C.P., Ghofrani, M.: A magnetic suspension system with a large angular range. Rev. Sci. Instrum. **64**(7), 1910–1917 (1993)
16. Sawada, H., Suda, S., Kunimasu, T.: NAL 60cm magnetic suspension and balance system. In: Congress of International Council of the Aeronautical Sciences, August 2004, pp. 2004-3.1.2
17. Berkelman, P., Dzadovsky, M.: Novel design, characterization, and control method for large motion range magnetic levitation. IEEE Magn. Lett. **1** (2010)

18. Zhang, X., Trakarnchaiyo, C., Zhang, H., Khamesee, M.B.: Magtable: a tabletop system for 6-dof large range and completely contactless operation using magnetic levitation. Mechatronics **77**, 102600 (2021)
19. Miyasaka, M., Berkelman, P.: Magnetic levitation with unlimited omnidirectional rotation range. Mechatronics **24**(3), 252–264 (2014)
20. Berkelman, P., Kang, S.: Multiple magnet independent levitation and motion control using a single coil array. In: IEEE/ASME International Conference on Advanced Intelligent Mechatronics, 2023, submitted
21. Chong, P., Commandeur, C., Davis, H., Whitehead, L.: Large-gap magnetic positioning system having advantageous configuration. In: International Symposium on Magnetic Suspension Technology, Part 2, vol. 2(NAS 1.55: 3152-PT-1), pp. 941–953 (1992)
22. Song, H., Lin, W., Zhou, M., Liu, G., Pan, H., Tong, M.: Robust h-infinity control for disturbance rejection in a magnetic levitation device. In: 2019 IEEE 28th International Symposium on Industrial Electronics (ISIE), pp. 2170–2174 (2019)

Chapter 8
Mathematical Simulation of Dynamics for Exoskeleton Including Variable-Length Links with Adjustable Stiffness

Alexander O. Blinov, Andrey V. Borisov, and Robert G. Mukharlyamov

Abstract A method of composing exoskeleton dynamics equations is presented in this paper. The exoskeleton is simulated with a system of rigid bodies including variable-length links with adjustable stiffness. The dependence of the force of resistance of the magnetic-rheological medium to the movement of the piston is obtained depending on the applied external magnetic field. The article sequentially considers models of one link of variable length with adjustable stiffness, two links and three links. For each model, a system of differential equations of motion is compiled. Then, based on the analysis of the structure of the obtained systems of differential equations of motion, a generalization is carried out for the case of a model consisting of an arbitrary finite number of links. The system dynamics is described by differential equations in generalized vector–matrix form. The damping forces, required for motion stabilization, are provided by magneto-rheological fluid controlled by magnetic field strength. The developed model of an exoskeleton with links of variable length with adjustable stiffness can be used in the creation of a human–machine mechatronic system as an auxiliary or main means for moving and performing various physical loads in order to reduce them on the human musculoskeletal system.

Keywords Exoskeleton · Dynamics · Equations · Variable-length link · Adjustable stiffness · Magneto-rheological fluid

---

A. O. Blinov (✉) · A. V. Borisov
The Branch of National Research University "Moscow Power Engineering Institute" in Smolensk, Energetichesky Passage, House 1, Smolensk 214013, Russian Federation
e-mail: alex-blinov67@yandex.ru

A. V. Borisov
e-mail: BorisowAndrej@yandex.ru

R. G. Mukharlyamov
Institute of Physical Research and Technology Peoples, Friendship University of Russia (RUDN University), 6 Miklukho-Maklaya Street, Moscow 117198, Russian Federation
e-mail: robgar@mail.ru

D. Azimov (ed.), *Proceedings of the IUTAM Symposium on Optimal Guidance and Control for Autonomous Systems 2023*, IUTAM Bookseries 40,
https://doi.org/10.1007/978-3-031-39303-7_8

8.1 Introduction

The problems of developing mechanical controlled systems with variable geometry and constraints become in demand for the implementation of the motion of solid body systems under the influence of internal energy sources and imposed internal and external constraints. This requires kinematic chains with hinges and connections, as they provide the transformation of internal energy into purposeful movement. Such mechanical systems are a wide and practically important class of new machines: manipulators, snake-like robots, anthropomorphic robots, exoskeletons, biological bodies. Their modeling and practical creation requires the use of many achievements of scientific and technological progress, including the use of magnetic rheological media.

Designing exoskeletons requires selecting a mechanical model with an optimal number of links specified by their purpose. An exoskeleton can be used for various purposes. It can be, for example, a muscle training aftermath of a traumatic event or a stroke to restore patient's mobility, a maintaining uncomfortable posture of a worker with raised arms, or carrying heavy loads. For each purpose, an exoskeleton including a corresponding number of links is required. Thus, for patient rehabilitation purposes an exoskeleton having from two to four mobile links is sufficient. Keeping a static posture of human body can be implemented by a system including up to eight links. Implementing movements relating to carrying heavy loads can require a system heaving ten or more links. Designing of a general exoskeleton model requires development of a mathematical model which should be appropriate for solving multidimensional system control problems. Shocks and abrupt jolts can be damped by including into the system the variable-length links with adjustable stiffness. This approach provides comfortable operation for exoskeleton user.

In the future, soft exoskeletons using magnetic rheological fluids will play a special role in the field of robotics.

A magnetic rheological fluid is a fluid that can change its rheological behavior under the influence of an applied magnetic field. It consists of polarizable and magnetizable particles that line up when a magnetic field is applied, resulting in a change in viscosity within a few milliseconds. Depending on the physicochemical composition of the magnetic rheological fluid, the yield stress varies in the range of 0–100 kPa [1]. Magnetic rheological fluid is colloidal solutions, which, in turn, consist of dispersed and ultrafine magnetic particles stabilized by surfactants.

We present the division of magnetic rheological fluids depending on the dispersion of particles and the structure of the magnet: (1) magnetic rheological suspensions, (2) magnetically controlled microcarriers, (3) ferromagnetic fluids [2].

The viscosity of magnetic rheological suspensions, unlike "ferrofluids", can increase up to 100 times. Magnetic rheological fluid has a high aggregative stability in a magnetic field. It is also capable of transmitting high shear stresses [3].

Examples of applications for magnetic rheological fluid are magnetic fluid seals for shaft sealing, hybrid drives with magnetic rheological clutches, magnetic rheological polishing, magnetic rheological shock absorbers, etc. So, hybrid drives work by rectifying magnetostrictive oscillations using magneto-rheological flow rectification. The operation of magnetic rheological couplings is influenced by factors such as the properties and response of the magnetic rheological fluid, structural rigidity, performance, control scheme, and others. Devices based on magnetic rheological fluid have the advantages of high torque, low power consumption and safety.

Based on the foregoing, it follows that magnetic rheological media have broad prospects for application in the creation of such complex mechatronic devices as exoskeletons.

As follows from the analysis of publications available in the open press and regular monitoring of this topic, exoskeleton models that have links of variable length with adjustable stiffness have not yet been identified. Therefore, we will consider the works in which the simulation of magnetic-rheological media and their application in various technical devices is carried out. An actuator with variable stiffness was proposed in [4]. The use of magneto-rheological materials in exoskeleton drives is discussed in [1, 5–14]. The concept of a magnetic rheological fluid, as a certain category of intelligent fluids that change their viscosity in the presence of a magnetic field, is introduced in [14–22]. The viscoelasticity of these liquids can be increased by increasing the magnetic field applied to them. One of the created applications of magnetic rheological fluid is the drive in the knee joint of the exoskeleton [7, 11, 12], since the torque in this joint has a damping effect on most of the gait cycle [23]. Magnetic rheological drives have been studied in [14, 24, 25]. Prospects for the use of magnetic rheological fluids in the creation of exoskeletons are considered in [5]. However, the creation of a full-fledged exoskeleton or an anthropomorphic robot with soft links of variable length and rigidity is not yet discussed. Thus, this work is ahead of the existing models and is the theoretical foundation for the further practical implementation of a new class of exoskeletons.

8.2 Model of One Link of Variable Length with Adjustable Stiffness in Space

Consider a spatial model of a link with a payload (Fig. 8.1). Let us introduce a fixed Cartesian coordinate system connected with the Earth $A_0x_0y_0z_0$ (Fig. 8.1) with the origin in the fixed spherical hinge A_0.

The model uses the angles between the links, which corresponds to the mode of operation of the exoskeleton, in which the actuators change the relative angles. Let us introduce a moving local coordinate system $A_0x_1y_1z_1$, rigidly connected to the spherical hinge A_0, in which the rotations of the link are realized (Fig. 8.1). On the section of the link B_1C_1 there is a cylinder with a magnetic rheological fluid and coils for controlling its rigidity by means of an applied magnetic field. At point A_1, there is a concentrated mass that simulates the payload: the weight of the body, head, arms and carried leg. The Lagrange equations of the second kind are compiled for

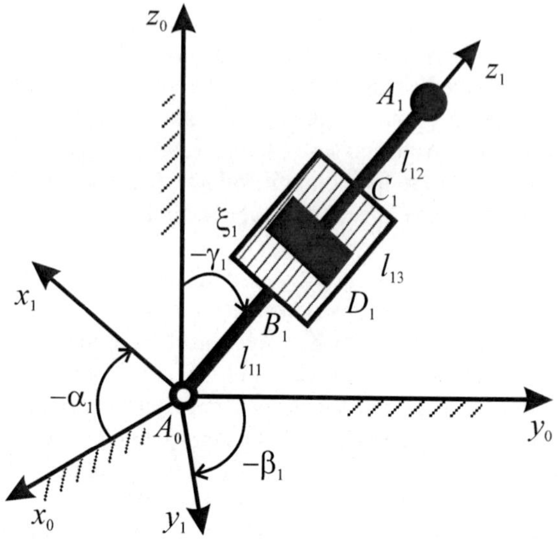

Fig. 8.1 Variable length link model with controlled stiffness and payload in space

the considered spatial model of a variable-length exoskeleton link with controlled stiffness.

The control of the change in the viscosity of the magnetic rheological fluid within the framework of the proposed model is implemented using external magnetic fields, which leads to a change in the force of resistance to the movement of the piston inside the cylinder. The dependence of the resistance force implemented by the magnetic rheological fluid in each link has the form:

$$F = \pi r^2 \eta_f \dot{\gamma}\left[1 + \phi_L\left(\alpha_n + \frac{(\zeta_n + \beta_n \lambda_n + \beta_n \cos(2\theta))}{2} + (\chi_n - 2\beta_n \lambda_n)\frac{\sin^2(2\theta)}{4}\right)\right] + \frac{9\pi r^2 \phi_L \mu_0 H^2 (n-1)(\chi_f + 1)\sin(2\theta)}{16 n_c}, \tag{8.1}$$

where ϕ_L—is the volume fraction of micron particles, n—is the quantity of particles in a chain, n_c is the maximum number of particles in a chain, $\alpha_n, \ldots, \lambda_n$—are the kinetic coefficients, $\dot{\gamma}$—is the shear rate, η_f—is the viscosity of the magnetic rheological fluid, θ—is the angle between the vector directed along the straight line connecting the centers of two micron particles and the vector of the external magnetic field [2, 14–16, 20, 22].

We use a mathematical model that takes into account the properties of the magnetic rheological fluid to simulate the control of the resistance force in the link. Knowing the dependence of the longitudinal force, using the obtained dependence (8.1), it will be possible to determine the strength of the external magnetic field that controls the properties of the magnetic rheological fluid.

8.3 Model of Two Links of Variable Length with Adjustable Rigidity in Space

Consider a 3D model of an exoskeleton fragment.consisting of two fixed links of variable length with adjustable stiffness. Let us introduce a fixed Cartesian coordinate system associated with the Earth $A_0x_0y_0z_0$ (Fig. 8.2) with the origin in the fixed spherical hinge A_0. The model uses the angles between the links, which corresponds to the mode of operation of the robotic mechanism, in which the actuators change the relative angles. We use moving local coordinate systems $A_0x_1y_1z_1$ and $A_1x_2y_2z_2$, rigidly connected with spherical hinges A_0 and A_1, in which the rotations of the links are realized. We direct the coordinate axes z_1 and z_2 along the moving links A_0A_1 and A_1A_2 (Fig. 8.2). Figure 8.2a shows the state of the magnetic rheological fluid without the application of an external magnetic field. In this case, the magnetic particles are arranged randomly. Figure 8.2b shows the case of an oriented arrangement of magnetic particles under the action of an external magnetic field.

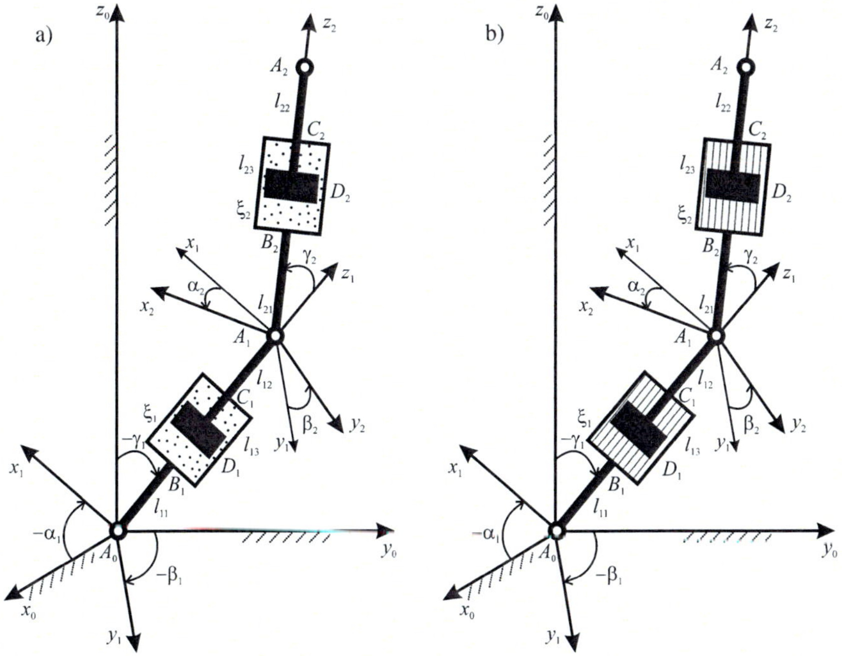

Fig. 8.2 Spatial model of two links of the exoskeleton of variable length with adjustable stiffness

Links A_0A_1 and A_1A_2 have the same design, consisting of rods A_1D_1 and A_2D_2 with pistons D_1 and D_2 located inside housings B_1C_1 and B_2C_2. The housings are cylinders filled with magnetic rheological fluid, on which electromagnetic coils are wound from the outside. Absolutely rigid rods A_0B_1 and A_1B_2 are rigidly attached to them from below, on the opposite side of which spherical hinges are located. All elements of the link in the first approximation are modeled by thin rods. Spherical hinges are installed at the ends of the links. In the hinges, it is possible to create the necessary control moments. The issues of creating technical devices that implement control moments are beyond the scope of this study and are not considered here.

Each link has two weighty absolutely rigid rods: $A_0B_1 = l_{11}, D_1A_1 = l_{12}, A_1B_2 = l_{21}, D_2A_2 = l_{22}$, where the first index is equal to the link number, the second is the section number on the link. Lengths of cylinders with magnetic rheological fluid: $B_1C_1 = l_{13}, B_2C_2 = l_{23}$. The masses and moments of inertia of the individual elements of the links are: $m_{11}, m_{12}, m_{13}, I_{11}, I_{12}, I_{13}$, respectively, and are constant values within the proposed model. As a first approximation, let us assume that the pistons D_1 and D_2 themselves have a negligible mass and moments of inertia in comparison with the rods D_1A_1 and D_2A_2. The change in the length of the link is carried out due to the relative movement along the direction of the links A_0A_1 and A_1A_2 of the pistons with rods D_1A_1 and D_2A_2 inside the cylinders with magnetic rheological fluid B_1C_1 and B_2C_2. The magnetic rheological fluid realizes the resistance forces acting along the rods with pistons, which we denote by F_1 and F_2, respectively. As a result of the action of an external magnetic field, the resistance force provides the necessary change in the length of the link in accordance with the phase of walking and the load from neighboring links.

The states of the links depend on four parameters and are uniquely determined by the angles $\alpha_1(t)$, $\beta_1(t)$, $\gamma_1(t)$ and $\alpha_2(t)$, $\beta_2(t)$, $\gamma_2(t)$, variable lengths of link sections B1D1 $= \xi_1(t)$ and $B_2D_2 = \xi_2(t)$. The model of two links of the supporting leg of the exoskeleton with variable length and adjustable rigidity has eight degrees of freedom. We denote the control moments developed in the spherical hinges A_0 and A_1 with ideal constraints $M_{1\alpha}, M_{1\beta}, M_{1\gamma}$ and $M_{2\alpha}, M_{2\beta}, M_{2\gamma}$, considered as instantly developing the required effort.

The kinetic energy of the considered mechanism is calculated by integrating over all weighty areas $A_0B_1, B_1C_1, D_1A_1, A_1B_2, B_2C_2, D_2A_2$

$$T = \sum_{i=1}^{2} \sum_{j=1}^{3} \int_{0}^{l_{ij}} \frac{\rho_{ij} V_{ij}^2}{2} dx_{ij} \tag{8.2}$$

where: i—number of link ($i = 1, 2$); j—section number on the link, and $j = 1$ corresponds to the lower sections A0B1 and A1B2, $j = 2$ matches rods with pistons D_1A_1 and D_2A_2, $j = 3$ matches the cylinders B_1C_1 and B_2C_2; x_{ij}—coordinate of an infinitely small particle of the j-th inertial section of the i-th link, ρ_{ij}—density of the j-th section of the i-th link, while $m_{ij} = \rho_{ij}l_{ij}$, $I_{ij} = \rho_{ij}{l_{ij}}^3/3$, l_{ij}, m_{ij}, I_{ij}—length, mass, moment of inertia of the j-th section of the i-th link, v_{ij}^2—the square of the speed of an infinitely small particle of the j-th section of the i-th link.

The kinetic energy of the entire mechanism under consideration is very cumbersome, since the second (upper) link performs a complex movement around the pole A_1, so we give the expression only for the kinetic energy of the first link

$$
\begin{aligned}
T_1 = \frac{1}{2}\big[\big(I_{11} + I_{12} + I_{13} + \big(l_{11}^2 + l_{11}l_{12} + 2l_{11}\xi_1 + l_{12}\xi_1 + \xi_1^2\big)m_{12} \\
+ \ \big(l_{11}^2 + l_{11}l_{13}\big)m_{13}\big)\big(\dot{\alpha}_1^2\cos^2\beta_1 + \dot{\beta}_1^2\big) + m_{12}\dot{\xi}_1^2\big].
\end{aligned}
\tag{8.3}
$$

Using the Lagrange equations of the second kind, a system of differential equations of motion was compiled for the mathematical model of the mechanism proposed in (Fig. 8.2), which requires supplementing with equations describing the behavior of the magnetic rheological fluid.

Dynamic equations of the exoskeleton supporting leg model, represented by two links of variable length with adjustable stiffness, are compiled in the form of Lagrange equations. To represent the structure of differential equations, we present the least cumbersome equation corresponding to the generalized coordinate ξ_1:

$$
\begin{aligned}
& -\eta_1 C_1^\beta\left(C_2^\beta C_1^\gamma S_2^\alpha - S_2^\beta S_1^\gamma\right)\ddot{\alpha}_1 - \eta_1 C_2^\beta S_2^\alpha \ddot{\alpha}_2 - \eta_1\left(C_1^\gamma S_2^\beta + C_2^\beta S_2^\alpha S_1^\gamma\right)\ddot{\beta}_1 \\
& -\eta_1 C_2^\alpha S_2^\beta \ddot{\beta}_2 + \eta_5\ddot{\xi}_1 + \eta_7 C_2^\alpha C_2^\beta \ddot{\xi}_2 \\
& +\left[\left(\eta_2 + \eta_3 C_2^\alpha C_2^\beta\right)\left(\left(S_1^\beta\right)^2 - \left(C_1^\beta\right)^2 - 1\right) + \eta_3\left(C_1^\gamma S_2^\beta + S_2^\alpha C_2^\beta S_1^\gamma\right)S_1^{2\beta}\right. \\
& \left. -\left(\eta_6\xi_1 + \eta_4 + (1/2)\xi_2\eta_7 C_2^\alpha C_2^\beta\right)\left(C_1^\beta\right)^2 + (1/4)\xi_2\eta_7\left(C_1^\gamma S_2^\beta + S_2^\alpha C_2^\beta S_1^\gamma\right)S_1^{2\beta}\right]\dot{\alpha}_1^2 \\
& -\eta_1 C_2^\alpha C_2^\beta \dot{\alpha}_2^2 - 2\left(\eta_4 + 2\eta_6\xi_1 + \eta_2 + \eta_1 C_2^\alpha C_2^\beta\right)\dot{\beta}_1^2 - \eta_1 C_2^\alpha C_2^\beta \dot{\beta}_2^2 \\
& -2\eta_7 C_2^\alpha S_2^\beta \dot{\beta}_2\dot{\xi}_2 - 2\eta_1\left(C_2^\beta C_1^\gamma + S_2^\alpha S_2^\beta S_1^\gamma\right)\dot{\beta}_1\dot{\beta}_2 \\
& -2\eta_1\left(C_2^\beta C_1^\gamma S_2^\alpha - S_2^\beta S_1^\gamma\right)\dot{\beta}_1\dot{\gamma}_1 - 2\eta_7\left(S_2^\beta C_1^\gamma + S_2^\alpha C_2^\beta S_1^\gamma\right)\dot{\beta}_1\dot{\xi}_2 \\
& -2\eta_1 C_1^\beta C_2^\alpha C_2^\beta C_1^\gamma \dot{\alpha}_1\dot{\alpha}_2 + 2\eta_1 C_1^\beta\left(C_1^\gamma S_2^\alpha S_2^\beta + C_2^\beta S_1^\gamma\right)\dot{\alpha}_1\dot{\beta}_2 \\
& +2\eta_1 C_1^\beta\left(C_1^\gamma S_2^\beta + C_2^\beta S_2^\alpha S_1^\gamma\right)\dot{\alpha}_1\dot{\gamma}_1 - 2\eta_7 C_1^\beta\left(C_2^\beta S_2^\alpha C_1^\gamma - S_2^\beta S_1^\gamma\right)\dot{\alpha}_1\dot{\xi}_2 \\
& -2\eta_1 C_2^\alpha C_2^\beta S_1^\gamma \dot{\alpha}_2\dot{\beta}_1 + 2\eta_1 S_2^\alpha S_2^\beta \dot{\alpha}_2\dot{\beta}_2 - 2\eta_7 C_2^\beta \dot{\alpha}_2\dot{\xi}_2 + \eta_5 g C_1^\alpha C_1^\beta = F_1
\end{aligned}
\tag{8.4}
$$

where: $C_1^\alpha = \cos\alpha_1$, $C_1^\beta = \cos\beta_1$, $C_1^\gamma = \cos\gamma_1$, $C_2^\alpha = \cos\alpha_2$, $C_2^\beta = \cos\beta_2$, $C_2^\gamma = \cos\gamma_2$, $S_1^\alpha = \sin\alpha_1$, $S_1^\beta = \sin\beta_1$, $S_1^\gamma = \sin\gamma_1$, $S_2^\alpha = \sin\alpha_2$, $S_2^\beta = \sin\beta_2$, $S_2^\gamma = \sin\gamma_2$, $S_1^{2\beta} = \sin(2\beta_1)$, $\left(C_1^\beta\right)^2 = \cos^2\beta_1$, $\left(S_1^\beta\right)^2 = \sin^2\beta_1$, $\eta_1 = m_{21}l_{21}/2 + m_{23}(l_{21} + l_{23}/2) + m_{22}(l_{21} + \xi_2 + l_{22}/2)$, $\eta_2 = (m_{21} + m_{22} + m_{23})(l_{11} + l_{12})/2$, $\eta_3 = (m_{21}l_{21} + 2l_{21}m_{22} + l_{22}m_{22})/4$, $\eta_4 = m_{12}(2l_{11} + l_{12} + 2\xi_1)/4$, $\eta_5 = m_{12} + m_{21} + m_{22} + m_{23}$, $\eta_6 = (m_{21} + m_{22} + m_{23})/2$, $\eta_7 = m_{22}$.

The magnetic rheological fluid control model is represented by expression (8.1). Due to the application of forces from the adjacent rods, the action of the reaction from the side of the fixed support and the force of gravity, the relative movement of the pistons D_1 and D_2 occurs in the cylinders B_1C_1 and B_2C_2 containing a magnetic rheological fluid, which, due to a change in viscosity, provides different resistance to the movement of the piston. The upper section in (Fig. 8.2) shows the chaotic behavior of magneto-rheological fluid particles in the absence of an external magnetic field. The lower section in (Fig. 8.2) shows oriented magnetic particles in the case of an external magnetic field. The change in the viscosity of the magnetic rheological fluid is controlled by an external magnetic field implemented using coils wound around the cylinder. In the absence of a magnetic field, the particles have a chaotic arrangement (upper section in Fig. 8.2). In this case, the magnetic rheological fluid does not show significant resistance to the movement of the piston. When an external magnetic field is applied, the liquid resists the movement of the piston, which is the greater, the more oriented particles it contains. It becomes possible to adjust the stiffness of the A_0A_1 and A_1A_2 links when changing their length, which will positively affect the comfort of the exoskeleton. The control of the change in the viscosity of the magnetic rheological fluid within the framework of the proposed model is implemented using external magnetic fields, which leads to a change in the force of resistance to the movement of the piston inside the cylinder. The dependence of the resistance force implemented by the magnetic rheological fluid in each link has the form (8.1).

We use a mathematical model that takes into account the properties of a magnetic rheological fluid for further modeling the movement of a mechanism in space by setting generalized coordinates in the form of differentiable functions that provide the synthesis of an anthropomorphic periodic movement of a link. The purpose of the simulation is to obtain the dependences of the control moments, longitudinal forces, strengths of external magnetic fields that generally control the movement of the proposed model of the two links of the supporting leg of the exoskeleton with variable length and adjustable stiffness.

To determine the control moment and longitudinal force that control the movement of the exoskeleton support leg model, which consists of links of variable length with adjustable stiffness, we set the angles of rotation and change in the lengths of the links in the form given in [5, 6].

8.4 Exoskeleton Model with Three Links of Variable Length and Adjustable Rigidity in Space

Let us consider a 3D model of an exoskeleton in the form of a supporting leg and a body consisting of two fixed links of variable length with adjustable stiffness and two masses simulating a portable leg and head. Let us introduce a fixed Cartesian coordinate system connected with the Earth $A_0x_0y_0z_0$ (Fig. 8.3) with the origin at

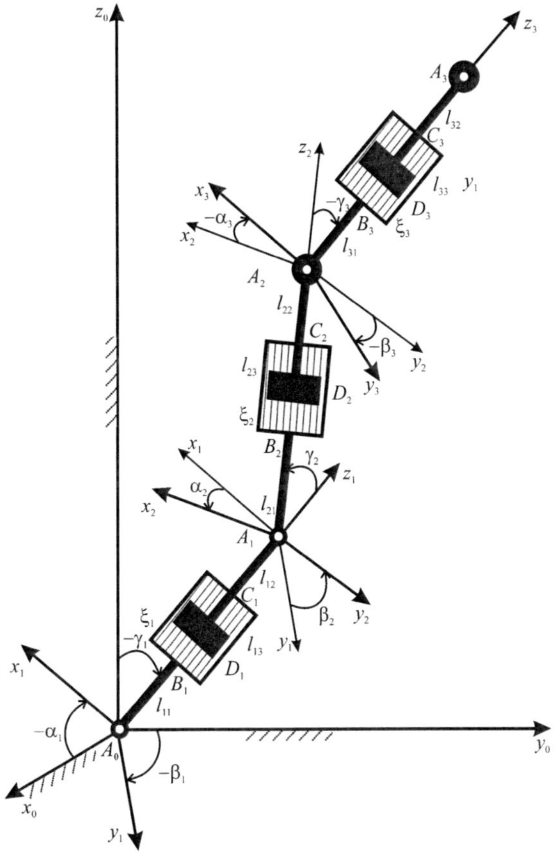

Fig. 8.3 Spatial model of three links of the exoskeleton of variable length with adjustable stiffness

the fixed hinge A_0, which is a combination of two cylindrical hinges with mutually orthogonal axes of rotation.

This design is due to the biomechanics of human walking, in which there is practically no rotation of the lower leg around its axis in the ankle joint. Therefore, the model of the ankle joint does not use a spherical hinge, which makes it possible to realize in excess the possible degrees of mobility that are practically absent in the real human ankle joint. At point A_1, there is a cylindrical hinge that adequately models the human knee joint. The human hip joint is a spherical joint, therefore, at point A_2 there is a spherical joint that models all available real degrees of mobility. To compile a system of differential equations of motion, we use moving local coordinate systems $A_0x_1y_1z_1$ and $A_1x_2y_2z_2$, rigidly connected to the hinges A_0, A_1 and A_2, in which the rotations of the links are realized. This way of description uses the angles between the links corresponding to the mode of operation of the real drives of the robotic exoskeleton, which change the relative angles. Axes z_1, z_2 and z_3 will be directed along the moving links A_0A_1, A_1A_2 and A_2A_3 (Fig. 8.3).

Links of the same design A_0A_1, A_1A_2 and A_2A_3 consist of absolutely solid rods A_1D_1, A_2D_2 and A_3D_3 with pistons D_1, D_2 and D_3 located inside the housings B_1C_1, B_2C_2 and B_3C_3. The cases are thin-walled cylinders with electromagnetic coils wound on them. Inside the cylinders are filled with magnetic rheological fluid. Absolutely rigid rods A_0B_1, A_1B_2 and A_2B_3 are rigidly attached to the cylinders from below, at the opposite end of which there are hinges. The hinges assume the possibility of creating the necessary control moments. At the points of the mechanism A_2 and A_3 there are concentrated masses m_P, and m_H, which approximately model the transferred leg and head, respectively.

The length of each link changes due to the relative movement along the direction of the links A_0A_1, A_1A_2 and A_2A_3 of pistons with rods D_1A_1, D_2A_2 and D_3A_3 inside magnetic rheological fluid cylinders B_1C_1, B_2C_2 and B_3C_3. The magnetic rheological fluid realizes the resistance forces F_1, F_2 and F_3 acting along the piston rods. As a result of the action of an external magnetic field, the magnetic particles are partially oriented along the magnetic field lines, and with an increase in the magnetic field strength, the proportion of oriented particles increases, thereby increasing the resistance force of the magnetic rheological fluid to the piston movement. Therefore, a controlled change in the length of the link is provided in accordance with the phase of walking and the load from neighboring links.

The positions of the links are uniquely determined by the angles $\alpha_1(t)$, $\beta_1(t)$, $\alpha_2(t)$ and $\alpha_3(t)$, $\beta_3(t)$, $\gamma_3(t)$, and variable lengths of sections of links between the piston and the bottom of the cylinder $B_1D_1 = \xi_1(t)$, $B_2D_2 = \xi_2(t)$ and $B_3D_3 = \xi_3(t)$ (Fig. 8.3). Therefore, the model has nine degrees of freedom. We denote the control moments developed in the hinges A_0, A_1 and A_2 with ideal constraints $M_{1\alpha}$, $M_{1\beta}$, $M_{2\alpha}$, and $M_{3\alpha}$, $M_{3\beta}$, $M_{3\gamma}$, considered as instantly developing the required effort.

The kinetic energy of the considered mechanism is calculated by integrating over all weighty areas A_0B_1, B_1C_1, D_1A_1, A_1B_2, B_2C_2, D_2A_2, A_2B_3, B_3C_3, D_3A_3

$$T = \frac{1}{2}\left(m_P V_{A_2}^2 + m_H V_{A_3}^2 + \sum_{i=1}^{3}\sum_{j=1}^{3} \int_0^{l_{ij}} \rho_{ij} V_{ij}^2 d\zeta_{ij} \right), \quad (8.5)$$

where i—link number ($i = 1, 2, 3$); j—section number on the link, and $j = 1$ corresponds to the lower sections A_0B_1, A_1B_2 and A_2B_3, $j = 2$ matches rods with pistons D_1A_1, D_2A_2 and D_3A_3, $j = 3$ matches the cylinders B_1C_1, B_2C_2 and B_3C_3; ζ_{ij}—coordinate of an infinitely small particle of the j-th inertial section of the i-th link, ρ_{ij}—density of the j-th section of the i-th link, while $m_{ij} = \rho_{ij}l_{ij}$, $I_{ij} = \rho_{ij}l_{ij}{}^3/3$, l_{ij}, m_{ij}, I_{ij}—length, mass, moment of inertia of the j-th section of the i-th link, m_P is the mass of the transferred leg located at the point A_2, $V_{A_2}^2$—square of the speed of the point A_2, simulating the hip joint, m_H is the mass of the head located at the point A_3, $V_{A_3}^2$—the square of the speed of point A_3, simulating the human head, V_{ij}^2—the square of the speed of an infinitely small particle of the j-th section of the i-th link.

A system of differential equations of Lagrange motion of the second kind was compiled for the exoskeleton model proposed in (Fig. 8.3) in the form of three links of variable length with adjustable stiffness. Further, based on the analysis of the structure of the obtained differential equations of motion of the mechanisms considered above, we obtain generalizations for a model containing a finite number of links of variable length with adjustable stiffness.

8.5 Generalization of the Exoskeleton Model to the Case of an Arbitrary Finite Number of Links of Variable Length with Adjustable Stiffness

It is assumed that the state of a 3-D exoskeleton model including n variable-length links with adjustable stiffness is described by the angles calculated between the links (Fig. 8.4). Let's introduce an immobile system of coordinates $A_0x_0y_0z_0$ and local mobile systems of coordinates $A_ix_iy_iz_i$ $(i = 1, 2,\ldots, n)$. It is considered, that sections A0B1, C1A1, …, $A_{n\text{-}1}B_n$, C_nA_n of the links are absolutely rigid bodies, and sections B1D1 $= \xi_1(t)$, …, $B_nD_n = \xi_n(t)$ have variable lengths.

The generalized coordinates describing the mechanism position in space are the angles between the coordinate axes and the lengths of variable-length links (Fig. 8.4): α_i, β_i, γ_i, ξ_i $(i = 1, 2,\ldots, n)$. Therefore, the model has $4n$ degrees of freedom. The link A_0A_1 has a stem A_1D_1 with a piston D_1 in a housing B_1C_1, which has a form of a cylinder with magneto-rheological fluid and an electromagnetic coil, with an absolutely rigid element A_0B_1 fixed below it. The ends of the links feature spherical hinges that allow applying the required control torques.

The exoskeleton model pieces having one, two, and three variable-length links have been considered in the research. The differential equations of motion have been composed for each model piece. The analysis of the composed systems has been made, a pattern has been identified, and the dynamics equations have been presented in the generalized vector–matrix form.

The subscripts of the symbols denoting a matrix point to the corresponding generalized coordinate: $\kappa = 1, 2, 3, 4$, index 1 corresponds to the vector of generalized coordinates $\alpha = (\alpha_1,\ldots, \alpha_n)^T$, index 2 corresponds to the vector $\beta = (\beta_1, \ldots, \beta_n)^T$, index 3 corresponds to the vector $\gamma = (\gamma_1, \ldots, \gamma_n)^T$, index 4 corresponds to the link lengths vector $\xi = (\xi_1, \ldots, \xi_n)^T$. The general form of dynamics equations is as follows:

$$\begin{aligned}
&A_\kappa(\alpha,\beta,\gamma,\xi)\ddot{\alpha} + B_\kappa(\alpha,\beta,\gamma,\xi)\ddot{\beta} + C_\kappa(\alpha,\beta,\gamma,\xi)\beta\ddot{\gamma} + D_\kappa(\alpha,\beta,\gamma,\xi)\ddot{\xi}\\
&+ E_\kappa(\alpha,\beta,\gamma,\xi)\dot{\Lambda}\dot{\alpha} + F_\kappa(\alpha,\beta,\gamma,\xi)\dot{\mathrm{B}}\dot{\beta} + G_\kappa(\alpha,\beta,\gamma,\xi)\dot{\Gamma}\dot{\gamma}\\
&+ 2H_\kappa(\alpha,\beta,\gamma,\xi)\dot{\mathrm{A}}\dot{\xi} + 2K_\kappa(\alpha,\beta,\gamma,\xi)\dot{\mathrm{B}}\dot{\xi}\\
&+ 2L_\kappa(\alpha,\beta,\gamma,\xi)\dot{\Gamma}\dot{\xi} + 2N_\kappa(\alpha,\beta,\gamma,\xi)\dot{\mathrm{A}}\dot{\beta} + 2P_\kappa(\alpha,\beta,\gamma,\xi)\dot{\mathrm{A}}\dot{\gamma}\\
&+ 2Q_\kappa(\alpha,\beta,\gamma,\xi)\dot{\mathrm{B}}\dot{\gamma} + gR_\kappa(\alpha,\beta,\gamma,\xi) = M_\kappa(\alpha,\beta,\gamma,\xi).
\end{aligned} \tag{8.6}$$

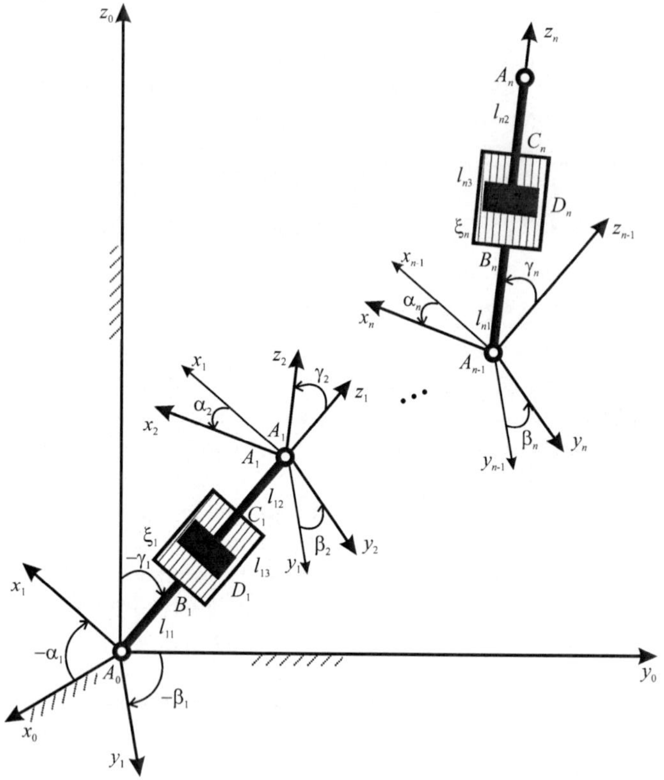

Fig. 8.4 3D model of a variable-length exoskeleton with adjustable stiffness

In this equation, $\dot{\alpha}$, $\dot{\beta}$, $\dot{\gamma}$—the vectors of angular velocities; $\ddot{\alpha}$, $\ddot{\beta}$, $\ddot{\gamma}$—the vectors of angular accelerations; $\dot{\mathrm{A}} = \mathrm{diag}(\dot{\alpha}_1, \ldots, \dot{\alpha}_n)$, $\dot{\mathrm{B}} = \mathrm{diag}(\dot{\beta}_1, \ldots, \dot{\beta}_n)$, $\dot{\Gamma} = \mathrm{diag}(\dot{\gamma}_1, \ldots, \dot{\gamma}_n)$—the diagonal matrices; M_κ—the vectors of generalized forces; $A_\kappa(\alpha, \beta, \gamma, \xi)$, $B_\kappa(\alpha, \beta, \gamma, \xi)$, $C_\kappa(\alpha, \beta, \gamma, \xi)$, $D_\kappa(\alpha, \beta, \gamma, \xi)$, $E_\kappa(\alpha, \beta, \gamma, \xi)$, $F_\kappa(\alpha, \beta, \gamma, \xi)$, $G_\kappa(\alpha, \beta, \gamma, \xi)$, $H_\kappa(\alpha, \beta, \gamma, \xi)$, $K_\kappa(\alpha, \beta, \gamma, \xi)$, $L_\kappa(\alpha, \beta, \gamma, \xi)$, $N_\kappa(\alpha, \beta, \gamma, \xi)$, $P_\kappa(\alpha, \beta, \gamma, \xi)$, $Q_\kappa(\alpha, \beta, \gamma, \xi)$—the matrices containing information about mechanism structure; $R_\kappa(\alpha, \beta, \gamma, \xi)$—the matrices defined by the moments of gravity force. The right parts of the Eqs. (8.1) are supplemented with the corresponding drag forces $M_\kappa(\alpha, \beta, \gamma, \xi)$, which are implemented by applying magnetic field strength to magneto-rheological fluid [1].

As a result, a closed system of dynamics equations in the generalized vector–matrix form has been composed. The obtained system can be used for solving control problems of exoskeletons made up of n variable-length links with adjustable stiffness.

8.6 Conclusion

As a result of the analysis of scientific publications devoted to magnetic rheological fluids, analytical relationships have been obtained for estimating the strength of a magnetic rheological fluid. As the analysis showed [2, 15–22], the direction and magnitude of the applied external magnetic field have a significant effect on the value of the viscosity of the magnetic rheological fluid. An analytical dependence of the drag force implemented by the magnetic rheological fluid in the link on the external magnetic field and empirical constants is obtained, a method is proposed for controlling the rigidity of the exoskeleton link by changing the external magnetic field.

A three-dimensional model of a link of variable length with adjustable stiffness is developed, models with two and three links are considered, and on their basis a generalization to the case of n links is obtained. To determine the positions of the links in the model, the angles between the links are used, which corresponds to the mode of operation of the robotic mechanism, in which the drives change the relative angles in cylindrical or spherical hinges. To compile differential equations of motion, mobile local coordinate systems are used, rigidly connected with the links. A mathematical model that takes into account the properties of a magnetic rheological fluid can be used in the future to model the movement of a mechanism in space by setting generalized coordinates in the form of differentiable time functions that provide the synthesis of an anthropomorphic periodic motion of a link similar to [9] to obtain the dependences of control moments, longitudinal forces, intensities of external magnetic fields that generally control the movement of the proposed models of exoskeletons, composed of links with variable length and adjustable stiffness.

The developed models of exoskeletons can be used for practical implementation in hardware. The results obtained can be used in practice when creating a human–machine system in the form of a spatial mechanism with controlled changes in the lengths of links as an auxiliary or main means for moving and performing various physical loads in order to reduce them on the human musculoskeletal system. In addition, the obtained models of mechanisms are applicable in the compensation of movements by people with diseases of the musculoskeletal system, restoring or completely replacing the motor functions of the human musculoskeletal system. It can be used in medicine during the rehabilitation of patients, for muscle training, restoration of broken neural connections, for example, in post-stroke patients. The model can be used as an auxiliary device that enhances the power capabilities of a person at work when carrying heavy loads, increasing the payload, increasing the endurance of the worker and protecting the musculoskeletal system from injury. Mechanisms with adjustable link lengths are applicable in medical institutions when caring for bedridden patients, for surgeons during long-term operations, in industry, agriculture, in the army, wherever a person has to endure heavy loads or stay in one position for a long time. It is possible to use the obtained models when creating exoskeletons for children with diseases or injuries of the musculoskeletal system, for example, spinal muscular atrophy, cerebral palsy, etc.

Acknowledgements This work was supported by the Russian Science Foundation No. 22-21-00491, https://rscf.ru/project/22-21-00491/.

References

1. Jinzhou, C., Liao, W.H.: Design and control of a Magnetorheological actuator for leg exoskeleton. In: IEEE International Conference on Robotics and Biomimetics (ROBIO) (2007)
2. Belyaev, E.S.: In: Belyaev, E.S., Ermolaev, A.I., Titov, E.Y., Tumakov, S.F., Plekhov, A.S. (eds.). Magnetorheological Fluids: Technologies of Creation and Application, 94 pp. Nizhny Novgorod State Technical University named R.E. Alekseeva (2017)
3. Simon, L.: Online Learning Algorithm for Structural Control Using Magnetorheological Actuatorsm, 88 pp. Massachusetts Institute of Technology (2007). https://dspace.mit.edu/bitstream/handle/1721.1/39271/170931934-MIT.pdf?sequence=2
4. Psomopoulou, E., Doulgeri, Z., Rovithakis, G., Tsagarakis, N.: A simple controller for a variable stiffness joint with uncertain dynamics and prescribed performance guarantees. In: IEEE/RSJ International Conference on Intelligent Robots and Systems, pp. 5071–5076. Vilamoura, Algarve (2012). https://doi.org/10.1109/BioRob.2012.6290757
5. Blinov, A.O., Borisov, A.V., Mukharlyamov, R.G.: Variable length exoskeleton link model with adjustable stiffness. In: All-Russia Conference on Problems in Dynamics, Particle Physics, Plasma Physics and Optoelectronics. Proceedings, pp. 165–170. RUDN University, Moscow (2021)
6. Blinov, A., Borisov, A., Konchina, L., Novikova, M.: Applying the models of magneto- rheological substances in the study of exoskeleton variable-length link with adjustable stiffness. J. Appl. Inform. **98**(2), 133–142 (2022). https://doi.org/10.37791/2687-0649-2022-17-2-133-142
7. Ahmadkhanlou, F., Zite, J.L., Washington, G.N.: A magnetorheological fluid-based controllable active knee brace. In: Proceedings SPIE, vol. 6527, pp. 652700 (2007). https://doi.org/10.1117/12.715902
8. Andrade, R.M., Fabriz Ulhoa, P.H., Fragoso Dias, E.A.: Design and testing a highly backdrivable and kinematic compatible magneto-rheological knee exoskeleton. J. Intell. Mater. Syst. Struct. **5** (2022). https://doi.org/10.1177/1045389X221117496
9. Blinov, A., Borisov, A., Konchina, L., Novikova, M.: Applying the models of magneto-rheological substances in the study of exoskeleton variable-length link with adjustable stiffness. J. Appl. Inform. **17**(2), 133–142 (2022). https://doi.org/10.37791/2687-0649-2022-17-2-133-142
10. Bougrinat, Y.: Design and Development of a Lightweight Ankle Exoskeleton for Human Walking Augmentation, 114 pp. Université de Montréal, Avril (2018). https://publications.polymtl.ca/3076/1/2018_YacineBougrinat.pdf
11. Carlson, J.: Magnetorheological fluid actuators. In: Adaptronics and Smart Structures: Basics, Materials, Design, and Applications, pp. 1808 (2013)
12. Chen, J., Liao, W.: Design, testing and control of a magnetorheological actuator for assistive knee braces. Smart Mater. Struct. **19**, 035029 (2010)
13. Laflamme, S.: Online Learning Algorithm for Structural Control Using Magnetorheological Actuators, 88 pp. Massachusetts Institute of Technology (2007). https://dspace.mit.edu/bitstream/handle/1721.1/39271/170931934-MIT.pdf?sequence=2
14. Blinov, A.O., Borisov, A.V., Konchina, L.V., Novikova M.A., Chigarev, A.V.: Development of properties control methods for magneto-rheological substance in order to regulate the stiffness of exoskeleton variable-length link. Adv. Eng. Res. **22**(4), pp. 296–305 (2022). https://doi.org/10.23947/2687-1653-2022-22-4-296-305
15. Zubarev, A.Y., Iskakova, L.Y., Chirikov, D.N.: On the nonlinear rheology of magnetic fluids. Colloid J. **73**(3), 320–333 (2011)

16. Zubarev, A.Y., Chirikov, D.N.: Magnetorheological properties of ferrofluids with cluster particles. Colloid J. **75**(5), pp. 567–576 (2013)
17. Konovalova, N.I., Martynov, S.I.: Dynamics of magnetic particles in a viscous fluid. Izv. Volga region. Physical and mathematical sciences. Mathematics. **3**(11), 3–11 (2009)
18. Makarova, L.A.: Study of the magnetic and electrical properties of composite rheological materials based on ferromagnetic and ferroelectric fillers. In: Abstract of the Dissertation of the Candidate of Physical and Mathematical Sciences, 27 pp. Moscow (2018)
19. Naigert, K.V., Tselishchev, V.A.: Applied Properties of Magnetorheological Fluids: Textbook [Electronic Resource]. Ufa State Aviation Technical University (USATU), Ufa (2021). https://www.ugatu.su/media/uploads/MainSite/Ob%20universitete/Izdateli/El_izd/2021-81
20. Chirikov, D.N.: Theoretical Study of the Rheological Properties of Bidisperse Magnetic Fluids, 17 pp. Research report. FGAOU HPE "Ural Federal University named after the first President of Russia B.N. Yeltsin. Ekaterinburg (2013). https://elar.urfu.ru/bitstream/10995/20813/1/chirikov_2.1.1.1-50.pdf
21. Yusupbekov, K.A., Sobirov, M.M., Yuldashev, A.R.: Active vehicle suspensions with shock absorbers of variable stiffness. Sci. Technol. Educ. **2**(66), 18–27 (2020)
22. Lebedev, A.V.: Viscosity of magnetic fluids must be modified in calculations of dynamic susceptibility. J. Magn. Magn. Mater. **431**, 30–32 (2017). https://doi.org/10.1016/j.jmmm.2016.09.110
23. Dollar, A.M., Herr, H.: Lower extremity exoskeletons and active orthoses: challenges and state-of-the-art. IEEE Trans. Rob. **24**, 144–158 (2008)
24. Glowinski, S., Krzyzynski, T., Bryndal, A., Maciejewski, I.: A kinematic model of a humanoid lower limb exoskeleton with hydraulic actuators. Sensors. **20**, 6116 (2020). https://doi.org/10.3390/s20216116
25. Lee, T., Lee, D., Song, B., Baek, Y.S.: Design and control of a polycentric knee exoskeleton using an electro-hydraulic actuator. Sensors **20**, 211 (2020). https://doi.org/10.3390/s20010211

Part III
Stochastic Systems and Game Theory

Chapter 9
Theoretical Guarantees for Satisfaction of Terminal State Constraints for Nonlinear Stochastic Systems

Ali Pakniyat

Abstract In several engineering applications, it is desired to bring a system from an initial configuration to a specific terminal configuration. A motivational example is the vertical landing of reusable rockets which are required to come to full stop at an exact location on the landing platform in an upright configuration with all linear and angular velocities coming to zero. While in a deterministic setting, one can study these problems and provide theoretical guarantees for the satisfaction of the terminal state requirements, e.g., by employing the Pontryagin Minimum Principle (PMP), no such guarantees can be provided for exact satisfaction of terminal state constraints in a stochastic setting and, inevitably, one needs to seek alternative expressions of the desired requirements and establish guarantees for those alternatives. This article presents two novel approaches, each with an alternative expression of the terminal state requirement, and each providing theoretical guarantees for optimality and the satisfaction of the associated terminal state constraints. The first approach is to impose a constraint on the conditional expectations of the terminal state at all future times in which case the associated optimality conditions are expressed in the form of the Terminally Constrained Stochastic Minimum Principle (TC-SMP). The second approach is to impose a terminal state constraint as the matching of the probability distribution of the terminal state with a desired probability distribution in which case the associated optimality conditions are expressed using Hamilton-Jacobi (HJ) type equations. Numerical examples are provided to illustrate the results.

Keywords Hamilton-Jacobi inequalities · Necessary conditions · Nonlinear systems · Optimal control · Probability constraints · Probability distribution control · Steering theory · Stochastic maximum principle · Stochastic processes · Stochastic systems

A. Pakniyat (✉)
Department of Mechanical Engineering, University of Alabama, Tuscaloosa, AL, USA
e-mail: apakniyat@ua.edu

D. Azimov (ed.), *Proceedings of the IUTAM Symposium on Optimal Guidance and Control for Autonomous Systems 2023*, IUTAM Bookseries 40,
https://doi.org/10.1007/978-3-031-39303-7_9

9.1 Introduction

In several missions for aerial and underwater vehicles, the system's state is required to be pinpoint delivered to a desired destination state. A motivational example is the landing of a reusable rocket, e.g., the booster rocket of SpaceX Falcon 9, which is required to come to full stop conditions at an exact location on the landing platform in an upright configuration with all linear and angular velocities coming to zero. In the absence of dynamic uncertainty, one can formulate the problem as a deterministic optimal control problem with a fixed terminal state and invoke powerful theoretical results such as the Pontryagin Minimum Principle (PMP) to obtain the optimal input signal together with the associated optimal trajectory. It is worth remarking that the celebrated Hamilton Jacobi Bellman (HJB) equation is not applicable for this problem since the terminal (boundary) condition for the HJB equation is not well defined (i.e., it turns into a singular function).

In the presence of a stochastic diffusion, these state steering problems are more challenging and have been the subject of a limited number of studies. More precisely, the majority of studies assume linearity of the dynamics and quadratic forms for the cost, so that the associated probabilities take the form of Gaussian distributions. In this case, and in the absence of any additional state constraints, the dynamics of the mean state process and the covariance state process can be shown to be decoupled. Within an infinite time horizon setting, the problem has been formulated as the association of a steady-state distribution with its mean being at the desired terminal location, and a comprehensive study over the assignable covariances for the infinite horizon problem is presented in [16, 21, 52, 53]. For linear stochastic systems over finite time horizons, a similar philosophy is taken in both continuous time and discrete time settings, and the associated distribution assignment methodologies are studied by [1–3, 9–11, 15, 17, 31, 32]. The accommodation of input constraints is considered in [2], and convex relaxations for linear systems subject to chance constraints, which are probabilistic constraints that impose a maximum probability of constraint violation, are studied in [31, 32]. Within the same class of linear quadratic systems, the accommodation of information obtained as the time progresses is presented in the form of a model predictive control (MPC) based approach in [25, 28, 30, 45].

Extensions of the probability distribution assignment to nonlinear systems has been presented for feedback-linearizable systems [6], and implementation through iterative linearization is proposed in [43] and via differential dynamic programming approximations [54]. More recently, new results have emerged via reformulations of the probability assignment problem as nonlinear Schrödinger bridge problem [7, 29]. However, a key limitation of the current nonlinear Schrödinger bridge results is that the Itô differential equation governing the dynamics of the nonlinear stochastic systems must take the special form of a gradient flow. In contrast, for stochastic systems whose dynamics are governed by a general class of nonlinear Itô differential equation, whose costs take general nonlinear forms, and whose desired probability distributions are permitted to take general (not necessarily Gaussian) distributions, the necessary optimality conditions are established in [35] in the form of Hamilton-

Jacobi problems where the optimization objective, in addition to the value function evaluation at the initial conditions, includes an extra term which is the integral of the product of the value function at the terminal time and the desired probability distribution. The methodology used in [35] is based upon the accommodation of terminal distribution constraints on the convex duality method for optimal control problems which was initiated by Vinter and Lewis [50, 51] for deterministic control systems and, later, by Fleming and Vermes for piecewise deterministic [47] and stochastic [14] processes. The fundamental idea of this approach is the introduction of a weak formulation that embeds the original (strong) problem into a convex linear program over the space of Radon measures. Upon establishing the equivalence of the two problems, new necessary and sufficient optimality condition are obtained by invoking the Fenchel-Rockafellar duality theorem. This approach is particularly useful in characterization of optimal policies in certain desirable classes of controls by investigating the extreme points of the set of Hamilton-Jacobi problems (see e.g. [5, 13, 22, 57]). For deterministic control systems, convex duality based numerical algorithms are established in [12, 23, 40] for continuous systems, and in [27, 46, 56] for hybrid systems.

A fundamental limit of methodologies based on the assignment of probability distributions is that the studied probabilities are conditioned on the filtration at the initial time. In contrast, as proposed in [38, 39] the employment of the Stochastic Minimum Principle (SMP) yields a natural accommodation of filtration-adapted updates because the same adaptation requirement must be provided for the adjoint process. In other words, the optimal input expressed in terms of the adjoint process is adapted to the current time filtration, since the solution of the backward stochastic differential equation (BSDE) for the adjoint process must remain adapted to the same forward filtration. This important characteristic provides an opportunity to impose terminal state constraints at all times, as apposed to the current literature where constraints are imposed on probability distributions as viewed at the initial time. In order to solve the associated problem, we invoke the Stochastic Maximum Principle (SMP) presented in [41] and, in particular, the version with terminal state constraints [41, Theorem 5], henceforth called the Terminally Constrained Stochastic Minimum Principle (TC-SMP). While, in general, obtaining numerical solutions to the BSDEs of the adjoint process are computationally expensive, for a class of linear stochastic systems with quadratic costs, we derive analytical solutions to the adjoint equation in terms of the system's state transition matrix, its controllability Gramian and the solution of a differential matrix Riccati equation. Moreover, the accommodation of various information structures on the TC-SMP is studied in [38] and, further, in this article.

The objective of this article is the presentation of a general framework within which theoretical guarantees are presented for the satisfaction of a large class of terminal state constraints for nonlinear stochastic systems, aw well as the illustration of the two key methodologies of the convex duality based HJ inequalities and the TC-SMP with analytic examples. The structure of the article is as follows.

Section 9.2 discusses the characterization of optimal solutions to the deterministic problem of steering the state towards a desired value, and Sect. 9.3 elaborates the discussion to the case with dynamic uncertainties and discusses how one can impose constraint on the state of stochastic systems without violating causality requirements. Section 9.4 presents the problem of constraining the family of conditional expectations of the terminal state and presents the Terminally Constrained Stochastic Minimum Principle (TC-SMP), its associated numerical algorithm and its specialization to linear stochastic systems with quadratic costs. Three examples are presented in this section to further illustrate the results. Section 9.5 presents the problem of constraining the probability distribution of the terminal state and establishes the associated optimality conditions, identifying the corresponding value function as the optimal solution to a family of Hamilton-Jacobi problems where the optimization objective, in addition to the value function evaluation at the initial conditions, includes an extra term which is the integral of the product of the value function at the terminal time and the desired probability distribution. A numerical algorithm based upon this methodology is presented and the results are illustrated via two numerical examples. Concluding remarks are presented in Sect. 9.6.

9.2 Deterministic Case Revisited

Let us first recall the fixed end point optimal control problem for the deterministic case, i.e., in the absence of dynamic uncertainties. For the nonlinear control system with the dynamics

$$\dot{x}_t = f(x_t, u_t, t) \tag{9.1}$$

subject to a given initial state $x_{t_0} = x_0$, and given a fixed finite time horizon $[t_0, t_f]$, we would like to find an input signal input signal $[u] := \{u_s : t_0 \le s \le t_f\}$ which brings the state to a desired value $x_{t_f} = x_f \in \mathbb{R}^n$. Furthermore, it is often the case that among all such inputs, one would like to find an optimal input which minimizes the cost

$$J(t_0, x_0, [u]; x_f) = \int_{t_0}^{t_f} \ell(x_s, u_s, s)\mathrm{d}s \tag{9.2}$$

For these problems, one can identify optimal inputs from the necessary optimality conditions of the Minimum Principle [42] which states that for the optimal input process $[u^*] \equiv \{u_s^* : t_0 \le s \le t_f\}$ and along the corresponding optimal trajectory $[x^*] = \{x_t^* : t_0 \le t \le t_f, \text{ where } x_t^* = x_0 + \int_{t_0}^t f(x_s^*, u_s^*, s)\mathrm{d}s\}$ there exist a constant $\gamma \in \{0, 1\}$ and an adjoint process $[\lambda^*] = \{\lambda_s^* : t_0 \le s \le t_f\}$ such that the Hamiltonian defined by

$$H(x, u, \lambda, \gamma, t) := \gamma\, \ell(x, u, t) + \lambda^\top f(x, u, t) \tag{9.3}$$

is minimized with respect to u, i.e.,

$$H(x_t^*, u_t^*, \lambda_t^*, \gamma, t) \leq H(x_t^*, u, \lambda_t^*, \gamma, t), \qquad \text{for all } u \in U \subset \mathbb{R}^m \tag{9.4}$$

where

$$\dot{x}_t^* = \left.\frac{\partial H}{\partial \lambda}\right|_{(x_t^*, u_t^*, \lambda_t^*, \gamma, t)} \equiv f(x_t^*, u_t^*, t) \tag{9.5}$$

$$\dot{\lambda}_t^* = \left.\frac{-\partial H}{\partial x}\right|_{(x_t^*, u_t^*, \lambda_t^*, \gamma, t)} \equiv -\gamma \frac{\partial \ell(x_t^*, u_t^*, t)}{\partial x} - \frac{\partial f(x_t^*, u_t^*, t)}{\partial x}^{\top} \lambda_t^* \tag{9.6}$$

subject to $x_{t_0}^* = x_0$ and $x_{t_f}^* = x_f$. If the problem is normal (see, e.g., [24]), it is possible to satisfy the necessary conditions with the constant γ taken to be 1.

In contrast to the Minimum Principle which is well suited for fixed terminal state problems, Dynamic Programming [4] and the associated Hamilton Jacobi Bellman (HJB) equation (see, e.g., [55]) is not able to handle terminal state constraints. As noted by [48] for terminally constrained problems of the type $x_{t_f} \in S \subset \mathbb{R}^n$, "the value function has at most only a subsidiary role. This is because, unless stringent conditions are imposed on the data, we cannot expect any longer that the value function will be defined on a sufficiently large subset of $\mathbb{R} \times \mathbb{R}^n$, or be sufficiently regular, for it to serve as a Carathéodory function". When the set S becomes a singleton, i.e., $x_{t_f} \in \{x_f\}$, then the HJB equation

$$\frac{\partial V(t,x)}{\partial t} + \inf_{u \in U} \left[\left(\frac{\partial V(t,x)}{\partial x} \right)^{\top} f(x,u,t) + \ell(x,u,t) \right] = 0 \tag{9.7}$$

becomes subject to the singular terminal condition

$$V(t_f, x) = \begin{cases} 0, & x = x_f, \\ \infty, & x \neq x_f. \end{cases} \tag{9.8}$$

However, by invoking convex duality relation between the space of measures and that of continuous functions, it is possible [48–51] that the value function can be identified as the upper envelope (i.e., supremum) of the smooth subsolutions of the Hamilton-Jacobi inequalities [48]

$$\begin{aligned} V(t_0, x_0) = \sup_{v \in C^1([t_0, t_f] \times \mathbb{R}^n)} \Bigg\{ & v(t_0, x_0) : \\ & \frac{\partial v(t,x)}{\partial t} + \left(\frac{\partial v(t,x)}{\partial x} \right)^{\top} f(x,u,t) + \ell(x,u,t) \geq 0, \\ & v(t_f, x) \leq 0, \text{ for } x = x_f. \Bigg\} \end{aligned} \tag{9.9}$$

While the above result is promising in the sense that it shows the ability of convex duality method to provide a solution to the deterministic fixed endpoint problem, its special form in its deterministic setting hinders the true nature of this identification and it cannot be directly extended for its stochastic equivalent problem. As a matter of fact, if instead of the above identification, [48] presented their results as

$$V(t_0, x_0) = \sup_{v \in C^1([t_0,t_f]\times\mathbb{R}^n)} \Bigg\{ v(t_0, x_0) - v(t_f, x_f) : \frac{\partial v(t,x)}{\partial t} + \left(\frac{\partial v(t,x)}{\partial x}\right)^{\top} f(x,u,t) + \ell(x,u,t) \geq 0, \quad v(t_f, x) \leq 0, \ \text{for all } x \in \mathbb{R}^n. \Bigg\} \tag{9.10}$$

then the discovery of its stochastic counterpart would have been made much earlier than [35] and Theorem 3 in this article.

9.3 The Stochastic Versions of the "Fixed" Endpoint Problem

We now formulate the stochastic equivalent of the deterministic fixed endpoint problem.

Let $\left(\Omega, \mathscr{F}, \{\mathscr{F}_t\}_{t=t_0}^{t_f}, \mathsf{P}\right)$ be a filtered probability space with $\mathscr{F}_t$ being an increasing family of sub σ-algebras of $\mathscr{F}$ such that $\mathscr{F}_{t_0}$ contains all the P-null sets, and $\mathscr{F}_{t_f} = \mathscr{F}$ for a fixed terminal time $t_f < \infty$. Consider a nonlinear stochastic systems governed by the controlled Itô differential equation

$$\mathrm{d}x_s = f(s, x_s, u_s)\,\mathrm{d}s + g(s, x_s)\,\mathrm{d}w_s, \tag{9.11}$$

where, at each $s \in [t_0, t_f]$, the system's state is denoted by $x_s \in \mathbb{R}^n$, its input is denoted by $u_s \in U \subset \mathbb{R}^m$, and the realization of a standard Wiener process is denoted by $w_s \in \mathbb{R}^d$. The input value set U is assumed to be convex and compact and the functions f and g are considered to be Lipschitz continuous functions over, respectively, $[t_0, t_f] \times \mathbb{R}^n \times U$ and $[t_0, t_f] \times \mathbb{R}^n$, which satisfy either boundedness or linear growth conditions.

Let $[u] := \{u_s : t_0 \leq s \leq t_f\}$ denote a nonanticipative, U-valued, input process such that $u_s \in U$ is progressively measurable with respect to $\mathscr{F}_s$ for all $s \in [t_0, t_f]$. We denote by $\mathcal{U}$ the set of all such inputs. We remark that the underlying policy for the determination of the input process $[u]$ can take any form, as long as the policy remains causal, that is, u_s does not depend on future values of the noise or the state. For instance, within a comprehensive closed loop policy, u_s is permitted to depend on s, $[x]_{t_0}^s$, $[w]_{t_0}^s$, $[u]_{t_0}^s$ and expectations of their future values under the $\mathscr{F}_s$ filtration.

However, it can be shown that there is no loss of optimality in feedback structures where u_s depends only on s, x_s and expectations of the cost gradient under the $\mathscr{F}_s$ filtration.

In this paper, we consider only the case with complete and accurate observations of the state. Thus, for time instances t and s such that $t_0 \le t \le s \le t_f$, within the interval $[t_0, t_f]$, and under the filtration $\mathscr{F}_t$, the variable x_s is treated as a *deterministic* variable whenever $s \le t$, and is treated as a *random* variable whenever $s > t$. However, at all future instances, $s \in (t, t_f]$, the state value x_s remains a random variable under the filtration $\mathscr{F}_t$. We define the notation

$$\mathbb{E}^{[u]}_{\mathscr{F}_t}[x_s] := \mathbb{E}\big[x_s \big| \mathscr{F}_t; [u]_t^{t_f}\big] \equiv \mathbb{E}\big[x_s \big| \mathscr{F}_t; [u]_t^{s}\big], \tag{9.12}$$

for the expected value[1] of x_s at $s \in [t, t_f]$, under the filtration $\mathscr{F}_t$ and given the input process $[u]_t^{t_f}$, where the last equality (conditioning on $[u]_t^s$ instead of $[u]_t^{t_f}$) is a consequence of the causality of the controlled process in (9.11).

The associated optimal control problem corresponds to the minimization of the cost

$$J\big(t, x_t; [u]_t^{t_f}\big) := \mathbb{E}^{[u]}_{\mathscr{F}_t}\left[\int_t^{t_f} \ell(x_s, u_s)\mathrm{d}s + L\big(x_{t_f}\big)\right]. \tag{9.13}$$

subject to appropriate terminal state constraint where, in the above cost, ℓ is a continuous function with polynomial growth.

A naïve approach for the consideration of terminal state constraints is to impose the constraint $x_{t_f} = x_f$ on the stochastic system (9.11). However, such a constraint violates causality as under each filtration $\mathscr{F}_t$ at time $t \in [t_0, t_f]$, future values of the state, i.e., $\{x_s : s \in (t, t_f]\}$ are random variable and, hence, it is not possible (i.e., there does not exists any nonanticipative input process $[u]$ such that the random variable x_{t_f} becomes deterministic so that it then satisfies a constraint such as $x_{t_f} = x_f$.

Another approach for the consideration of terminal state constraints is to impose the constraint $x_{t_f} \overset{a.s.}{=} x_f$ on the stochastic system (9.11). This constraint can be equivalently expressed as $\mathsf{P}(x_{t_f} = x_f) = 1$ or $\mathsf{P}(x_{t_f} \ne x_f) = 0$. While this constraint is both mathematically well-posed and practically desirable, a major challenge is the absence of theoretical guarantees of this time for stochastic processes.

In this article, we presents two novel approaches for the consideration of terminal state constraints on nonlinear stochastic systems, and each providing theoretical guarantees for optimality and the satisfaction of the associated terminal state constraints.

The first method is to impose a terminal state constraint as

$$\mathbb{E}^{[u]}_{\mathscr{F}_t}[x_{t_f}] = x_f, \tag{9.14}$$

[1] Since x_s becomes deterministic for $s = t$, and thus, $\mathbb{E}^{[u]}_{\mathscr{F}_t}[x_s] \equiv \mathbb{E}^{[u]}_{\mathscr{F}_t}[x_t] = x_t$ whenever $s = t$.

at all time instances $t \in [t_0, t_f]$, in order for the state to be steered to a desired value $x_f \in \mathbb{R}^n$

The second method is to impose a terminal state constraint as $x_{t_f}^{[u]} \sim \mathsf{p}_d$, i.e., we require the probability distribution of the terminal state to take the desired form p_d. This, by definition, signifies that for every Borel set $B_x \in \mathbb{R}^n$,

$$\mathsf{P}^{[u]}\big(x_{t_f} \in B_x\big) = \int_{B_x} \mathsf{p}_d(\mathrm{d}x), \tag{9.15}$$

where $\mathsf{P}^{[u]}(\cdot)$ denotes the probability of an event given the input $[u]$.

9.4 Constraining the Family of Conditional Expectations of the Terminal State

As mentioned earlier, in this method we impose a family terminal state constraints in the form of

$$\mathbb{E}^{[u]}_{\mathscr{F}_t}[x_{t_f}] = x_f, \tag{9.16}$$

at all time instances $t \in [t_0, t_f]$, in order for the state to be steered to a desired value $x_f \in \mathbb{R}^n$. Our strategy to solve this problem hinges on the restriction of the class of controllers to those yielding the expected value of the terminal state matching the desired value *under filtrations at all future times*. The novelty of this approach lies within its change of viewpoint, from the conventional conditioning the probability distributions on the information available at the design time, to the less explored, and mathematically more elaborate, approach of conditioning these probabilities on the family of σ-algebras of all possible scenarios for future uncertainties.

In the absence of the constraint (9.16), there are several versions of the Stochastic Maximum Principle (SMP), see, e.g., [55] for historical remarks on the development of the SMP. For the case of problems with terminal constraints, [41] first considered the case of a terminal constraint in the form of the total expectation of a nonlinear function of the state, including the terminal state constraint (9.16) consisting of n individual constraints $\mathbb{E}^{[u]}_{\mathscr{F}_t}\big[x^{(i)}_{t_f}\big] = \mu^{(i)}_f$, $(i = 1, 2, \cdots, n)$. However, the implementation of [41] still uses conditioning under the total expectation, which is equivalent to the filtration $\mathscr{F}_{t_0}$, instead of imposing the expectation on the σ-algebra of all potential realizations of the information at time t which are contained in $\mathscr{F}_t$.

9.4.1 Terminally Constrained Stochastic Minimum Principle (TC–SMP)

Theorem 1 [39] For the system (9.11), the optimal input for the cost (9.13) subject to the constraint (9.16) is determined from

$$u_s^* = \underset{u \in U \subset \mathbb{R}^m}{\operatorname{argmin}} \left\{ \ell(x_s, u) + \lambda_s^\top f(x_s, u) \right\}, \tag{9.17}$$

where the adjoint pair (λ_s, Λ_s), $s \in [t, t_f]$ are governed by the backward stochastic differential equation

$$\mathrm{d}\lambda_s = -\left(\frac{\partial f(x_s^*, u_s^*)}{\partial x} \lambda_s + \frac{\partial \ell(x_s^*, u_s^*)}{\partial x} \right) \mathrm{d}s + \Lambda_s \mathrm{d}w_s, \tag{9.18}$$

subject to the terminal condition

$$\lambda_{t_f} = \alpha \frac{\partial L(x_{t_f}^*)}{\partial x} + \beta, \tag{9.19}$$

where $\alpha \in \mathbb{R}$ and $\beta \in \mathbb{R}^n$ are constants which are not simultaneously zero. □

It shall be remarked that, as is conventional for backward stochastic processes, the second-order adjoint process Λ is implicitly defined as the (unique) process leading to the statisfaction of the terminal condition in (9.19). However, under sufficient smoothness of the functions, it has been shown (see e.g. [26]) that

$$\Lambda_s = g(x_s)^\top \frac{\partial}{\partial x} \lambda\big(s, x_s\big) \tag{9.20}$$

This is due to the fact that, whenever there exist a twice continuously differentiable function $\tilde{\lambda}$ such that $\lambda_s = \tilde{\lambda}(s, x_s)$, a direct application of Itô's formula yields

$$\begin{aligned} \mathrm{d}\tilde{\lambda}(s, x_s) = &\left(\frac{\partial \tilde{\lambda}(s, x_s)}{\partial s} + \frac{\partial \tilde{\lambda}(s, x_s)}{\partial x}^\top f(x_s, u_s) \right. \\ &\left. + \frac{1}{2} \mathrm{tr}\left(g(x_s)^\top g(x_s) \frac{\partial^2 \tilde{\lambda}(s, x_s)}{\partial x^2} \right) \right) \mathrm{d}s + g(x_s)^\top \frac{\partial \tilde{\lambda}(s, x_s)}{\partial x} \mathrm{d}w_s \\ &\overset{\lambda_s = \tilde{\lambda}(s, x_s)}{\overset{\downarrow}{=}} \mathrm{d}\lambda_s \overset{(9.18)}{=} -\left(\frac{\partial f(x_s, u_s)}{\partial x} \lambda_s + \frac{\partial l(x_s, u_s)}{\partial x} \right) \mathrm{d}s + \Lambda_s \mathrm{d}w_s \end{aligned} \tag{9.21}$$

which requires that

$$-\frac{\partial l(x_s,u_s)}{\partial x}-\frac{\partial f(x_s,u_s)}{\partial x}\tilde{\lambda}(s,x_s)=\frac{\partial \tilde{\lambda}(s,x_s)}{\partial s}$$
$$+\frac{\partial \tilde{\lambda}(s,x_s)}{\partial x}^{\top} f(x_s,u_s)+\frac{1}{2}\mathrm{tr}\Big(g(x_s)^{\top}g(x_s)\frac{\partial^2 \tilde{\lambda}(s,x_s)}{\partial x^2}\Big), \tag{9.22}$$
$$\Lambda_s=g(x_s)^{\top}\frac{\partial \tilde{\lambda}(s,x_s)}{\partial x}. \tag{9.23}$$

It shall be remarked that (9.22) is a partial differential equation (PDE).[2] Thus, it is possible to obtain the solution to the adjoint equation (9.18) from the solution of (9.22). However, it is possible (and in certain applications, it is more numerically efficient) to directly solve the adjoint equation (9.18) while invoking (9.23) to accelerate numerical integrations.

A general solution methodology for the TC-SMP is as follows.

- Step 1: Select $\epsilon > 0$, set the iteration counter to $k = 0$, and initiate the algorithm with an arbitrary control policy π^k, such that $u_s = \pi^k(s, x_s)$. Discretize the time interval $[t_0, t_f]$ into a set of discrete times $\{t_0, t_1, \dots, t_i, \dots, t_N = t_f\}$.
- Step 2: For M realizations of the Wiener process w, generate sample paths for the state forward in time using the policy π^k:

$$x_{t_{i+1}}=x_{t_i}+f\big(x_{t_i},\pi^k(t_i,x_{t_i})\big)\Delta t_i+g\big(x_{t_i}\big)\Delta w_i,\quad x_{t_0}=x_0.$$

- Step 3: For each sample path, compute the associated adjoint processes backward in time using

$$\lambda^k_{t_N}=\alpha\frac{\partial L(x_{t_N})}{\partial x}+\beta,$$
$$\Lambda^k_{t_N}=\alpha g\big(x_{t_i}\big)^{\top}\frac{\partial^2 L(x_{t_N})}{\partial x^2}$$
$$\lambda^k_{t_{i-1}}=\lambda^k_{t_i}-\mathbb{E}^{\pi^k}_{t_{i-1},x_{t_{i-1}}}\Bigg[\Bigg(\frac{\partial f\big(x_{t_i},\pi^k(t_i,x_{t_i})\big)}{\partial x}\lambda^k_{t_i}$$
$$+\frac{\partial \ell\big(x_{t_i},\pi^k(t_i,x_{t_i})\big)}{\partial x}\Bigg)\Delta t_i-\Lambda^k_{t_i}\Delta w_i\Bigg],$$
$$\Lambda^k_{t_{i-1}}\approx\frac{1}{\Delta t_{i-1}}\mathbb{E}^{\pi^k}_{t_{i-1},x_{t_{i-1}}}\big[\Delta w_{i-1}\lambda^k_{t_i}\big],$$

[2] Notice that in the classical (free end-point) LQG case where $f(x,u^*)=Ax+Bu^*=Ax-BR^{-1}B^{\top}\lambda$ and $l(x,u)=\frac{1}{2}x^{\top}Qx+\frac{1}{2}u^{\top}Ru=x^{\top}Qx+\lambda^{\top}BR^{-1}B^{\top}\lambda$, the conjecture $\lambda=\Pi x$ leads to the celebrated Riccati equation. More specifically, $\frac{\partial l}{\partial x}=Qx$, $\frac{\partial f}{\partial x}=A^{\top}$, $\frac{\partial \lambda}{\partial s}=\dot{\Pi}x$, $\frac{\partial \lambda}{\partial x}=\Pi$ and $\frac{\partial^2\lambda}{\partial x^2}=0$. Substitution of these expressions into (9.22) yields

$$-Qx-A^{\top}\Pi x=\dot{\Pi}x+\Pi^{\top}(Ax-BR^{-1}B^{\top}\Pi x)+0=\big(\dot{\Pi}+\Pi A-\Pi BR^{-1}B^{\top}\Pi\big)x,$$

which, after cancelling out the x-factor, is the Riccati equation.

- Step 4: Update the policy according to

$$\pi^{k+1}(t_i, x) = \underset{u \in \mathbb{R}^m}{\operatorname{argmin}} \{ \ell(x, u) + f(x, u)^\top \lambda_{t_i}^k \}.$$

- Step 5: If $\left| J(\pi^k) - J(\pi^{k-1}) \right| < \epsilon$, stop. Otherwise, increment k by and 1 and go to Step 1.

9.4.2 TC-SMP for Linear Quadratic Problems

In this section, we present the analytical solutions to the the TC-SMP in Theorem 1 for linear stochastic systems with quadratic cost. To this end, let the dynamics (9.11) be of the form

$$\mathrm{d}x_s = \big(A_s x_s + B_s u_s\big)\mathrm{d}s + D_s \mathrm{d}w_s, \tag{9.24}$$

where the time varying parameters in the system dynamics $A \in L^\infty([t_0, t_f]; \mathbb{R}^{n \times n})$, $B \in L^\infty([t_0, t_f]; \mathbb{R}^{n \times m})$, $D \in L^\infty([t_0, t_f]; \mathbb{R}^{n \times k})$, are essentially bounded measurable matrix functions of time.

For simplicity, we assume that the cost (9.13) is a quadratic function of the input and the terminal state, that is,

$$J\big(t, x_t, [u]_t^{t_f}\big) := \frac{1}{2}\mathbb{E}_{\mathscr{F}^t}^{[u]}\left[\int_t^{t_f} u_s^\top R_s u_s \mathrm{d}s + (x_{t_f} - \mu_f)^\top H_f\, (x_{t_f} - \mu_f) \right], \tag{9.25}$$

with $R \in L^\infty([t_0, t_f]; \mathcal{S}^{m \times m})$, $R_s > 0$, for all $s \in [t_0, t_f]$, and $H_f \in \mathcal{S}^{n \times n}$, $H_f \geq 0$, where $\mathcal{S}^{m \times m}$ denotes the space of $m \times m$-dimensional symmetric matrices.

We assume that the system (A_s, B_s) is controllable,[3] and that the system is noise controllable,[4] equivalently, $\mathrm{Im}(D_s) \subset \mathrm{Im}(B_s)$, for all $s \in [t_0, t_f]$, that is,

$$\forall w \in \mathbb{R}^k, \exists u \in \mathbb{R}^m \text{ s.t. } B_s u = D_s w. \tag{9.26}$$

Theorem 2 [38, 39] For the system (9.24) and the cost (9.25) subject to the constraint (9.16), the optimal input is determined by

[3] Hence, the Gramian (9.29) is full rank.

[4] As a requirement for solvability of the Riccati equations (9.30) and (9.31).

$$
\begin{aligned}
u_s^* = & -R_s^{-1}B_s^\top \Phi(t_f,s)^\top\big[\mathcal{G}(t,t_f)\big]^{-1}\left(\Phi(t_f,t)x_t-\mu_f\right)\\
& - R_s^{-1}B_s^\top \Pi(s;t_f)\Bigg(x_s-\Phi(t;s)x_t\\
& + \mathcal{G}(t,s)\Phi(s;t_f)^\top\big[\mathcal{G}(t,t_f)\big]^{-1}\Big(\Phi(t;t_f)x_t-\mu_f\Big)\Bigg),
\end{aligned} \tag{9.27}
$$

where $\Phi(t;s)\in\mathbb{R}^{n\times n}$ is the state transition matrix from t to s for the system (9.24), which is the solution of

$$
\dot{\Phi}\equiv\frac{\partial\Phi(t;s)}{\partial s}=A_s\Phi,\qquad\qquad \Phi(t;t)=I_{n\times n}, \tag{9.28}
$$

and where

$$
\mathcal{G}(\tau,t):=\int_t^\tau \Phi(s;\tau)B_sR_s^{-1}B_s^\top\Phi(s;\tau)^\top \mathsf{d}s, \tag{9.29}
$$

is the controllability Gramian (see e.g., [8, Theorem 6.1]) over the horizon $[t,\tau]\subset[t_0,t_f]$, and $\Pi(s;t_f)$ is the solution of the following Riccati equation

$$
\dot{\Pi}_s\equiv\frac{\mathsf{d}}{\mathsf{d}s}\Pi(s;t_f)=\Pi_sB_sR_s^{-1}B_s^\top\Pi_s-\Pi_sA_s-A_s^\top\Pi_s, \tag{9.30}
$$

subject to the terminal condition

$$
\Pi(t_f;t_f)=H_f. \tag{9.31}
$$

□

9.4.3 Numerical Illustrations

In order to illustrate the results of Theorem 1 and its specialization to linear quadratic Gaussian problems, Theorem 2, let us consider the following examples.

Example 1 Consider the scalar case of a linear stochastic system with the dynamics

$$
\mathsf{d}x_s:=(ax_s-bu_s)\mathsf{d}s+d\,\mathsf{d}w_s, \tag{9.32}
$$

with a,b,d scalar constants, and consider the problem of steering the state to the desired value $\mu_f\in\mathbb{R}$ by enforcing

$$
\mathbb{E}^{[u]}_{\mathscr{F}_t}[x_{t_f}]=\mu_f, \quad (16)
$$

at all $t \in [t_0, t_f]$, with the cost

$$J\big(t, x_t, [u]_t^{t_f}\big) := \frac{1}{2}\mathbb{E}^{[u]}_{\mathscr{F}_t}\Bigg[\int_t^{t_f} r\, u_s^2 \mathrm{d}s + h(x_{t_f} - \mu_f)^2\Bigg]. \tag{9.33}$$

where $r > 0$ and $h \in \mathbb{R}_{\geq 0} \setminus \{2ar/b^2\}$.[5]

For this problem, we can analytically represent Φ, $\mathcal{G}$ and Π as

$$\Phi(t; s) = e^{a(s-t)}, \tag{9.34}$$

$$\mathcal{G}(t, \tau) = \frac{b^2}{2a} e^{2a\,t_f}\left(e^{-2a\,t} - e^{-2a\,\tau}\right), \tag{9.35}$$

$$\Pi(s; t_f) = \frac{2ar}{b^2\left(1 - \frac{h}{h - \frac{2ar}{b^2}} e^{\frac{b^2}{r}(t_f - s)}\right)}. \tag{9.36}$$

and, therefore, the optimal input (9.27) becomes

$$\begin{aligned} u_s^* = &\frac{-2a}{br\left(e^{2a(t_f-t)} - 1\right)} e^{a(t_f-s)}\left(e^{a(t_f-t)} x_t - \mu_f\right) \\ &- \frac{2a}{b\left(1 - \frac{h}{h - \frac{2ar}{b^2}} e^{\frac{b^2}{r}(t_f - s)}\right)} \Bigg(x_s - e^{a(s-t)} x_t \\ &+ e^{a(t_f-s)} \frac{e^{-2at} - e^{-2as}}{e^{-2at} - e^{-2at_f}}\left(e^{a(t_f-t)} x_t - \mu_f\right)\Bigg). \end{aligned} \tag{9.37}$$

Let $a = b = d = r = h = 1$, and the time horizon be $[t_0, t_f] = [0, 2]$. For the steering towards the desired stated $\mu_f = 5$, from the initial condition $x_0 = -3$, the optimal input satisfying (9.16) for all $t \in [0, 2]$, and the associated trajectories for 50 sample paths are illustrated in Fig. 9.1. As can be seen from the figure, all trajectories are almost surely driven to the required terminal state at $t = t_f$. The associated probability distributions for the state and the input processes from the implementation of the TC-SMP on this system are illustrated in Fig. 9.2. It can be observed that the state distribution, starting from a lumped delta distribution at the initial condition, is steered towards a delta distribution at the desired terminal state while in intermediate times the state distribution does not remain lumped. In contrast, the input distribution, starting from a lumped delta distribution, does not remain lumped at any later time, which shows that the satisfaction of the desired terminal state comes at the expense of input uncertainty at later times over a wide range of values.

[5] The exclusion of this value, which occurs only if $a > 0$, is due to the appearance of $h - 2ar/b^2$ as a denominator in (9.37).

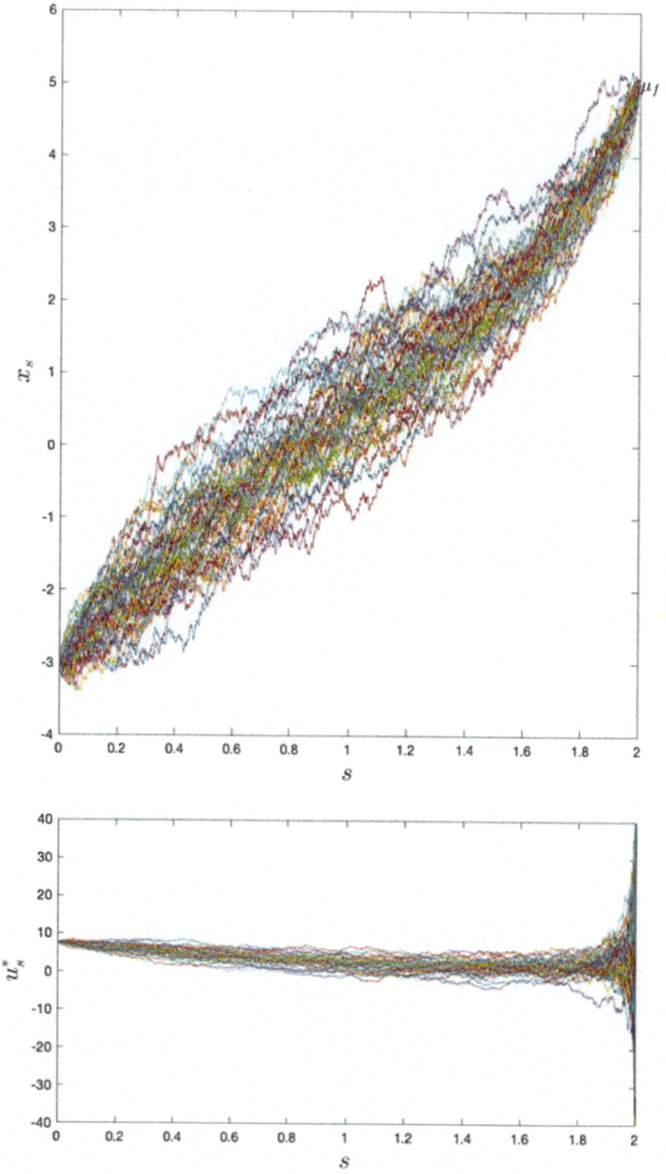

Fig. 9.1 Sample paths associated with the implementation of the terminally constrained stochastic minimum principle (TC-SMP) on the system in Example 1

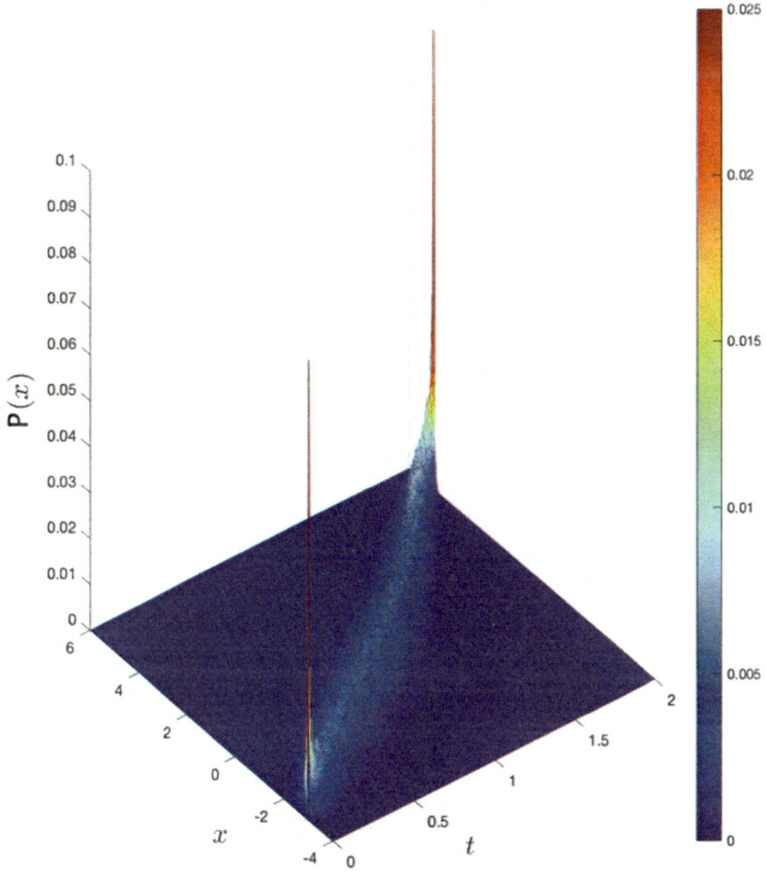

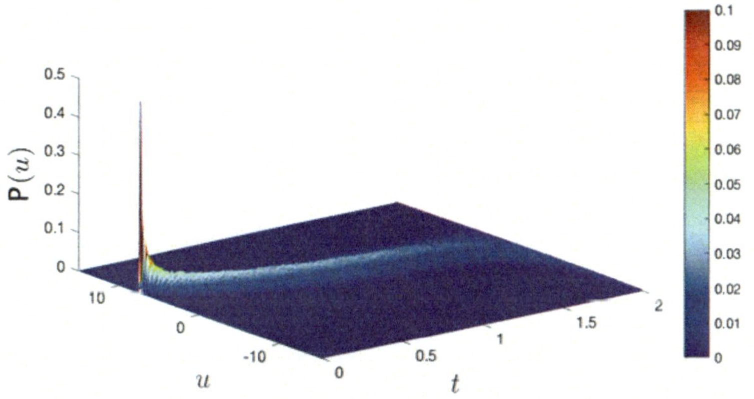

Fig. 9.2 Probability distributions associated with the implementation of the terminally constrained stochastic minimum principle (TC-SMP) on the system in Example 1

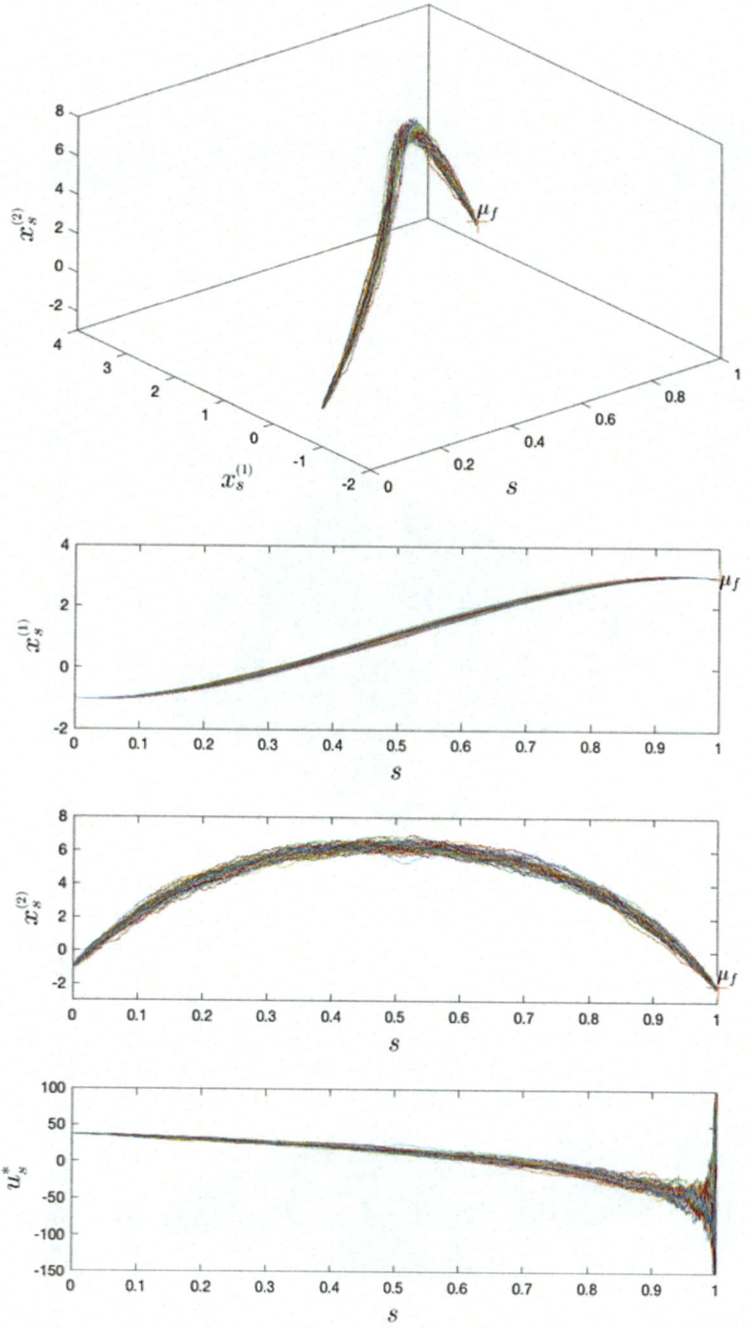

Fig. 9.3 The implementation of the terminally constrained stochastic minimum principle (TC-SMP) on the system in Example 2

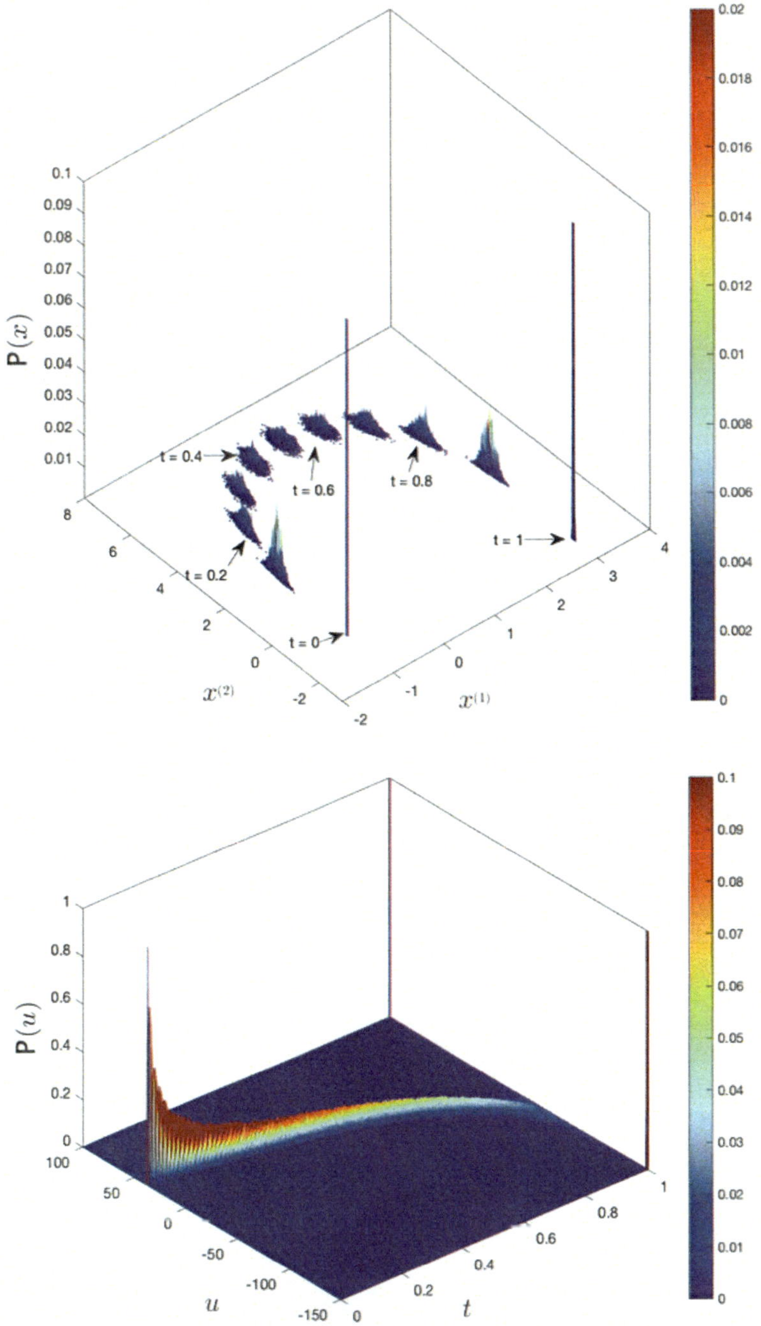

Fig. 9.4 The evolution of probability distributions of the state (top) and the input process (bottom) associated with the implementation of the terminally constrained stochastic minimum principle (TC-SMP) on the system in Example 2

Example 2 Consider the system governed by

$$\mathrm{d}x_s = \left(\begin{bmatrix} 0 & 1 \\ 2 & -3 \end{bmatrix} \begin{bmatrix} x_s^{(1)} \\ x_s^{(2)} \end{bmatrix} + \begin{bmatrix} 0 \\ 1 \end{bmatrix} u_s \right) \mathrm{d}s + \begin{bmatrix} 0 \\ 1 \end{bmatrix} \mathrm{d}w_s, \tag{9.38}$$

over the time horizon $[t_0, t_f] = [0, 1]$, starting from the initial condition $x_0 = [-1, -1]^\top$, and steered towards the desired terminal state by enforcing

$$\mathbb{E}^{[u]}_{\mathscr{F}_t}[x_1] = \begin{bmatrix} 3 \\ -2 \end{bmatrix}, \tag{9.39}$$

at all $t \in [0, 1]$, and consider the associated optimal control problem with the cost

$$J\big(t, x_t, [u]\big) := \mathbb{E}^{[u]}_{\mathscr{F}_t}\left[\int_t^{t_f} \frac{1}{2} u_s^2 \mathrm{d}s + \frac{1}{2} \left\| x_{t_f} - \mu_f \right\|^2 \right]. \tag{9.40}$$

The implementation of the TC-SMP for 50 Sample paths are illustrated in Fig. 9.3 and the evolution of the associated probability distributions for the state process and the input process are displayed in Fig. 9.4. In order to better illustrate the evolution of the probability distribution of the state process, the associated 3 dimensional snapshots of the state distribution at times $t \in \{0, 0.1, \ldots, 0.9, 1\}$ are displayed. As can be observed in these figures, the TC-SMP steers all trajectory realizations to the final desired state. It is worth remarking that the enforcement of the terminal state constraint (9.16) especially when $t \to t_f$, forces the controller to consume as much large values as required, that for x_t realizations away from μ_f, this requires the consumption of large input values. This can also be deduced from expression (9.27) for the optimal input, by noting that $\mathcal{G}(t, t_f) \to 0$ and therefore its inverse becomes as $t \to t_f$. It shall, however, be remarked that the associated singularity is isolated by noting that $t \leq s \leq t_f$

9.5 Constraining the Probability Distribution of the Terminal State

As mentioned earlier, in this method, we impose a terminal state constraint as $x_{t_f}^{[u]} \sim \mathsf{p}_d$, i.e., we require the probability distribution of the terminal state to take the desired form p_d. This, by definition, signifies that for every Borel set $B_x \in \mathbb{R}^n$,

$$\mathsf{P}^{[u]}\big(x_{t_f} \in B_x\big) = \int_{B_x} \mathsf{p}_d(\mathrm{d}x), \tag{9.41}$$

where $\mathsf{P}^{[u]}(\cdot)$ denotes the probability of an event given the input $[u]$. Hence, the value function at the initial time and state is defined as

$$V(t_0, x_0) = \inf_{[u]\in\mathcal{U}} \left\{ \mathbb{E}^{[u]}\left[\int_{t_0}^{t_f} \ell\,(x_s, u_s)\,\mathsf{d}s \right] \text{s.t. } x_T^{[u]} \sim \mathsf{p}_d \right\} \quad (P)$$

In the absence of the constraint (9.41), one can invoke the convex duality approach of [14] to identify the value function as the upper envelope of the smooth subsolutions of the Hamilton-Jacobi and the associated boundary value inequalities. However, the presence of the constraint requires a more elaborate version of the convex duality approach established in [35] which is presented below.

9.5.1 *Convex Duality and the Associated Hamilton–Jacobi (HJ) Inequalities*

Theorem 3 [35] For every $x_0 \in \mathbb{R}^n$ and given a desired terminal distribution p_d, the optimal cost (P) is obtained as

$$\begin{aligned} V(t_0, x_0) = \sup_{v\in C^2([t_0,t_f]\times\mathbb{R}^n)} &\left\{ v(t_0, x_0) - \int_{\mathbb{R}^n} v(t_f, x)\mathsf{p}_d(\mathsf{d}x), \right. \\ \text{s.t. } \frac{\partial v(t,x)}{\partial t} &+ \left[\frac{\partial v(t,x)}{\partial x}\right]^\top f(t,x,u) \\ &+ \frac{1}{2}\mathrm{tr}\Big(g(t,x)^\top g(t,x)\,\frac{\partial^2 v(t,x)}{\partial x^2}\Big) + \ell(t,x,u) \geq 0, \\ &\left. \text{for all } (t,x,u) \in [t_0, t_f] \times \mathbb{R}^n \times U \right\}. \end{aligned} \quad (9.42)$$

□

Based upon the Theorem 3, a general procedure for the numerical solution can be developed as follows.

- Step 1: Set the iteration counter to $k = 0$, and initiate the algorithm with an arbitrary terminal cost function $L^k(x)$.
- Step 2: Solve the HJB equation[6]

[6] If a classical solution does not exist, one needs to consider an additional supremization over subsolutions of the HJB, i.e., the family of functions (indexed by another iteration j, satisfying the HJ inequalities $\frac{\partial v_j^k(t,x)}{\partial t} + \min_{u\in U}\left\{\left(\frac{\partial v_j^k(t,x)}{\partial x}\right)^\top f(x,u,t) + \ell(x,u,t)\right\} \geq 0$, but subject to the equality conditions $v_j^k(t_f, x) = L^k(x)$ for all j. However, it can be shown that the suprimizing function (over all j) converges to the viscosity solution of the HJB equation (9.43).

$$\frac{\partial v^k(t,x)}{\partial t} + \min_{u\in U}\left\{\left(\frac{\partial v^k(t,x)}{\partial x}\right)^{\top} f(x,u,t) + \ell(x,u,t)\right\} = 0, \tag{9.43}$$

subject to $v^k(t_f, x) = L^k(x)$.

- Step 3: Evaluate $v^k(t_0, x_0) - \int_{\mathbb{R}^n} v^k(t_f, x)\mathsf{p}_d(\mathrm{d}x)$.
- Step 4: Update $L^{k+1}(x)$ using an ascent direction[7] for the cost $v^k(t_0, x_0) - \int_{\mathbb{R}^n} v^k(t_f, x)\mathsf{p}_d(\mathrm{d}x)$.

9.5.2 Numerical Illustration

In order to illustrate the results of Theorem 3 and the difficulties and challenges associated with its algorithmic implementation let us consider the following examples.

Example 3 First, consider the the scalar system

$$\mathrm{d}x_s = (x_s + u_s)\mathrm{d}s + \mathrm{d}w_s, \tag{9.44}$$

with the total cost

$$J(t_0, x_0, [u]) = \mathbb{E}\int_0^T \frac{1}{2}u_s^2 \mathrm{d}s. \tag{9.45}$$

and subject to the constraint $x_{t_f} \sim \mathsf{p}_d$ for the desired probability distribution $\mathsf{p}_d = \mathcal{N}(\mu_d, \sigma_d)$. In order to illustrate (9.42), consider the family of functions $\{v_\gamma\}_{\gamma\in\mathbb{R}_+\times\mathbb{R}}$, where for each $\gamma \equiv (h_\gamma, \mu_\gamma)$, the function

$$v_\gamma(t,x) = \frac{1}{2}\pi(t)x^2 + \beta(t)x + \alpha(t), \tag{9.46}$$

is constructed from the Riccati equations

$$\dot{\pi}(t) = \big(\pi(t)\big)^2 - 2\pi(t), \qquad \pi(T) = h_\gamma, \tag{9.47}$$

$$\dot{\beta}(t) = -\big(1 - \pi(t)\big)\beta(t), \qquad \beta(T) = -h_\gamma\mu_\gamma, \tag{9.48}$$

$$\dot{\alpha}(t) = \frac{1}{2}\big(\beta(t)\big)^2 - \frac{1}{2}\pi(t), \qquad \alpha(T) = \frac{1}{2}h_\gamma\mu_\gamma^2. \tag{9.49}$$

Then for every $\gamma\equiv(h_\gamma, \mu_\gamma) \in [0,\infty) \times \mathbb{R}$, the corresponding function $v_\gamma(t,x) \in C^\infty([t_0, t_f] \times \mathbb{R})$ satisfies

[7] Due to the computationally expensive nature of the cost, and the infinite dimensionality of the space of terminal costs, derivative-free methods such as Bayesian optimization shall be used in this general procedure.

$$\frac{\partial v_\gamma(t,x)}{\partial t} + \left[\frac{\partial v_\gamma(t,x)}{\partial x}\right](x+u) + \frac{1}{2}\frac{\partial^2 v_\gamma(t,x)}{\partial x^2} + \frac{1}{2}u^2 \geq 0, \tag{9.50}$$

for all $(t,x,u) \in [t_0, t_f] \times \mathbb{R} \times \mathbb{R}$.

For $x_0 = 0$, $T = 2$ and $\mathsf{p}_d = \mathcal{N}(3,1)$, the values of $v_\gamma(0,x_0) - \int_{\mathbb{R}^n} v_\gamma(T,x)$ are displayed over the region $(h_\gamma, \mu_\gamma) \in [0,5] \times [2.5, 3.5]$ in Fig. 9.5. As observed in the simulation, for this family of HJ-subsolutions, the maximum occurs at $(h_{\gamma^*}, \mu_{\gamma^*}) = (0.96, 3)$. As shown in [9], for LQG problems with Gaussian desired distributions, the value function is indeed quadratic and takes the Riccati form, and hence the desired value function V coincides with v_{γ^*} up to the constant $\int_{\mathbb{R}^n} v_{\gamma^*}(T,x)$.

Example 4 Now consider consider the the scalar system

$$\mathsf{d}x_s = (x_s + u_s)\mathsf{d}s + \mathsf{d}w_s, \tag{9.51}$$

with the total cost

$$J(t_0, x_0, [u]) = \mathbb{E}\int_0^T \frac{1}{2}u_s^2 \mathsf{d}s. \tag{9.52}$$

and subject to the constraint $x_{t_f} \sim \mathsf{p}_d$ where the desired probability distribution p_d is not Gaussian, e.g., when it's a mixture of two Gaussian distributions, e.g., $\mathsf{p}_d = \frac{1}{2}\mathcal{N}(\mu_1^d, \sigma_1^d) + \frac{1}{2}\mathcal{N}(\mu_2^d, \sigma_2^d)$. These problems, despite their LQ form of the dynamics and cost, cannot be solved by the conventional covariance control methodologies, e.g., [9]. In contrast, the results of Theorem 1 can be implemented in the following way to identify the value function and the corresponding optimal policy, as explained below.

Consider the family of functions $\{v_\gamma\}$ where for each $\gamma = (\eta_\gamma, \rho_\gamma, h_1^\gamma, \mu_1^\gamma, h_2^\gamma, \mu_2^\gamma)$, the function

$$v_\gamma(t,x) = \frac{-1}{\eta_\gamma}\ln\Bigg(e^{-\eta_\gamma\rho_\gamma\left(\frac{1}{2}\pi_1(t)x^2 + \beta_1(t)x + \alpha_1(t)\right)} + e^{-\eta_\gamma(1-\rho_\gamma)\left(\frac{1}{2}\pi_2(t)x^2 + \beta_2(t)x + \alpha_2(t)\right)}\Bigg) \tag{9.53}$$

where π_i, β_i, α_i, $i = 1, 2$, satisfy (9.47)–(9.49) with $(h_\gamma, \mu_\gamma) = (h_i^\gamma, \mu_i^\gamma)$. Then it can be verified (see, e.g., [44]) that (9.53) satisfies the HJ-inequality (9.50) for all $(t,x,u) \in [t_0, t_f] \times \mathbb{R} \times \mathbb{R}$.

In order to restrict the search [as the primary purpose of the example is to illustrate the characterization of the value function by (9.42)] we consider the symmetric case where $\mu_1^d = -\mu_2^d = \mu_d$, $\sigma_1^d = \sigma_2^d$, and $x_0 = (\mu_1^\gamma + \mu_2^\gamma)/2 = 0$, thus $\rho_\gamma = 1/2$, and $\mu_1^\gamma = \mu^d$, and $\mu_2^\gamma = -\mu^d$. In particular, we consider the case with $x_0 = 0$, $T = 2$ and the desired distribution $\mathsf{p}_d = \frac{1}{2}\mathcal{N}(3,1) + \frac{1}{2}\mathcal{N}(-3,1)$, and hence we restrict attention to the sequence of functions parameterized by $\gamma = (\eta_\gamma, \frac{1}{2}, h_\gamma, 3, h_\gamma, -3) \equiv$

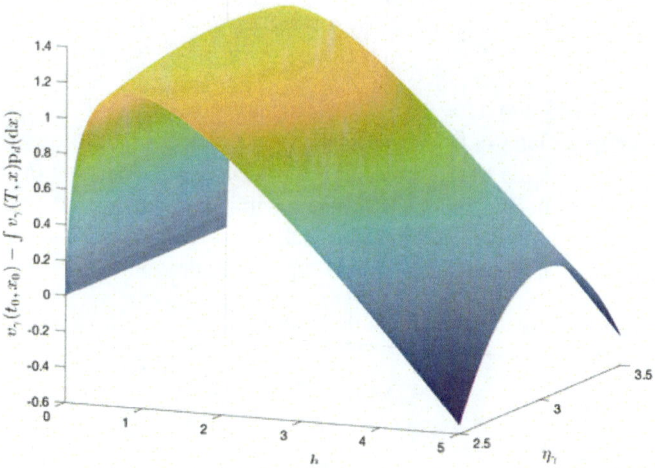

(a)

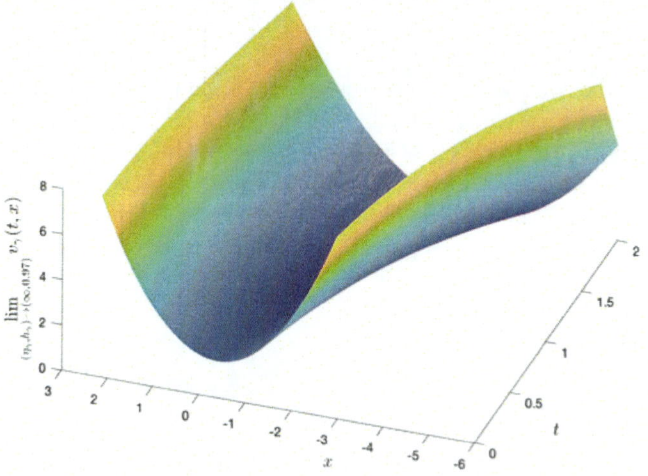

(b)

Fig. 9.5 The identification of the parameters of the value function for $x_0 = 0, T = 2$ and the desired distribution $\mathsf{p}_d = \mathcal{N}(3, 1)$ employing (9.42) and the class of functions $\{v_\gamma\}$ defined by (9.46). **a** Parameterization. **b** Associated limiting function. **c** Trajectories of 200 optimal sample paths. **d** Distribution pf x_T for 100,000 sample paths versus desired distribution

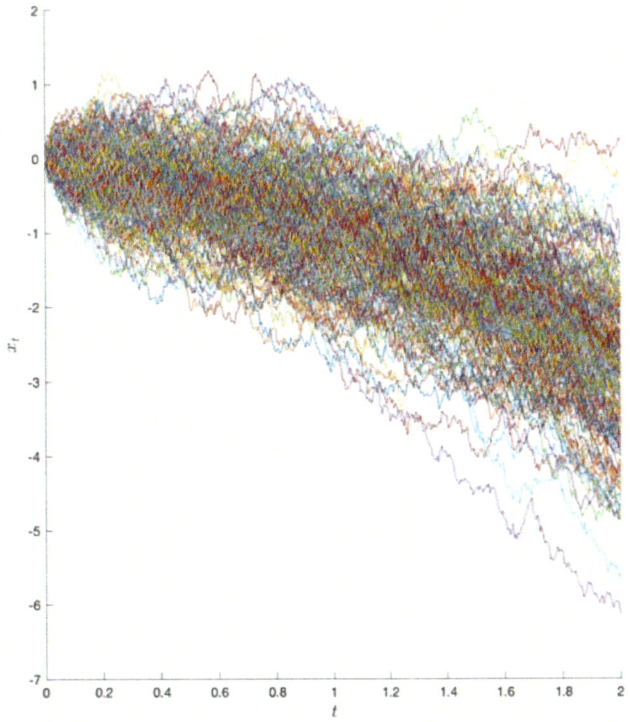

(c)

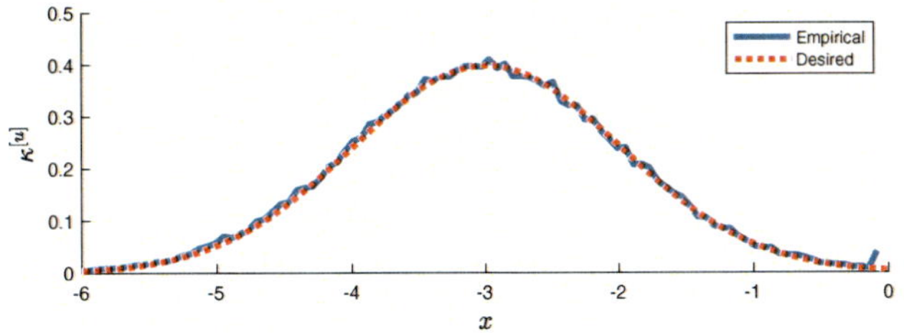

(d)

Fig. 9.5 (continued)

(η_γ, h_γ). The corresponding values of $v_\gamma(0, x_0) - \int_{\mathbb{R}^n} v_\gamma(T, x)$ are displayed over the region $(\eta_\gamma, h_\gamma) \in [1, 10^3] \times [0, 4]$ in Fig. 9.6.

As observed in Fig. 9.6a, for the family (9.53) of HJ-subsolutions, the supremum in not attained over the bounded domain and, while $h_{\gamma^*} = 0.97$, the suprimum requires $\eta_\gamma \to \infty$. Indeed, the value function in this case is nonsmooth and is required to be identified from $\lim_{(\eta_\gamma, h_\gamma) \to (\infty, 0.97)} v_\gamma(t, x)$ as displayed in Fig. 9.6b.

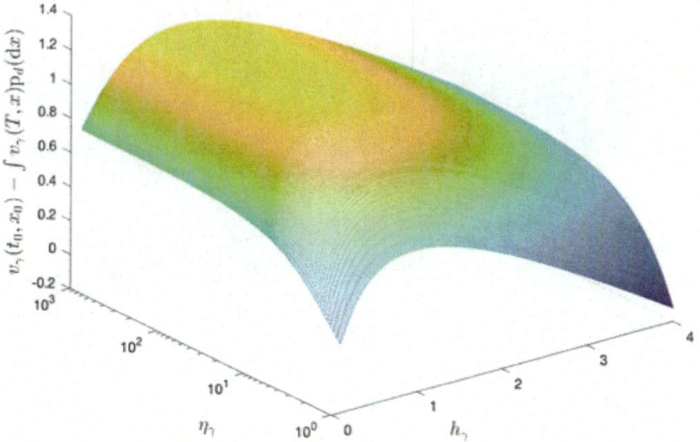

(a)

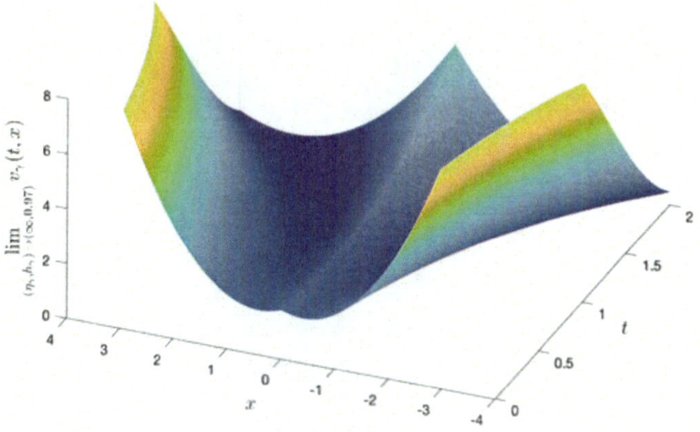

(b)

Fig. 9.6 The identification of parameters (**a**), the associated value function (**b**), sample paths (**c**), and the corresponding distribution (**d**) for $x_0 = 0$, $T = 2$ and $\mathsf{p}_d = \frac{1}{2}\mathcal{N}(3, 1) + \frac{1}{2}\mathcal{N}(-3, 1)$ employing (9.42) and the class of functions $\{v_\gamma\}$ defined by (9.53). **a** Parameterization. **b** Associated limiting function. **c** Trajectories of 200 optimal sample paths. **d** Distribution pf x_T for 100,000 sample paths versus desired distribution

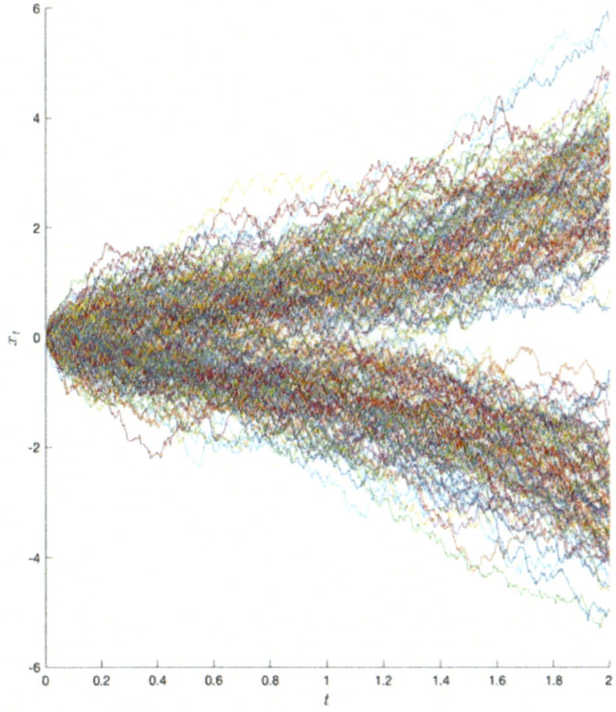

(c)

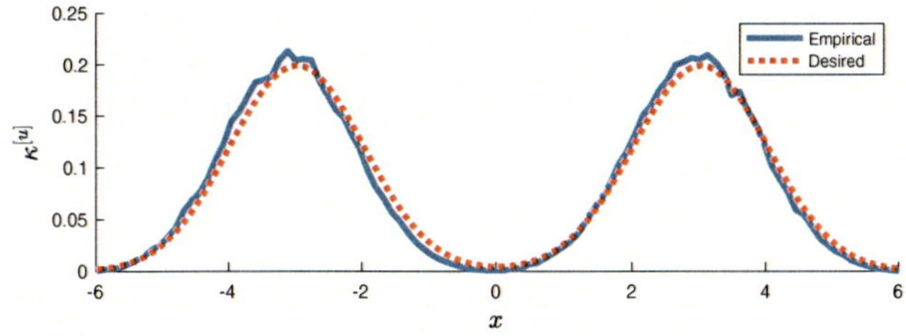

(d)

Fig. 9.6 (continued)

In order to illustrate that the desired probability distribution is attained, the optimal trajectories of 200 sample paths are displayed in Fig. 9.6c and the empirical distribution of these trajectories obtained from 100,000 sample paths are displayed in Fig. 9.6d.

9.6 Concluding Remarks

This article presents two novel approaches for steering the state of nonlinear stochastic systems towards a desired value. Since equating the random variable of the term state to the desired value violates causality requirements, each of the presented methods provide an alternative causal expression of the terminal state requirement, and for each of these alternatives, theoretical guarantees for optimality and the satisfaction of the associated terminal state constraints are provided. The first approach is to impose a constraint on the first moment (expected value) of of the terminal state and to reimpose this constraint under conditional expectations at all future times. For this case the associated optimality conditions are expressed in the form of the Terminally Constrained Stochastic Minimum Principle (TC-SMP). The second approach is to impose a terminal state constraint as the matching of the probability distribution of the terminal state with a desired probability distribution in which case the associated optimality conditions are expressed using Hamilton-Jacobi (HJ) type equations. While both the TC-SMP and the convex duality based HJ inequalities are power methods in establishing the necessary optimality conditions for their corresponding optimal control problems, it is worth comparing the two.

The TC-SMP performs very successfully in terms of steering of the state towards the desired location an almost all sample paths are being steered to the desired value, however, this is achieved at the expense of increased control effort at the final time. Numerical examples illustrate that the TC-SMP achieves it goal by exploiting the unboundedness of the input value set. In other words, the enforcement of the terminal state constraint (9.16) especially when $t \to t_f$, forces the controller to consume as much large values as required, that for x_t realizations away from μ_f, this requires the consumption of large input values. This can also be deduced from expression (9.27) for the optimal input of the LQG case, by noting that $\mathcal{G}(t, t_f) \to 0$ and therefore its inverse becomes as $t \to t_f$. It shall, however, be remarked that the associated singularity is isolated by noting that $t \leq s \leq t_f$.

The convex duality approach, in contrast, is by definition a methodology for problems subject to terminal constraints of distribution type and, hence, its accuracy in delivering the state to a desired value depends on the expression of the desired probability distribution. Moreover, due to the involvement of the second moment (covariance) and higher order moments of the state distribution, this approach is inevitably time-inconsistent as the desired probability itself is only expressed under the total (and not other conditional) probability measure. The convex duality approach characterizes the value function by a set of Hamilton-Jacobi type conditions. A benefit of this characterization is that it holds true despite potential nonsmoothness of the value function. However, special care must be taken as (i) the sequence of test functions might converge to a value smaller than the optimal cost which suggests that the family of functions does not contain a function characterizing the value function, or (ii) a maximum does not exist but the suprimum (and hence the value function) can be characterized from the limiting behavior of the associated family of functions.

In both of these cases, a notion of suboptimality is required to be developed when the solution is numerically constructed from the associated optimization problem.

Future research directions include the accommodation of chance constraints, which are probabilistic constraints that impose a maximum probability of constraint violation, as a nonlinear systems extension of those established for linear systems in [31, 32], and also the accommodation of hybrid systems features, in particular the presence of controlled and autonomous switchings with exact equality and almost surely equality constraints as switching manifolds as in [33] and nonlinear jump maps as in [34, 36], as well as the development of numerical algorithms for numerical solutions, including the stochastic version of the Hybrid Minimum Principle - Multiple Autonomous Switching Algorithm [37], and Feynman-Kac based algorithms as in [18–20].

Acknowledgements Discussions concerning this work with Panagiotis Tsiotras are gratefully acknowledged.

References

1. Bakolas, E.: Optimal covariance control for discrete-time stochastic linear systems subject to constraints. In: Proceedings of the 55th IEEE Conference on Decision and Control, 1153–1158 (2016)
2. Bakolas, E.: Finite-Horizon covariance control for discrete-time stochastic linear systems subject to input constraints. Automatica **91**, 61–68 (2018)
3. Beghi, A.: On the relative entropy of discrete-time Markov processes with given end-point densities. IEEE Trans. Inf. Theor. **42**(5), 1529–1535 (1996)
4. Bellman, R.: Dynamic programming. Science **153**(3731), 34–37 (1966)
5. Bhatt, A.G., Borkar, V.S.: Occupation measures for controlled Markov processes: characterization and optimality. Ann. Probab. 1531–1562 (1996)
6. Caluya, K.F., Halder, A.: Finite horizon density steering for multi-input state feedback linearizable systems. In: 2020 American Control Conference (ACC), 3577–3582. IEEE (2020)
7. Caluya, K.F., Halder, A.: Wasserstein proximal algorithms for the Schrödinger bridge problem: density control with nonlinear drift. IEEE Trans. Autom. Control **67**(3), 1163–1178 (2021)
8. Chen, C.T.: Linear System Theory and Design. Rinhart and Winston, Orlando (1984)
9. Chen, Y., Georgiou, T.T., Pavon, M.: Optimal steering of a linear stochastic system to a final probability distribution, part I. IEEE Trans. Autom. Control **61**(5), 1158–1169 (2015)
10. Chen, Y., Georgiou, T.T., Pavon, M.: Optimal steering of a linear stochastic system to a final probability distribution, part II. IEEE Trans. Autom. Control **61**(5), 1170–1180 (2015)
11. Chen, Y., Georgiou, T.T., Pavon, M.: Stochastic control liaisons: Richard Sinkhorn meets Gaspard Monge on a Schrodinger bridge. Siam Rev. **63**(2), 249–313 (2021)
12. Chen, Y., Ames, A.D.: Duality between Density Function and Value Function with Applications in Constrained Optimal Control and Markov Decision Process. arXiv:1902.09583 (2019)
13. Cho, M., Stockbridge, R.: Linear programming formulation for optimal stopping problems. SIAM J. Control Optim. **40**(6), 1965–1982 (2002)
14. Fleming, W.H., Vermes, D.: Convex duality approach to the optimal control of diffusions. SIAM J. Control Optim. **27**(5), 1136–1155 (1989)
15. Goldshtein, M., Tsiotras, P.: Finite-Horizon covariance control of linear time-varying systems. In: Proceedings of the 56th IEEE Conference on Decision and Control, 3606–3611 (2017)
16. Grigoriadis, K.M., Skelton, R.E.: Minimum-energy covariance controllers. Automatica **33**(4), 569–578 (1997)

17. Halder, A., Wendel, E.: Finite horizon linear quadratic gaussian density regulator with Wasserstein terminal cost. In: Proceedings of the American Control Conference, 7249–7254 (2016)
18. Hawkins, K.P., Pakniyat, A., Theodorou, E., Tsiotras, P.: Forward-backward rapidly-exploring random trees for stochastic optimal control. In: 2021 60th IEEE Conference on Decision and Control (CDC), 912–917. IEEE (2021a)
19. Hawkins, K.P., Pakniyat, A., Tsiotras, P.:. On the time discretization of the Feynman-Kac forward-backward stochastic differential equations for value function approximation. In: 2021 60th IEEE Conference on Decision and Control (CDC), 892–897. IEEE (2021b)
20. Hawkins, K.P., Pakniyat, A., Tsiotras, P.: Value function estimators for Feynman-Kac forward-backward SDEs in stochastic optimal control. Automatica (to appear in) (2023)
21. Hotz, A., Skelton, R.E.: Covariance control theory. Int. J. Control **46**(1), 13–32 (1987)
22. Kurtz, T., Stockbridge, R.: Existence of Markov controls and characterization of optimal Markov controls. SIAM J. Control Optim. **36**(2), 609–653 (1998)
23. Lasserre, J.B., Henrion, D., Prieur, C., Trélat, E.: Nonlinear optimal control via occupation measures and LMI-relaxations. SIAM J. Control Optim. **47**(4), 1643–1666 (2008)
24. Lewis, A.D.: The maximum principle of Pontryagin in control and in optimal control. In: Handouts for the course taught at the Universitat Politcnica de Catalunya. Department of Mathematics and Statistics Queen's University [online]. Available http://www.mast.queensu.ca/andrew/teaching/courses.shtml (May 5 2006)
25. Liu, C., Gray, A., Lee, C., Hedrick, J.K., Pan, J.: Nonlinear stochastic predictive control with unscented transformation for semi-autonomous vehicles. In: 2014 American Control Conference, 5574–5579. IEEE (2014)
26. Ma, J., Protter, P., Yong, J.: Solving forward-backward stochastic differential equations explicitly-a four step scheme. Probab. Theory Rel. Fields **98**(3), 339–359 (1994)
27. Majumdar, A., Vasudevan, R., Tobenkin, M.M., Tedrake, R.: Convex optimization of nonlinear feedback controllers via occupation measures. Int. J. Robot. Res. **33**(9), 1209–1230 (2014)
28. Mesbah, A., Streif, S., Findeisen, R., Braatz, R.D.: Stochastic nonlinear model predictive control with probabilistic constraints. In: Proceedings of the American Control Conference, 2413–2419 (2014)
29. Nodozi, I., Halder, A.: Schrödinger meets Kuramoto via Feynman-Kac: minimum effort distribution steering for noisy nonuniform Kuramoto oscillators. In: 2022 IEEE 61st Conference on Decision and Control (CDC), 2953–2960. IEEE (2022)
30. Okamoto, K., Tsiotras, P.: Stochastic model predictive control for constrained linear systems using optimal covariance steering (2019a). http://arxiv.org/abs/1905.13296
31. Okamoto, K., Goldshtein, M., Tsiotras, P.: Optimal covariance control for stochastic systems under chance constraints. IEEE Control Syst. Lett. **2**(2), 266–271 (2018)
32. Okamoto, K., Tsiotras, P.: Optimal stochastic vehicle path planning using covariance steering. IEEE Robot. Autom. Lett. **4**(3), 2276–2281 (2019)
33. Pakniyat, A., Caines, P.E.: On the stochastic minimum principle for hybrid systems. In: Proceedings of the 55th IEEE Conference on Decision and Control, Las Vegas, NV, USA, 1139–1144 (2016)
34. Pakniyat, A., Caines, P.E.: Hybrid optimal control of an electric vehicle with a dual-planetary transmission. Nonlin. Anal. Hybrid Syst. **25**, 263–282 (2017)
35. Pakniyat, A.: A convex duality approach for assigning probability distributions to the state of nonlinear stochastic systems. IEEE Control Syst. Lett. **6**, 3080–3085 (2022)
36. Pakniyat, A., Caines, P.E.: On the relation between the minimum principle and dynamic programming for classical and hybrid control systems. IEEE Trans. Autom. Control **62**(9), 4347–4362 (2017). https://doi.org/10.1109/TAC.2017.2667043
37. Pakniyat, A., Caines, P.E.: On the hybrid minimum principle: the Hamiltonian and adjoint boundary conditions. IEEE Trans. Autom. Control **66**(3), 1246–1253 (2020)
38. Pakniyat, A., Tsiotras, P.: Partially observed steering the state of linear stochastic systems. In: Proceedings of the 60th IEEE Conference on Decision and Control (CDC), 3780–3785. https://doi.org/10.1109/CDC45484.2021.9683692

39. Pakniyat, A., Tsiotras, P.: Steering the state of linear stochastic systems: a constrained minimum principle formulation. In: Proceedings of the 2021 IEEE American Control Conference (ACC), 1300–1305 (2021b)
40. Pauwels, E., Henrion, D., Lasserre, J.B.: Linear conic optimization for inverse optimal control. SIAM J. Control Optim. **54**(3), 1798–1825 (2016)
41. Peng, S.: A general stochastic maximum principle for optimal control problems. SIAM J. Control Optim. **28**(4), 966–979 (1990)
42. Pontryagin, L., Boltyanskii, V., Gamkrelidze, R., Mishchenko, E.: The Mathematical Theory of Optimal Processes, vol. 4. Wiley Interscience, New York (1962)
43. Ridderhof, J., Okamoto, K., Tsiotras, P.: Nonlinear uncertainty control with iterative covariance steering. In: 2019 IEEE 58th Conference on Decision and Control (CDC), 3484–3490. IEEE (2019)
44. Salhab, R., Malhamé, R.P., Le Ny, J.: Collective stochastic discrete choice problems: a min-LQG dynamic game formulation. IEEE Trans. Autom. Control **65**(8), 3302–3316 (2019)
45. Sehr, M.A., Bitmead, R.R.: Particle model predictive control: tractable stochastic nonlinear output-feedback MPC. IFAC-PapersOnLine **50**(1), 15361–15366 (2017)
46. Vasudevan, R., Gonzalez, H., Bajcsy, R., Sastry, S.S.: Consistent approximations for the optimal control of constrained switched systems-part 1: a conceptual algorithm. SIAM J. Control Optim. **51**(6), 4463–4483 (2013)
47. Vermes, D.: Optimal control of piecewise deterministic Markov processes. Stochast. Int. J. Probab. Stochast. Process. **14**(3), 165–207 (1985)
48. Vinter, R.B.: Dynamic programming for optimal control problems with terminal constraints. In: Recent Mathematical Methods in Dynamic Programming, pp. 190–202. Springer, Berlin Heidelberg (1985)
49. Vinter, R.: Convex duality and nonlinear optimal control. SIAM J. Control Optim. **31**(2), 518–538 (1993)
50. Vinter, R.B., Lewis, R.M.: A necessary and sufficient condition for optimality of dynamic programming type, making no a priori assumptions on the controls. SIAM J. Control Optim. **16**(4), 571–583 (1978)
51. Vinter, R.B., Lewis, R.M.: The equivalence of strong and weak formulations for certain problems in optimal control. SIAM J. Control Optim. **16**(4), 546–570 (1978)
52. Xu, J.H., Skelton, R.: An improved covariance assignment theory for discrete systems. IEEE Trans. Autom. Control **37**(10), 1588–1591 (1992)
53. Yasuda, K., Skelton, R.E., Grigoriadis, K.M.: Covariance controllers: a new parametrization of the class of all stabilizing controllers. Automatica **29**(3), 785–788 (1993)
54. Yi, Z., Cao, Z., Theodorou, E., Chen, Y.: Nonlinear covariance control via differential dynamic programming. In: 2020 American Control Conference (ACC), 3571–3576. IEEE (2020)
55. Yong, J., Zhou, X.Y.: Stochastic Controls: Hamiltonian Systems and HJB Equations. Springer-Verlag, New York (1999)
56. Zhao, P., Mohan, S., Vasudevan, R.: Optimal control of polynomial hybrid systems via convex relaxations. IEEE Trans. Autom. Control **65**(5), 2062–2077 (2019)
57. Zhu, H.: Convex duality for finite-fuel problems in singular stochastic control. J. Optim. Theory Appl. **75**(1), 155–181 (1992)

Chapter 10
Applications of the Π-Strategy When Players Move with Acceleration

Bahrom Samatov and Ulmasjon Soyibboev

Abstract In the present paper, we have carried out a fairly complete study of the pursuit problem when the players move by inertia, i.e. when the players carry out their movements with the help of acceleration vectors. The main tool for solving this pursuit problem is the parallel approach strategy (Π-strategy), which, under an arbitrary action of the evader, allows the best approach of the players. It is known that for the simple pursuit problem, the set of meeting points of the players is the ball of Apollonius. In the present paper, it is shown that in the case of inertial motions of players, the set of meeting points of the players is a linear combination of two such Apollonius balls, the first of which is built from the initial states, and the second from the initial states of the velocities.

Keywords Differential game · Pursuer · Evader · Strategy · Pursuit · Evasion · Acceleration · Attainability domain · Ball of apollonius · Life line

10.1 Introduction

Differential games are a special kind of problems for conflict-controlled dynamic systems described by differential equations. The concept of "Differential game" first appeared in a series of classified works by the American mathematician R. Isaacs on the project of the RAND Corporation (USA), completed in the early 50s of the 20th century. Research by R. Isaacs was published in 1965 in the form of a monograph [1], which considered a number of applied problems and proposed general ideas, which are mainly based on game-theoretic and variational methods of solution. Further, L. D. Bercovitz, W. H. Fleming, A. Friedman, Y. Ho, A. Bryson, S. Baron, O. Hajek,

B. Samatov (✉)
V.I. Romanovsky Institute of Mathematics, Academy of Sciences of the Republic of Uzbekistan, Tashkent, Uzbekistan 100174
e-mail: samatov57@gmail.com

U. Soyibboev
Namangan State University, Namangan, Uzbekistan 116019

D. Azimov (ed.), *Proceedings of the IUTAM Symposium on Optimal Guidance and Control for Autonomous Systems 2023*, IUTAM Bookseries 40,
https://doi.org/10.1007/978-3-031-39303-7_10

R. J. Elliott, N. J. Kalton, L. S. Pontryagin, N. N. Krasovskii, B. N. Pshenichnii, and many other followers developed the ideas of R. Isaacs.

The book by R. Isaacs remains today a pointer to the main path to solving many interesting problems in the theory of differential games. One of these problems is the Differential game with a "Life line" (Problem 9.5.1, [1]), where some subset in the space under consideration is given, which is called the "Life line". The goal of the escaping player is to cross this "Life line" before being caught by the pursuer. In the works [1–10] it is shown that for a simple pursuit game, i.e. when the players carry out their movements without inertia and the speed of the pursuer is greater than that of the evader, then the boundary of the meeting area of the players is the circle of Apollonius. Further, in the papers [2–4, 8, 9] it is shown that the evader can reach any interior point of this Apollonius ball and remain uncaught by the pursuer. However, if the pursuer uses a parallel approach strategy (Π-strategy), which intercepts the evader, the capture can be carried out inside or on the boundary of this ball with an arbitrary action of the evader. Of particular note is the use of the Π-strategy in games with many players in pursuit problems [2, 8, 11–13] and its ability to generalize to more general classes (for example, [5, 14–17] and others).

In the future, with the help of Π-strategies in the papers [4, 18–21] the Isaacs problem the Differential game with the "Life line" was studied for the case with various constraints on the players' controls. In these papers, interesting results were obtained for the boundaries of the maximum attainability domain of an evader. For example, in [20] it is shown that when the evader's control is selected from class L_∞, i.e. out of the set of all measurable functions, the value of which does not exceed a certain number, but the pursuer's control from the class L_2, i.e. from the space of quadratically integrable functions, then it turns out that the boundary of the reachable area of the evader is Descartes' Oval or Pascal's Snail's Loop. In the case when the controls of both players are chosen from the same class L_2, then in [4] it is shown that the maximum reachable area of the evader is also the Apollonian ball.

As early as in the book by R. Isaacs [1] it was noted that the Π-strategy of the pursuer provides the best convergence of players in the case of inertial-free motion of the players. However, the application of this effective strategy for the case with the inertial motions of the players remained unaffected. In this paper, we consider the pursuit problem when the players carry out their movements with the help of accelerating controls, and the Π-strategy is used to solve the problem, which allows the best convergence of the players. It is shown that here, using this strategy, the pursuer captures the evader in the set consisting of combinations of two Apollonius balls. The first ball of Apollonius is formed using the initial states of the players, and the second from their initial velocities. In turn, the above-mentioned Differential game with the "Life line" of R. Isaacs is also considered from the point of view of the pursuer. The work is a development of the works of [1–4, 10, 11, 22] and a continuation of the work [23] of the authors.

10.2 Formulation of the Problem

Let us assume that in the space $\mathbb{R}^n$ the controlled object **P**, called the Pursuer, is chasing another object **E**, called the Evader. Denote by x the location of the Pursuer, and by y the location of the Evader in $\mathbb{R}^n$. In this section, we consider the pursuit problem when movements of the objects are based on the equations

$$\ddot{x} = u, \quad x(0) = x_{10}, \quad \dot{x}(0) = x_{11}, \tag{10.1}$$

$$\ddot{y} = v, \quad y(0) = y_{10}, \quad \dot{y}(0) = y_{11}, \tag{10.2}$$

respectively, where $x, y, u, v \in \mathbb{R}^n$, $n \geq 2$; x_{10}, y_{10} are the initial positions, and x_{11}, y_{11} are the initial velocity vectors of the objects; it is assumed $x_{10} \neq y_{10}$; u, v are the acceleration vectors that serve as control parameters of the objects. In this case, the temporal change in the vector u must be a measurable function $u(\cdot)\colon [0, +\infty) \to \mathbb{R}^n$, for which imposes a geometrical constraint of the form

$$|u(t)| \leq \alpha \text{ for almost all } t \geq 0, \tag{10.3}$$

where α is a non-negative parametric number denoting the maximum value of Pursuer's acceleration.

Similarly, the temporal change of the vector v must be a measurable function $v(\cdot)\colon [0, +\infty) \to \mathbb{R}^n$, for which a geometrical constraint of the form

$$|v(t)| \leq \beta \text{ for almost all } t \geq 0, \tag{10.4}$$

where β is a non-negative parametric number, which means the maximum value of Evader's acceleration. Note that in (10.3) and (10.4) as the norms of the control vectors u and v we will consider the usual Euclidean norm, i.e. $|u| = \sqrt{u_1^2 + u_2^2 + \cdots + u_n^2}$, where $u_1, u_2, \ldots, u_n$ are the coordinates of the vector u in the space $\mathbb{R}^n$, and $|v| = \sqrt{v_1^2 + v_2^2 + \cdots + v_n^2}$, where $v_1, v_2, \ldots, v_n$ are the coordinates of the vector v in the same space $\mathbb{R}^n$.

A measurable function $u(\cdot)$ satisfying condition (10.3) will be called an admissible Pursuer's control, and the set of all such controls will be denoted by **U**. A measurable function $v(\cdot)$ satisfying condition (10.4) will be called admissible Evader's control, and the set of all such controls will be denoted by **V**.

By virtue of equations (10.1)–(10.2), every triple $(x_{10}, x_{11}, u(\cdot))$, where $u(\cdot) \in \mathbf{U}$, and $(y_{10}, y_{11}, v(\cdot))$, where $v(\cdot) \in \mathbf{V}$, generate the trajectories of the players **P** and **E**

$$x(t) = x_{10} + x_{11}t + \int_0^t (t-s)u(s)ds, \tag{10.5}$$

$$y(t) = y_{10} + y_{11}t + \int_0^t (t-s)v(s)ds, \tag{10.6}$$

respectively.

Assume that a closed subset **M** called the "Life line" is given in $\mathbb{R}^n$. The main target for Pursuer **P** is to catch Evader **E**, i.e. to obtain the equality $x(t_*) = y(t_*)$ at some $t_* > 0$ while Evader **E** stays in the zone $\mathbb{R}^n \setminus \mathbf{M}$. The main goal for Evader **E** is to reach the zone **M** before being caught by Pursuer **P** or to sustain the relation $x(t) \neq y(t)$ for each $t \in [0, +\infty)$, and if this is not possible, then maximize the capture time. We need to remark that the zone **M** does not restrict motion of the Pursuer. In addition, it is supposed that the initial positions x_{10}, y_{10} satisfy the conditions $x_{10} \neq y_{10}$ and $y_{10} \notin \mathbf{M}$ at the beginning of the game.

In the future, when solving the pursuit problem using the strategy of parallel approach (velocities, the objects themselves), the following denotations is introduced: $z = x - y$, $z_{10} = x_{10} - y_{10}$, $z_{11} = x_{11} - y_{11}$. Then system (10.1)–(10.2) reduces to the form

$$\ddot{z} = u - v, \quad z(0) = z_{10}, \quad \dot{z}(0) = z_{11}, \tag{10.7}$$

where z_{10} is the difference between the initial states, and z_{11} is the difference between the initial speeds of the players. In this case, when choosing admissible controls $u(\cdot) \in \mathbf{U}$ and $v(\cdot) \in \mathbf{V}$, the solution of the Eq. (10.7) will have form

$$z(t) = z_{10} + z_{11}t + \int_0^t (t-s)(u(s) - v(s))ds. \tag{10.8}$$

Now, with the new notation introduced above, we first consider only the pursuit problem, without phase constraints. Let the Pursuer's goal is to achieve the equality $z(t) = 0$ from the given initial states z_{10} and z_{11} in the shortest time t, and the goal of the Escaper is to achieve the inequality $z(t) \neq 0$ for all $t \geq 0$.

We will construct a solution to the pursuit problem based on the given initial states z_{10} and z_{11}. Note that only two cases are possible here:

Case 1 The vectors z_{10} and z_{11} are collinear, that is, there exists a finite number k such that $z_{11} = kz_{10}$.

Case 2 The vectors z_{10} and z_{11} are non-collinear, that is, are linearly independent.

Note 1 Note that in [23] the differential game (10.7) was studied only for Case 1, when $k = 0$. Here we present a full study for Case 1–2.

10.3 Applying the Π-Strategy in Case 1

10.3.1 Solution of the Pursuit Problem in Case 1

Definition 1 Let $\alpha \geq \beta$ and $z_{11} = kz_{10}$. Then in the game (10.7), the function

$$\mathbf{u}(v) = v - \lambda(v)\xi_{10} \tag{10.9}$$

is called the Π-*strategy of the Pursuer*, where

$$\lambda(v) = \langle v, \xi_{10}\rangle + \sqrt{\langle v, \xi_{10}\rangle^2 + \alpha^2 - |v|^2}, \quad \xi_{10} = z_{10}/|z_{10}|, \tag{10.10}$$

and $\mathbf{u}(v(t)), t \geq 0$ – *its realization* for any $v(\cdot) \in \mathbf{V}$, $\langle v, \xi_{10}\rangle$ denotes the inner product of vectors v, ξ_{10} in $\mathbb{R}^n$ and $|z_{10}|$ is the Euclidean norm of the vector z_{10} in the space $\mathbb{R}^n$.

It is easy to check that, for $\alpha \geq \beta$, the functions (10.9) and (10.10) are defined and continuous for all v, $|v| \leq \beta$, and moreover, the function (10.9) satisfies the equality $|\boldsymbol{u}(v)| = \alpha$.

Definition 2 The Π-strategy is called *winning for the Pursuer* in the time interval $[0, T_{10}]$ if, for any $v(\cdot) \in \mathbf{V}$:

(a) there exists such a time $t^* \in [0, T_{10}]$ such that $z(t^*) = 0$;
(b) $\boldsymbol{u}(v(\cdot)) \in \mathbf{U}$ in the time interval $[0, t^*]$.

In this case, the number T_{10} is usually called *a guaranteed pursuit or capture time* in Case 1.

Theorem 1 *Let in Case 1 the game (10.7) satisfy at least one of the following conditions:* (a) $\alpha > \beta$ *and* k *– arbitrary number;* (b) $\alpha = \beta$ *and* $k < 0$. *Then* Π-*strategy (10.9) is winning in the time interval* $[0, T_{10}]$, *where*

$$T_{10} = \begin{cases} \left(|z_{10}|k + \sqrt{|z_{10}|^2k^2 + 2|z_{10}|(\alpha - \beta)}\right)/(\alpha - \beta) & \text{if } \alpha > \beta, \\ -1/k & \text{if } \alpha = \beta, k < 0. \end{cases}$$

Proof Let the Evader choose an arbitrary control $v(\cdot) \in \mathbf{V}$, and let the Pursuer implement Π-strategy (10.9). Then by virtue of (10.8), we have

$$z(t) = z_{10} + z_{11}t - \int_0^t (t - s)\lambda(v(s))\xi_{10}ds.$$

From here and from the condition $z_{11} = kz_{10}$ we find

$$z(t) = z_{10}\Lambda(t, v_t(\cdot)), \tag{10.11}$$

where

$$\Lambda(t, v_t(\cdot)) = 1 + kt - \frac{1}{|z_{10}|}\int_0^t (t-s)\lambda(v(s))ds, \quad v_t(\cdot) = \{v(s) : 0 \le s \le t\}.$$

Let us estimate the function $\Lambda(t, v_t(\cdot))$ by $v_t(\cdot) \in \mathbf{V}$ from above. Then from the lemma on the minimum in the elementary optimal control problem (see [24], p. 360) we have

$$\Lambda(t, v_t(\cdot)) \le 1 + kt - \frac{1}{|z_{10}|}\min_{v(\cdot)\in\mathbf{V}}\int_0^t (t-s)\lambda(v(s))ds$$
$$\le 1 + kt - \frac{t^2}{2|z_{10}|}\min_{|v|\le\beta}\lambda(v).$$

Since, $\min_{|v|\le\beta}\lambda(v) = \alpha - \beta$, then from the last inequality we find

$$\Lambda(t, v_t(\cdot)) \le \Lambda(t), \tag{10.12}$$

where $\Lambda(t) = 1 + kt - \frac{t^2}{2|z_{10}|}(\alpha - \beta)$. It is clear that the function $\Lambda(t)$ increases monotonically as $0 \le t < \vartheta$, where $\vartheta = \frac{k|z_{10}|}{\alpha-\beta}$, and reaches its maximum at time ϑ, and decreases when $t > \vartheta$. Based on this property of the function $\Lambda(t)$ and from (10.11), (10.12) we can conclude that the distance $|z(t)|$ between players:

(a) can increase up to ϑ and reach its maximum value

$$|z(\vartheta)| = |z_{10}|\Lambda(\vartheta) = |z_{10}|\left(1 + \frac{k^2|z_{10}|}{2(\alpha-\beta)}\right);$$

(b) start decreasing at $t > \vartheta$ and vanish no later than in the time $t = T_{10}$. The latter means that there exists a moment $t^* \le T_{10}$ such that $\Lambda(t^*, v_{t^*}(\cdot)) = 0$. Therefore, from (10.11) implies that $z(t^*) = 0$, which completes the proof of Theorem 1. □

10.3.2 *The Attainability Domain of the Evader in Case 1*

By virtue of Theorem 1, if $\alpha > \beta$, then using Π-strategy (10.9) the Pursuer captures the Evader at some point in the space $\mathbb{R}^n$. For the case under consideration, we find a set of points of "meetings" of the players.

Let the triple $(y_{10}, y_{11}, v(\cdot))$, $v(\cdot) \in \mathbf{V}$ generate Evader's trajectory in the form (10.6), and let the triad $(x_{10}, x_{11}, \boldsymbol{u}(v(\cdot)))$, $\boldsymbol{u}(v(\cdot)) \in \mathbf{U}$ generate Pursuer's trajectory [see (10.5)] in the form

$$x(t) = x_{10} + x_{11}t + \int_0^t (t-s)\boldsymbol{u}(v(s))ds,$$

where $t \in [0, t^*]$, and t^* – a moment of the "meeting" of the players, that is, $x(t^*) = y(t^*)$. For each pair $(x(t), y(t))$ in the time interval $[0, t^*]$, we define a set of the form

$$\mathbf{A}(x(t), y(t)) = \{w\colon\ \beta|w - x(t)| \geq \alpha|w - y(t)|\}. \tag{10.13}$$

From Theorem 1 we have $|y(t) - x(t)| \geq 0$ in the interval $[0, t^*]$. From here and from the form of $\mathbf{A}(x(t), y(t))$ we obtain

$$y(t) \in \mathbf{A}(x(t), y(t)) \tag{10.14}$$

for all $t \in [0, t^*]$. It is easy to calculate that the boundary of the set $\mathbf{A}(x(t), y(t))$ is the Apollonian sphere centered at $C(z(t))$ and of radius at $R(z(t))$, where

$$C(z(t)) = x(t) - \frac{\alpha^2 z(t)}{\alpha^2 - \beta^2},\quad R(z(t)) = \frac{\alpha\beta|z(t)|}{\alpha^2 - \beta^2}.$$

Now consider a multi-valued mapping of the form

$$\mathbf{A}^*(x(t), y(t), t) = \mathbf{A}(x(t), y(t)) - t\mathbf{A}_{11}, \tag{10.15}$$

where

$$\mathbf{A}_{11} = \mathbf{A}(x_{11}, y_{11}) = \frac{\alpha^2 y_{11} - \beta^2 x_{11}}{\alpha^2 - \beta^2} + \frac{\alpha\beta|z_{11}|}{\alpha^2 - \beta^2}\mathbf{S}, \tag{10.16}$$

and here $\mathbf{S}$ is the unit ball centered at the origin of the space $\mathbb{R}^n$.

Similar to Lemma 5 (see [23], p. 101), and here the following property is satisfied.

Property 1 The multi-valued mapping $\mathbf{A}^*(x(t), y(t), t)$ is monotonically decreasing with respect to inclusion in time $t \in [0, t^*]$, that is, if $t_1, t_2 \in [0, t^*]$ and $t_1 < t_2$, then $\mathbf{A}^*(x(t_2), y(t_2), t_2) \subset \mathbf{A}^*(x(t_1), y(t_1), t_1)$.

From Property 1 and from (10.13), (10.15) implies the following statement.

Property 2 $\mathbf{A}^*(x(t), y(t), t) \subset \mathbf{A}_{10}$ for $t \in [0, t^*]$, where $\mathbf{A}_{10} = \mathbf{A}^*(x(0), y(0), 0) = \mathbf{A}(x_{10}, y_{10})$ and

$$\mathbf{A}_{10} = \frac{\alpha^2 y_{10} - \beta^2 x_{10}}{\alpha^2 - \beta^2} + \frac{\alpha\beta|z_{10}|}{\alpha^2 - \beta^2}\mathbf{S}. \tag{10.17}$$

Theorem 2 *If $\alpha > \beta$, then in Case 1, for all $t \in [0, t^*]$ the following inclusion is performed:*

$$y(t) \in \mathbf{A}_{10} + t\mathbf{A}_{11}, \tag{10.18}$$

where $\mathbf{A}_{11}$ and $\mathbf{A}_{10}$ are Apollonius balls of the types (10.16) and (10.17) respectively.

Proof From Property 2 and from the form (10.15) we have $\mathbf{A}(x(t), y(t)) \subset \mathbf{A}_{10} + t\mathbf{A}_{11}$ for all $t \in [0, t^*]$. From here and from (10.14) the validity of relation (10.18) follows for each $t \in [0, t^*]$. The proof is complete. □

Property 3 In Case 1, the attainability domain of the Evader is the set $\boldsymbol{W}_1$, where

$$\boldsymbol{W}_1 = \mathbf{A}_{10} + \bigcup_{t=0}^{T_{10}} t\mathbf{A}_{11},$$

i.e. the trajectory of the Evader $y_{T_{10}}(\cdot) = \{y(s): 0 \leq s \leq T_{10}\}$ does not leave the closed convex set $\boldsymbol{W}_1$.

Proof The proof of Property 3 follows from Theorem 2. □

From Property 3 we achieve the following theorem.

Theorem 3 *If the set $\boldsymbol{W}_1$ doesn't intersect with the $\boldsymbol{M}$, then the Pursuer wins in Case 1 in the game (10.1)–(10.2) with the "Life line".*

Note 2 We note some important special cases for the attainability domain of the Evader. In the space $\mathbb{R}^n$, the boundary of the capture domain $\boldsymbol{w}_1$ of the Evader is:

(a) the Apollonian sphere bounding ball $\mathbf{A}_{10}$ if $x_{11} = y_{11} = 0$ (see [1–4, 23]);
(b) the boundary of the set $\mathbf{A}_{10} + \bigcup_{t=0}^{T_{10}} tx_{11}$ if $x_{11} = y_{11} \neq 0$ (see [23]).

10.4 Applying the Π-Strategy in Case 2

Now let the original non-zero vectors z_{10}, z_{11} be non-collinear, and let $\alpha > \beta$, i.e. Case 2 from Sect. 10.2 is considered. First, we solve the following auxiliary problem.

Problem 1 For an arbitrary control $v(\cdot) \in \mathbf{V}$ of Evader $\mathbf{E}$, it is required to construct a Π-strategy of Pursuer $\mathbf{P}$, with the help of which equality of the velocities

$$\dot{x}(\theta) = \dot{y}(\theta) \tag{10.19}$$

is achieved, where θ is some finite time.

10.4.1 Solution of the Problem 1

To solve the Problem 1, the second-order differential game (10.1)–(10.2) is reduced to a first-order differential game. To do this, we introduce new variables $x_1 = \dot{x}$, $y_1 = \dot{y}$. Then from (10.1)–(10.2) we have

$$\dot{x}_1 = u, \quad x_1(0) = x_{11}, \tag{10.20}$$

$$\dot{y}_1 = v, \quad y_1(0) = y_{11}, \tag{10.21}$$

respectively, where u, v are the rates of change of the vectors x_1, y_1 depending on time t. It's clear that $x_{11} \neq y_{11}$, otherwise we return to Case 1, where x_{11}, y_{11} are the initial states of the vectors x_1, y_1.

Definition 3 For the Problem 1, the function

$$\boldsymbol{u}_1(z_{11}, v) = v - \lambda_1(z_{11}, v)\xi_{11} \tag{10.22}$$

is called the Π_1-*strategy of the Pursuer*, where

$$\lambda_1(z_{11}, v) = \langle v, \xi_{11}\rangle + \sqrt{\langle v, \xi_{11}\rangle^2 + \alpha^2 - |v|^2},$$
$$\xi_{11} = z_{11}/|z_{11}|, \quad z_{11} = x_{11} - y_{11},$$

and $\boldsymbol{u}_1(z_{11}, v(t))$, $t \geq 0$ – *its realization* for each $v(\cdot) \in \mathbf{V}$.

Theorem 4 *If $\alpha > \beta$ in the Problem 1, and the Pursuer uses Π_1-strategy (10.22), then for an arbitrary control $v(\cdot) \in \mathbf{V}$ of the Evader the equality (10.19) can be realized in some time $\theta \in [T_*, T^*]$, where*

$$T^* = \frac{|z_{11}|}{\alpha - \beta}, \quad T_* = \frac{|z_{11}|}{\alpha + \beta}. \tag{10.23}$$

Proof Let the Evader choose some control $v(\cdot) \in \mathbf{V}$, and the Pursuer implement the Π_1-strategy (10.22). Introducing the notation $z_1 = x_1 - y_1$, from (10.20)–(10.22) we find the Cauchy problem of the form

$$\dot{z}_1 = -\lambda_1(z_{11}, v(t))\xi_{11}, \quad z_1(0) = z_{11}.$$

From here we obtain the function of convergence of the velocities of the players

$$z_1(t) = z_{11}\Lambda_1(z_{11}, t, v_t(\cdot)), \tag{10.24}$$

where

$$\Lambda_1(z_{11}, t, v_t(\cdot)) = 1 - \frac{1}{|z_{11}|}\int_0^t \lambda_1(z_{11}, v(s))ds. \tag{10.25}$$

Since, $\min_{|v|\le\beta}\lambda_1(z_{11}, v) = \alpha - \beta$ and $\max_{|v|\le\beta}\lambda_1(z_{11}, v) = \alpha + \beta$, then for the function (10.25) we have the relations

$$\Lambda_*(z_{11}, t) \le \Lambda_1(z_{11}, t, v_t(\cdot)) \le \Lambda^*(z_{11}, t),$$

where $\Lambda^*(z_{11}, t) = 1 - \frac{t}{|z_{11}|}(\alpha - \beta)$ and $\Lambda_*(z_{11}, t) = 1 - \frac{t}{|z_{11}|}(\alpha + \beta)$. Hence we get that the function $\Lambda_1(z_{11}, t, v_t(\cdot))$ vanishes on the time interval $[T_*, T^*]$, i.e. there exists $\theta \in [T_*, T^*]$, such that $\Lambda_1(z_{11}, \theta, v_\theta(\cdot)) = 0$. Here T_* is the root of the equation $\Lambda_*(z_{11}, t) = 0$, and T^* is the root of the equation $\Lambda^*(z_{11}, t) = 0$, which have the form (10.23). Therefore, from (10.24) we get $z_1(\theta) = 0$ or $x_1(\theta) = y_1(\theta)$, which completes the proof of the theorem 4. □

Note 3 It was shown in [2–4] that for games of simple pursuit of the form (10.20)–(10.21), a Π-strategy of the type (10.22) ensures the optimal approach of the objects and the optimal time of their coincidence. Therefore, here too, the strategy (10.22) guarantees the optimal convergence of velocity trajectories and their coincidence.

10.4.2 Estimating the Distance Between the Players in $[0, \theta]$

In this subsection, we will show that in case 2 the players cannot coincide with each other on the time interval $[0, \theta]$, i.e. no capture is possible.

Let $v(\cdot) \in \mathbf{V}$ be a control chosen by the Evader, and let the Pursuer implement Π_1-strategy (10.22) in the time interval $[0, \theta]$. Then by virtue of (10.8) and (10.22), we obtain

$$z(t) = z_{10} + z_{11}t - \int_0^t (t - s)\lambda_1(z_{11}, v(s))\xi_{11}ds.$$

Whence we find that

$$z(t) = z_{10} + z_{11}\Lambda_2(z_{11}, t, v_t(\cdot)), \tag{10.26}$$

where

$$\Lambda_2(z_{11}, t, v_t(\cdot)) = t - \frac{1}{|z_{11}|}\int_0^t (t - s)\lambda_1(z_{11}, v(s))ds. \tag{10.27}$$

Therefore, from (10.25) and (10.27) we find

$$\frac{d\Lambda_2(z_{11}, t, v_t(\cdot))}{dt} = \Lambda_1(z_{11}, t, v_t(\cdot)).$$

Since $\Lambda_1(z_{11}, t, v_t(\cdot)) \ge 0$ for $t \in [0, \theta]$, then from the last equality implies that $\Lambda_2(z_{11}, t, v(\cdot))$ is a monotonically increasing function in $[0, \theta]$. From this and from the fact that $\lambda_1(z_{11}, v) \le \alpha - \beta$ for $|v| \le \beta$ we find the following relations

$$\begin{aligned} &\Lambda_2(z_{11}, t, v_t(\cdot)) \le \Lambda_2(z_{11}, \theta, v_\theta(\cdot)) \le \\ &\le \theta - \tfrac{1}{|z_{11}|} \int_0^\theta (\theta - s)(\alpha - \beta) ds = \theta - \tfrac{\alpha-\beta}{2|z_{11}|}\theta^2. \end{aligned} \tag{10.28}$$

Since $\theta \in [T_*, T^*]$ (see Theorem 2), it is easy to check that the right side of the latter in (10.28) reaches its maximum value at $\theta = T^*$. Therefore, for an arbitrary $v(\cdot) \in \mathbf{V}$ in the interval $[0, \theta]$, the following estimate holds:

$$0 \le \Lambda_2(z_{11}, t, v_t(\cdot)) \le \frac{|z_{11}|}{2(\alpha - \beta)}. \tag{10.29}$$

Now, consider the function $w(\tau) = z_{10} + \tau z_{11}$ for $\tau \in \Delta$, where $\Delta = [0, \frac{|z_{11}|}{2(\alpha-\beta)}]$, when the vectors z_{10} and z_{11} are non-collinear between yourself. It is easy to see that the function $w(\tau)$ for $\tau \in \Delta$ in the space $\mathbb{R}^n$ represents some finite line segment with the direction vector z_{11}. By the form of this function, it is easy to verify the following important property.

Property 4 (a) If $\langle z_{10}, z_{11}\rangle > 0$, then the function $|w(\tau)|$ is increasing in $\tau \in \Delta$, and the following estimate holds:

$$|z_{10}| \le |w(\tau)| \le \left| z_{10} + \frac{|z_{11}|}{2(\alpha - \beta)} z_{11} \right|;$$

(b) if $\langle z_{10}, z_{11}\rangle < 0$, then

$$0 < \left| z_{10} - \frac{\langle z_{10}, z_{11}\rangle}{|z_{11}|^2} z_{11} \right| \le |w(\tau)| \le \max\left\{ |z_{10}|, \left| z_{10} + \frac{|z_{11}|}{2(\alpha - \beta)} z_{11} \right| \right\}$$

is true on the interval Δ.

So, from Property 4 and from (10.26) and (10.29) for the distance between the players, we obtain the following estimate in the time interval $[0, \theta]$:

$$L_{10} \le |z(t)| \le L_{11}, \tag{10.30}$$

where

$$L_{10} = \begin{cases} |z_{10}| & \text{if } \langle z_{10}, z_{11}\rangle > 0, \\ \left| z_{10} - \frac{\langle z_{10}, z_{11}\rangle}{|z_{11}|^2} z_{11} \right| & \text{if } \langle z_{10}, z_{11}\rangle < 0, \end{cases}$$

$$L_{11} = \begin{cases} \left| z_{10} + \frac{|z_{11}|}{2(\alpha-\beta)} z_{11} \right| & \text{if } \langle z_{10}, z_{11}\rangle > 0, \\ \max\left\{ |z_{10}|, \left| z_{10} + \frac{|z_{11}|}{2(\alpha-\beta)} z_{11} \right| \right\} & \text{if } \langle z_{10}, z_{11}\rangle < 0. \end{cases}$$

Theorem 5 *If $\alpha > \beta$ and the Pursuer uses Π_1-strategy (10.22), then for an arbitrary control $v(\cdot) \in \mathbf{V}$ of the Evader on the time interval $[0, \theta]$ for the distance between the players, the estimate (10.30) is valid.*

Proof The proof of Theorem 5 follows directly from Property 4. □

10.4.3 Solution of the Pursuit Problem for the Case $t \geq \theta$

Thus, by virtue of Theorem 5, the players cannot coincide in the time interval $[0, \theta]$. However, at the time θ their speeds coincide. Now, starting from this time θ, we will consider the differential game

$$\ddot{x} = u, \quad x(\theta) = x_*, \quad \dot{x}(\theta) = x_{**}, \tag{10.31}$$

$$\ddot{y} = v, \quad y(\theta) = y_*, \quad \dot{y}(\theta) = y_{**}, \tag{10.32}$$

where x_*, y_* are the positions of the players, and x_{**}, y_{**} are the positions of their velocities at the time θ. Note that, by virtue of Theorem 5, the positions x_* and y_* cannot coincide. However, in system (10.31)–(10.32), the equality (10.19) is satisfied in the form $x_{**} = y_{**}$, i.e. the velocities at the time θ coincide, and the game reduces to Case 1, when $k = 0$.

Definition 4 Let $\alpha > \beta$ and $t \geq \theta$. Then in the game (10.31)–(10.32), the function

$$\mathbf{u}_2(z_*, v) = v - \lambda_2(z_*, v)\xi_*, \tag{10.33}$$

is called the Π_1-*strategy of the Pursuer*, where

$$\lambda_2(z_*, v) = \langle v, \xi_* \rangle + \sqrt{\langle v, \xi_* \rangle^2 + \alpha^2 - |v|^2},$$
$$\xi_* = z_*/|z_*|, \quad z_* = x_* - y_*,$$

and $\boldsymbol{u}_2(z_*, v(t))$, $t \geq \theta$ – *its realization* for each $v(\cdot) \in \mathbf{V}$.

Theorem 6 *If $\alpha > \beta$ and $t \geq \theta$, then in the game (10.31)–(10.32), Π_2-strategy (10.33) is winning in the time interval $[\theta, T_2]$, where*

$$T_2 = \theta + \sqrt{\frac{2|z_*|}{\alpha - \beta}}.$$

Proof The proof of Theorem 6 is carried out similarly to Theorem 1. Only here the strategy (10.33) is implemented, and it is taken into account that $k = 0$. □

Theorem 7 *If $\alpha > \beta$ and the vectors z_{10} and z_{11} are non-collinear, then in the game (10.1)–(10.2) for an arbitrary control $v(\cdot) \in \mathbf{V}$ of the Evader, the Pursuer using Π_1*

and Π_2*-strategies, catches the Evader in the time interval* $[T_{21}, T_{22}]$*, where*

$$T_{21} = \frac{|z_{11}|}{\alpha + \beta} + \sqrt{\frac{2}{\alpha + \beta} L_{10}},$$

$$T_{22} = \frac{|z_{11}|}{\alpha - \beta} + \sqrt{\frac{2}{\alpha - \beta} L_{11}}.$$

Proof The proof of Theorem 7 follows directly from Theorems 4–6. □

10.4.4 The Attainability Domain of the Evader in Case 2

In this subsection, the following problem is solved.

Problem 2 For the differential game (10.1)–(10.2) in Case 2 in the space $\mathbb{R}^n$, it is required to find the attainability domain:

(1) of the speed $\dot{y}(t)$ in the time interval $[0, \theta]$;
(2) of the trajectory $y(t)$ in the time interval $[0, \theta]$;
(3) of the trajectory $y(t)$ in the time interval $[0, T_{22}]$ for the Evader.

Solutions of the Parts (1) and (2) of the Problem 2. As noted in Sect. 10.1, it was shown in [1–10] that when the players carry out their movements without inertia and the speed of the pursuer is greater than the speed of the evader, then the meeting area of the players is the ball of Apollonius. This effect can also be applied to the case of constructing the meeting area of the players' velocities trajectories, i.e. for Problem 2 in part (1).

Property 5 In the time interval $[0, \theta]$, Evader's velocity trajectory satisfies the relation

$$\dot{y}(t) \in \mathbf{A}_{11}, \tag{10.34}$$

where $\mathbf{A}_{11}$ has the form (10.16).

The following property easily follows from Property 5.

Property 6 In the time interval $[0, \theta]$, the state of the Evader satisfies the relation

$$y(t) \in y_{10} + t\mathbf{A}_{11}. \tag{10.35}$$

Solution of the Part (3) of the Problem 2. To solve part (3) of the Problem 2, we first find the Evader's the attainability domain for the time interval $[\theta, T_{22}]$. To do this, for the differential game (10.31)–(10.32) we find the reachable area for the Evader in the same way as in Sect. 10.3.2. However, here we take into account that the initial speeds of the players coincide and the initial time of the game is considered as $t = \theta$. Then, similar to Theorem 2, we find the following assertion.

Property 7 If $\alpha > \beta$, then in Case 2 for the game (10.1)–(10.2) (or for (10.31)–(10.32)) for all $t \in [\theta, T_{22}]$ the following inclusion

$$y(t) \in \mathbf{A}(x(\theta), y(\theta)) + (t - \theta)\dot{y}(\theta), \tag{10.36}$$

where

$$\mathbf{A}(x(\theta), y(\theta)) = \frac{\alpha^2 y(\theta) - \beta^2 x(\theta)}{\alpha^2 - \beta^2} + \frac{\alpha\beta|z(\theta)|}{\alpha^2 - \beta^2}\mathbf{S}. \tag{10.37}$$

Proof The proof is carried out in a similar way to Theorem 2. In this case, it must be taken into account that for $\dot{x}(\theta) = \dot{y}(\theta)$ it has the equality

$$\mathbf{A}(\dot{x}(\theta), \dot{y}(\theta)) = \frac{\alpha^2 \dot{y}(\theta) - \beta^2 \dot{x}(\theta)}{\alpha^2 - \beta^2} + \frac{\alpha\beta|\dot{z}(\theta)|}{\alpha^2 - \beta^2}\mathbf{S} = \dot{x}(\theta) = \dot{y}(\theta).$$

□

Now the set $\mathbf{A}(x(\theta), y(\theta)) + (t - \theta)\dot{y}(\theta)$ is estimated from above. To do this, taking into account the inequality (10.30), we have that $|z(\theta)| \le L_{11}$ and hence from (10.37) we obtain a relation of the form

$$\mathbf{A}(x(\theta), y(\theta)) \subset \frac{\alpha^2 y(\theta) - \beta^2 x(\theta)}{\alpha^2 - \beta^2} + \frac{\alpha\beta L_{11}}{\alpha^2 - \beta^2}\mathbf{S}.$$

From (10.35) and $z(\theta) \in L_{11}\mathbf{S}$ we have

$$\frac{\alpha^2 y(\theta) - \beta^2 x(\theta)}{\alpha^2 - \beta^2} = y(\theta) - \frac{\beta^2}{\alpha^2 - \beta^2} z(\theta) \in y_{10} + \theta\mathbf{A}_{11} + \frac{\beta^2}{\alpha^2 - \beta^2} L_{11}\mathbf{S}.$$

Hence we get

$$\mathbf{A}(x(\theta), y(\theta)) \subset y_{10} + \theta\mathbf{A}_{11} + \frac{\beta}{\alpha - \beta} L_{11}\mathbf{S}.$$

From here and from (10.34), (10.36) and $\theta \in [T_*, T^*]$ [see (10.23)] we obtain

$$\begin{aligned}
y(t) &\in y_{10} + \theta\mathbf{A}_{11} + \frac{\beta}{\alpha - \beta} L_{11}\mathbf{S} + (t - \theta)\dot{y}(\theta) \subset \\
&\subset y_{10} + \theta\mathbf{A}_{11} + \frac{\beta}{\alpha - \beta} L_{11}\mathbf{S} + (t - \theta)\mathbf{A}_{11} \\
&= y_{10} + t\mathbf{A}_{11} + \frac{\beta}{\alpha - \beta} L_{11}\mathbf{S} + \theta\frac{2\alpha\beta|z_{11}|}{\alpha^2 - \beta^2}\mathbf{S} \subset y_{10} + t\mathbf{A}_{11} + \mathbf{B}_{10},
\end{aligned}$$

where $\mathbf{B}_{10} = \frac{\beta}{\alpha-\beta}\left(L_{11} + \frac{2\alpha|z_{11}|^2}{\alpha^2-\beta^2}\right)\mathbf{S}$.

Property 8 In the time interval $[\theta, T_{22}]$, the state of the Evader satisfies the relation

$$y(t) \in y_{10} + t\mathbf{A}_{11} + \mathbf{B}_{10}. \tag{10.38}$$

Consequently, from here and from Property 6 we derive the following result.

Theorem 8 *If $\alpha > \beta$, then in Case 2 for the game (10.1)–(10.2), the inclusion (10.38) is performed for all $t \in [0, T_{22}]$ and for any $v(\cdot) \in \mathbf{V}$.*

Property 9 If $\alpha > \beta$ and $\beta|x_{11}| \geq \alpha|y_{11}|$, then the multi-valued mapping $\mathbf{W}(t)$ is monotonically increasing with respect to the inclusion, i.e. from $t_2 > t_1 \Rightarrow \mathbf{W}(t_1) \subset \mathbf{W}(t_2)$ for $t \in [0, T_{22}]$, where $\mathbf{W}(t) = y_{10} + t\mathbf{A}_{11} + \mathbf{B}_{10}$.

Proof From the form (10.16) we obtain

$$\mathbf{W}(t) = y_{10} + t\frac{\alpha^2 y_{11} - \beta^2 x_{11}}{\alpha^2 - \beta^2} + t\frac{\alpha\beta|z_{11}|}{\alpha^2 - \beta^2}\mathbf{S} + \mathbf{B}_{10}. \tag{10.39}$$

Let $F(\mathbf{W}, \psi) = \max_{\omega \in \mathbf{W}}\langle \omega, \psi\rangle$ is the support function of the closed set $\mathbf{W}$ for each $\psi \in \mathbb{R}^n$, $|\psi| = 1$. We calculate the derivative of the support function of the set (10.39). Since $F(\mathbf{S}, \psi) = 1$ (see [25]), then from form (10.39) we get

$$\frac{d}{dt}F(\mathbf{W}(t), \psi) = \left\langle \frac{\alpha^2 y_{11} - \beta^2 x_{11}}{\alpha^2 - \beta^2}, \psi \right\rangle + \frac{\alpha\beta|z_{11}|}{\alpha^2 - \beta^2}$$

for any $\psi \in \mathbb{R}^n$, $|\psi| = 1$. Since $\langle \alpha^2 y_{11} - \beta^2 x_{11}, \psi\rangle$ reaches its smallest value at $\psi = -\frac{\alpha^2 y_{11} - \beta^2 x_{11}}{|\alpha^2 y_{11} - \beta^2 x_{11}|}$, then we have the inequality

$$\frac{d}{dt}F(\mathbf{W}(t), \psi) \geq -\frac{|\alpha^2 y_{11} - \beta^2 x_{11}|}{\alpha^2 - \beta^2} + \frac{\alpha\beta|z_{11}|}{\alpha^2 - \beta^2}. \tag{10.40}$$

It is easy to check from the condition $\beta|x_{11}| \geq \alpha|y_{11}|$ that the right side of inequality (10.40) is non-negative. Whence follows the validity of Property 8.

Definition 5 For the game (10.1)–(10.2) in Case 2, the set

$$\mathbf{W}_2 = y_{10} + \mathbf{B}_{10} + \bigcup_{t=0}^{T_{22}} t\mathbf{A}_{11} \tag{10.41}$$

is called *the attainability domain of the Evader*.

By Property 9 and Definition 5, we find the following property.

Property 10 If $\alpha > \beta$ and $\beta|x_{11}| \geq \alpha|y_{11}|$, then the attainability domain of the Evader $\mathbf{W}_2$ has the form of a ball $\mathbf{W}_2 = y_{10} + T_{22}\mathbf{A}_{11} + \mathbf{B}_{10}$.

From Theorem 8 we can assert the following theorem.

Theorem 9 *If the set W_2 doesn't intersect with M, then the Pursuer wins in Case 2 in the game (10.1)–(10.2) with the "Life line".*

Proof The proof of Theorem 9 follows from Theorem 8 and from the form of the set (10.41). □

References

1. Isaacs, R.: Differential Games. Wiley, New York (1965)
2. Petrosjan, L.A.: Differential Games of Pursuit. Series on Optimization, Vol. 2. World Scientific Poblishing, Singapore (1993). https://doi.org/10.1114/1670
3. Azamov, A.: On the quality problem for simple pursuit games with constraint. Publ. Sofia: Serdica Bulgariacae Math. **12**(1), 38–43 (1986)
4. Azamov, A.A., Samatov, B.T.: The Π-strategy: analogies and applications. Contrib. Game Theor. Manage. **4**, 33–46 (2011)
5. Sun, W., Tsiotras, P., Lolla, T., Subramani, D.N., Lermusiaux, P.F.J.: Multiple-pursuit/one-evader pursuit-evasion game in in dynamical flow fields. J. Guid. Control Dyn. **40**(7), 1627–1637 (2017). https://doi.org/10.2514/1.G002125
6. Garcia, E., Casbeer, D.W., Pachter, M.: Optimal strategies of the differential game in a circular region. IEEE Control Syst. Lett. **4**(2), 492–497 (2019). https://doi.org/10.1109/LCSYS.2019.2963173
7. Liang, L., Deng, F., Peng, Z., Li, X., Zha, W.: A differential game for cooperative target defense. Automatica **102**(April), 58–71 (2019). https://doi.org/10.1016/j.automatica.2018.12.034
8. Weintraub, I.E., Pachter, M., Garcia, E.: An introduction to pursuit–evasion differential games. In: American Control Conference (ACC), July 01–03 (2020). https://doi.org/10.23919/ACC45564.2020.9147205
9. Dorothy, M., Maity, D., Shishika, D., Von Moll, A.: One Apollonius Circle is Enough for Many Pursuit-Evasion Games. https://doi.org/10.48550/arXiv.2111.09205
10. Samatov, B.T., Horilov, M.A., Akbarov, AKh.: Differential game: "life line" for non-stationary geometric constraints on controls. Lobachevskii J. Math. **43**(1), 237–248 (2022). https://doi.org/10.1134/S1995080222040187
11. Pshenichnii, B.N.: Simple pursuit by several objects. Cybern. Syst. Anal. **12**(5), 484–485 (1976). https://doi.org/10.1007/BF01070036
12. Satimov, NYu.: Methods of Solving the Pursuit Problems in the Theory of Differential Games. Izd-vo NBRUz, Tashkent (2019)
13. Samatov, B.T.: Problems of group pursuit with integral constraints on controls of the players II. Cybern. Syst. Anal. **49**(6), 907–921 (2013). https://doi.org/10.1007/s10559-013-9581-5
14. Grigorenko, N.L.: Mathematical Methods of Control for Several Dynamic Processes. Izdat. Gos. Univ, Moscow (1990)
15. Chikrii, A.A.: Conflict-Controlled Processes. Kluwer Academic Publishers, Dordrecht (1997). https://doi.org/10.1007/978-94-017-1135-7
16. Munts, N.V., Kumkov, S.S.: On the coincidence of the minimax solution and the value function in a time-optimal game with a lifeline. Proc. Steklov Inst. Math. **305**, S125–S139 (2019). https://doi.org/10.1134/S0081543819040138
17. Samatov, B.T., Sotvoldiyev, A.I.: Intercept problem in dynamic flow field. Uzbek Math. J. **2**, 103–112 (2019). https://doi.org/10.29229/uzmj.2019-2-12

18. Samatov, B.T., Ibragimov, G.I., Hodjibayeva, I.V.: Pursuit-evasion differential games with the Grönwall type constraints on controls. Ural Math. J. **6**(2), 95–107 (2020). https://doi.org/10.15826/umj.2020.2.010
19. Samatov, B.T., Umaraliyeva, N.T., Uralova, S.I.: Differential games with the Langenhop type constraints on controls. Lobachevskii J. Math. **42**(12), 2942–2951 (2021). https://doi.org/10.1134/S1995080221120295
20. Samatov, B.T.: The pursuit-evasion problem under integral-geometric constraints on pursuer controls. Autom. Remote Control **74**(7), 1072–1081 (2013). https://doi.org/10.1134/S0005117913070023
21. Samatov, B.T.: The Π-strategy in a differential game with linear control constraints. J. Appl. Math. Mech. **78**(3), 258–263 (2014). https://doi.org/10.1016/j.jappmathmech.2014.09.008
22. Bakolas, E.: Optimal guidance of the isotropic rocket in the presence of wind. J. Optim. Theor. Appl. **162**(3), 954–974 (2014). https://doi.org/10.1007/s10957-013-0504-4
23. Samatov, B.T., Soyibboev, U.B.: Differential game with a lifeline for the inertial movements of players. Ural Math. J. **7**(2), 94–109 (2021). https://doi.org/10.15826/umj.2021.2.007
24. Alekseev, V.M., Tikhomirov, V.M., Fomin, S.V.: Optimal'noye upravleniye [Optimal Control]. Nauka, Moscow (1979)
25. Blagodatskikh, V.I.: Introduction to Optimal Control (Linear Theory). Visshaya Shkola, Moscow (2001)

Chapter 11
Optimal Number of Pursuers in the Game on the 1-Skeleton of 4D Cube

Gafurjan Ibragimov and Zahriddin Muminov

Abstract In the paper, we study pursuit and evasion differential games within a four-dimensional cube, where all the players move along the edges. The problem is to find the optimal number of pursuers in the game, to construct strategies for the pursuers in pursuit game, and evasion strategy in evasion game.

Keywords Graph of 4-cube · Differential game · Pursuer · Evader · Strategy

11.1 Introduction

Though the notion of differential game was introduced by Rufus Isaacs [1] for the first time, in fact, it was studied by some researchers. The theory of differential games was further developed by many researchers such as Petrosyan [2], Satimov [3] and others.

One of the most significant current discussions in differential games is multi player differential games. In recent years, there has been an increasing interest in differential games of several players with or several evaders (see for example [4–13])

However, if exhaustable resources such as energy, fuel, resources etc. are limited for the modeling control processes, then control functions are necessarily restricted by integral constraints. The method of resolving functions for the games with integral constraints on control functions was developed by Belousov [14] to obtain a sufficient condition for the pursuit differential game. The solution was further extended to the case of convex integral constraints [15]. Other works in differential game in $\mathbb{R}^n$ of integral constraints include [9, 10, 16, 17].

Some games with either geometric or integral constraint, restrict the movement of players to some specific state constraints. For examples, differential games in a convex subset of $\mathbb{R}^n$ were studied by [18–21].

Furthermore, games within a geometrical structure in the form of abstract graphs as its state constraint, are of increasing interest. These types of games are models

---

G. Ibragimov · Z. Muminov (✉)
V.I.Romanovskiy Institute of Mathematics, University street 9, 100174 Tashkent, Uzbekistan
e-mail: zimuminov@gmail.com

D. Azimov (ed.), *Proceedings of the IUTAM Symposium on Optimal Guidance and Control for Autonomous Systems 2023*, IUTAM Bookseries 40,
https://doi.org/10.1007/978-3-031-39303-7_11

for the search problems of a moving object and have minimax forms as mentioned in [22, 23]. It could be also called multi-move games as in the work of [24, 25]. In such type of games, players move from one vertex to its adjacent vertex by jumping.

Another type of games on abstract graphs are differential games, where players move along the edges of a given graph embedded in a Euclidean space, as studied in [5, 23, 26–29]. The works [5, 26, 28, 29] are devoted to the differential games on 1-skeleton of regular polyhedrons. The main research questions in these works are: What is pursuit strategy? What is evasion strategy? What is the minimum number of pursuers to complete the game?

In the recent work [30] a differential game of many pursuers and one evader within 1-skeleton graph of an orthoplex of dimension $d+1$ was discussed. Both pursuit and evasion games were considered on the edge graph K_{d+1} of the orthoplex $\Sigma^{2(d+1)}$ in the Euclidean space $\mathbb{R}^{d+1}$. It was proven that pursuit can be completed in the case of $n = k = d+1$, or when $n \neq k$ and $n + k \geq 2d$, where k is the number of pursuers whose speed belongs to the interval $[1/2, 1)$. If one of these conditions is not satisfied, then it was established that evasion is possible.

The current paper intends to study both pursuit and evasion differential games within a four-dimensional cube (Fig. 11.1). All the players move along the edges of the cube. We search the optimal number of slow pursuers to ensure that pursuit can be completed. We find the optimal number of pursuers, construct pursuers' and evader's strategies.

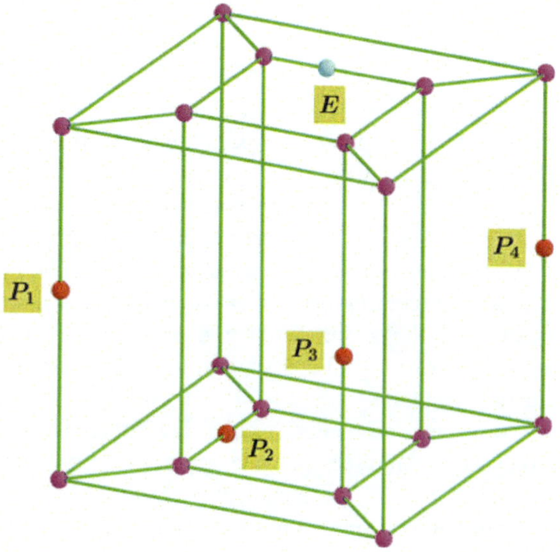

Fig. 11.1 The graph of four dimensional cube Q

11.2 Statement of Problem

We consider a differential game of n pursuers $\xi_1, \xi_2, ..., \xi_n$, $n \geq 2$, and one evader η whose dynamics are given by the following equations

$$\begin{aligned} \dot{\xi}_i &= v_i, \ \xi_i(0) = \xi_{i0}, \quad i = 1, ..., n, \\ \dot{\eta} &= v, \quad \eta(0) = \eta_0, \end{aligned} \tag{11.1}$$

where $\xi_{i0}, \eta_0 \in Q$, $\xi_{i0} \neq \eta_0$, $i = 1, ..., n$; v_i is the control parameter of i-th pursuer, and v is the control parameter of the evader. All the players move along the edges of four-dimensional cube Q. The maximal speeds of the pursuers $\xi_1, \xi_2, ..., \xi_n$ are $\rho_1, \rho_2, ..., \rho_n$, respectively, and that of evader η is 1, i.e., $|v_i| \leq \rho_i$, $i = 1, ..., n$, $|v| \leq 1$. It is assumed that $1/3 \leq \rho_i < 1$.

We let $H(r)$ denote the ball of radius r and centered at the origin of the Euclidean space $\mathbb{R}^{d+1}$.

Definition 1 A measurable function $v_i(\cdot)$, $v_i\colon [0, \infty) \to H(\rho_i)$ is called admissible control of the i-th pursuer, $i \in \{1, ..., n\}$, if for the solution $\xi_i(\cdot)$ of the equation $\dot{\xi}_i = v_i$, $\xi_i(0) = \xi_{i0}$, we have $\xi_i(t) \in Q$, $t \geq 0$.

Definition 2 A measurable function $v(\cdot)$, $v\colon [0, \infty) \to H(\sigma)$ is called admissible control of the evader, if for the solution $\eta(\cdot)$ of the equation $\dot{\eta} = v$, $\eta(0) = \eta_0$, we have $\eta(t) \in Q$, $t \geq 0$.

In the present paper, we consider pursuit and evasion differential games. In the pursuit differential game pursuers apply some strategies and the behavior of the evader is arbitrary, that is, the evader during the game uses an arbitrary admissible control. Let us give now definition for strategies of pursuers.

Definition 3 The functions $(t, \xi_1, ..., \xi_n, \eta, v) \to V_i(t, \xi_1, ..., \xi_n, \eta, v)$, $i=1, 2, ..., n$, are called strategies of pursuers ξ_i, $i = 1, 2, ..., n$, if the initial value problem (11.1) has a unique solution $\xi_1(t), ..., \xi_n(t), \eta(t) \in Q$, $t \geq 0$, for $v_i = V_i(t, \xi_1, ..., \xi_n, \eta, v)$, $i = 1, 2, ..., n$, and for arbitrary admissible control $v = v(t)$ of the evader.

According to this definition the information about the current time t, the current states $\xi_1(t), ..., \xi_n(t)$ of the pursuers and the current state $\eta(t)$ of the evader, and the evader's velocity vector is available for each pursuer.

Definition 4 If, for some number $\Theta > 0$, there exist strategies of pursuers such that $\xi_i(\theta) = \eta(\theta)$ at some θ, $0 < \theta \leq \Theta$, and $i \in \{1, ..., n\}$, then pursuit is said to be completed. The pursuers are interested in completing the pursuit as earlier as possible.

Definition 5 A function $(t, \xi_1, ..., \xi_n, \eta) \to V(t, \xi_1, ..., \xi_n, \eta)$ is called a strategy of the evader η if the initial value problem (11.1) has a unique solution $\xi_1(t), ..., \xi_n(t)$, $\eta(t) \in Q$, $t \geq 0$, for $v = V(t, \xi_1, ..., \xi_n, \eta)$ and for any admissible controls of pursuers $v_i = v_i(t)$, $i = 1, 2, ..., n$.

Hence, the evader is allowed to know information about the current time, the current states $\xi_1(t)$, ..., $\xi_n(t)$, $\eta(t)$ of all players.

Definition 6 If, for some initial states of players $\xi_{10}, ..., \xi_{n0}, \eta_0 \in Q$, there exists a strategy of evader such that $\xi_i(t) \neq \eta(t)$ for all $t \geq 0$, and $i = 1, 2, ..., n$, then we say that evasion is possible in the game in Q.

The evader tries to maintain the inequality $\xi_i(t) \neq \eta(t)$ as long as possible. Note that in this definition the phrase "for some initial states of players $\xi_{10}, ..., \xi_{n0}, \eta_0 \in Q$" is important. Because, for some initial states of the players, the evader may be trapped by some pursuers and therefore pursuit can be completed easily.

We call the number $N = N(Q)$ optimal number of pursuers for the differential game on the cube Q if, (i) for any initial states of players, pursuit can be completed in the differential game with N pursuers, and (ii) evasion is possible in the differential game with $N - 1$ pursuers.

The problem is to find the optimal number of pursuers N in the differential game on the cube Q, to construct pursuit strategies for the pursuers in pursuit differential game, and an evasion strategy in evasion differential game.

11.3 Main Result

Without any loss of generality we assume that the lengths of edges of the cube Q is equal to 1.

11.3.1 Pursuit Differential Game

We prove the following statement.

Theorem 1 *Four pursuers $\xi_1, \xi_2, \xi_3, \xi_4$ can complete the pursuit in the differential game on 1-skeleton of the four dimensional cube Q.*

Proof Let the points D' and C' divide the edge AB into three equal segments: $AD' = D'C' = C'B$ (Fig. 11.2). To construct strategies of pursuers, we define the shadow $E' \in AB$ of the evader $\bar{E} \notin AB$ on the edge AB as follows.

1. If $\bar{E} \in AE$ or AD or AA_1 (these edges are highlighted in green in Fig. 11.2), then $A\bar{E} = 3AE'$.

2. If $\bar{E} \in BC$ or BF or BB_1 (these edges are highlighted in green), then $B\bar{E} = 3BE'$.

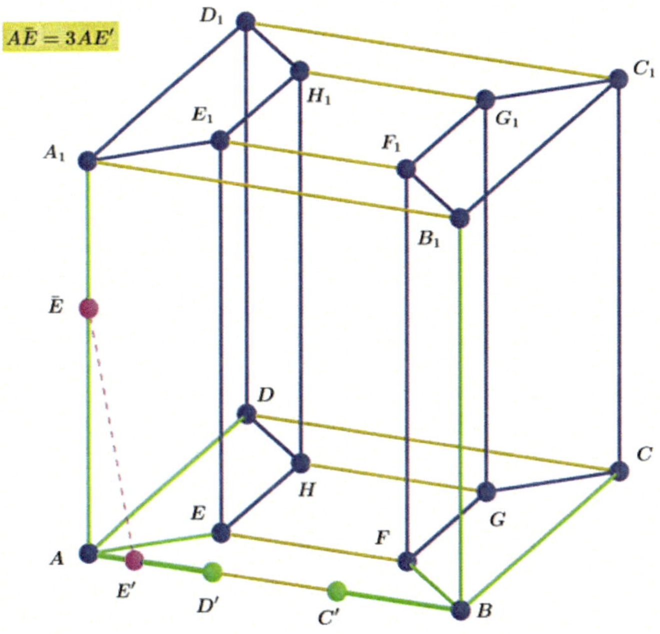

Fig. 11.2 The shadow E' of the point $\bar{E} \in AA_1$: $A\bar{E} = 3AE'$

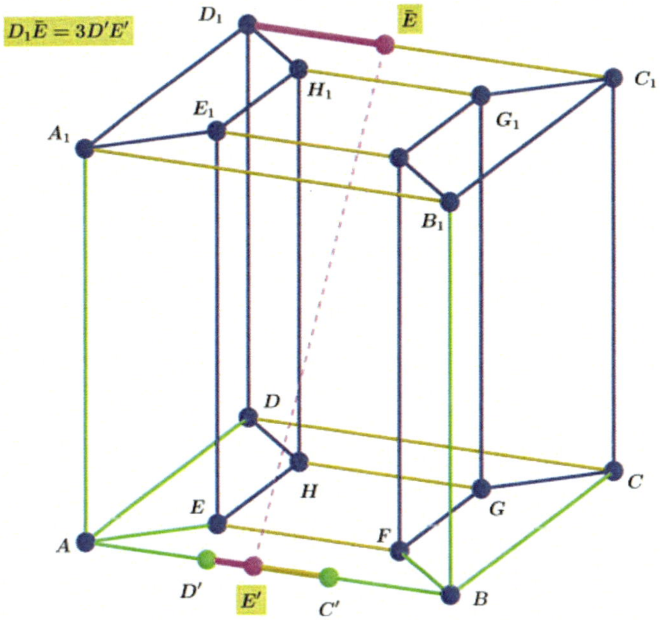

Fig. 11.3 The shadow E' of the point $\bar{E}$: $D_1\bar{E} = 3D'E'$

3. If $\bar{E} \in HE$ or HD or HH_1 or DD_1 or EE_1 or A_1D_1 or D_1H_1 or H_1E_1 or A_1E_1 (these edges are highlighted in blue), then $E' = D'$.

4. If $\bar{E} \in GF$ or GC or GG_1 or FF_1 or CC_1 or B_1C_1 or C_1G_1 or G_1F_1 or F_1B_1 (these edges are highlighted in blue), then $E' = C'$.

5. If $\bar{E}$ is on the edge parallel to AB, that is, $\bar{E} \in EF$ or HG or DC or A_1B_1 or E_1F_1 or H_1G_1 or D_1C_1 (these edges are highlighted in gold), then $E' \in C'D'$ is defined from the condition that the distance of $\bar{E}$ from the left end point of the edge that contains $\bar{E}$ is equal to $3D'E'$ (Fig. 11.3).

Since the maximum speed of the evader is 1, the speed of the point E' doesn't exceed 1/3. If the pursuer P_1 moves from the vertex A to the vertex B along the edge AB, then P_1 coincides with either the real evader $\bar{E}$ or its shadow E'. If P_1 coincides with the real evader $\bar{E}$, then pursuit is completed. If P_1 coincides with the shadow of evader E', then P_1 can further move on the point E' holding this point. Then, as the evader $\bar{E}$ reaches one of the vertices A and B at some time, we have $P_1 = E' = \bar{E}$ at that time, that is, the evader is captured at that time. Thus, starting from the time when $P_1 = E'$ pursuer P_1 can guard the edge AB from the evader.

We construct now strategies for the pursuers. Let pursuers ξ_1, ξ_2, ξ_3, and ξ_4 come to the vertices A, H, E_1, and D_1, respectively. Next, the pursuers ξ_1, ξ_2, ξ_3, and ξ_4 move along the edges AB, HG, E_1F_1, and D_1C_1, respectively, and catch the shadows of the evader on these edges, respectively. Each pursuer starting from the time when he catches the shadow of the evader moves holding the shadow of the evader on that edge.

Let all the pursuers $\xi_1, \xi_2, \xi_3, \xi_4$ catch the shadows of the evader on the edges AB, HG, E_1F_1, D_1C_1, respectively, by the time T (Fig. 11.4).

Then at the time T the evader is on one of the edges colored in Green or Blue or Cyan or Magenta (Fig. 11.4). In each case, the evader is trapped by three pursuers and cannot walk from one edge to another edge of distinct colors.

Without any loss of generality, we assume that the evader is on a green edge. Then it is trapped by the pursuers ξ_1, ξ_3, ξ_4. Then, we let the pursuers ξ_1, ξ_3, ξ_4 control the edges AB, E_1F_1, D_1C_1, respectively, holding the evader's shadow and let the pursuer ξ_2 move towards the evader. Since the green edges form a tree, therefore the pursuer ξ_2 catches the evader or forces it to reach one of the edges AB, E_1F_1, D_1C_1. In the latter case, the evader will be caught by one of the pursuers ξ_1, ξ_3, ξ_4. The proof of the theorem is complete.

11.3.2 Evasion Differential Game

We prove now the following statement.

Theorem 2 *Evasion from three pursuers ξ_1, ξ_2, ξ_3 is possible in the differential game on 1-skeleton of the four dimensional cube Q.*

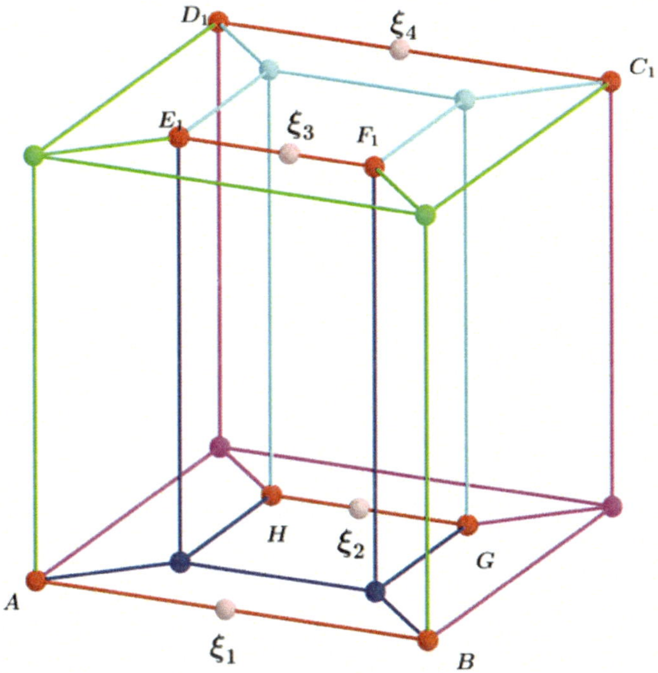

Fig. 11.4 The edges controlled by the pursuers of the 4D cube

Proof To prove this theorem, we show that, for some initial states of players, there exist a strategy of the evader such that evasion is possible.

Let the evader is at some vertex of the cube say at the vertex A and any pursuer is not at the vertex A. We show that from such initial positions of players evasion is possible. We construct a strategy for the evader as follows. The evader stays at the point A until the distance between the point A and closest to this point pursuer becomes less than or equal to 1/3 at some time t_1. Note that it is possible that $t_1 = 0$. For example, if the distance of a pursuer from the point A is less than or equal to 1/3 at the initial time, then, clearly, $t_1 = 0$.

For the definiteness, we assume that the neighboring to A vertices of the cube are B, C, D, E, and $AP_1 = 1/3$ at some time $t_1 \geq 0$ and that the pursuer P_1 is on the edge AB (Fig. 11.5). Since the distance between any two of the points C, D, E along the 1-skeleton of the cube is greater or equal to 2, and the speeds of pursuers P_2 and P_3 are less than or equal to 1, therefore these pursuers can reach only one of the vertices C, D, E for the unit time. Clearly, the pursuer P_1 cannot reach these vertices for the unit time. Hence, the evader can reach one of these vertices, say the vertex C for the unit time not being captured by the pursuers. Thus, the evader is at the vertex C at the time $t_1 + 1$. We repeatedly use this reasoning, to conclude that evasion is possible on the infinite time interval $[0, \infty)$. The proof of the theorem is complete.

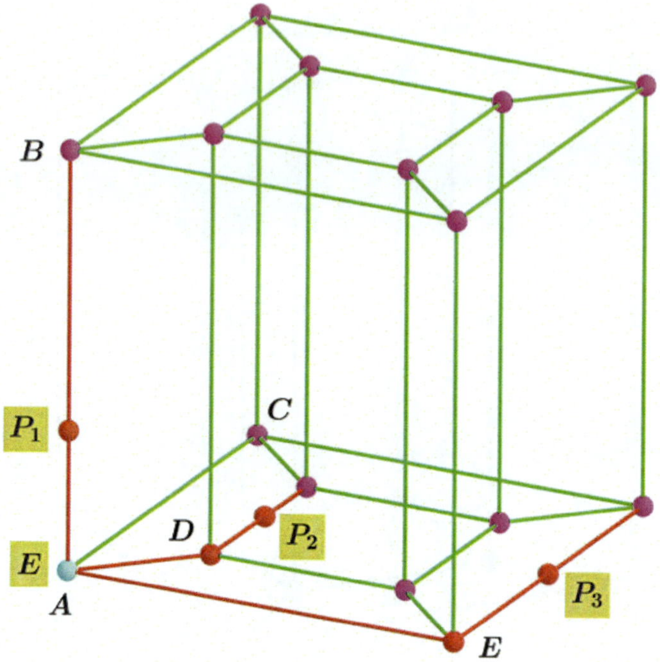

Fig. 11.5 The pursuer P_1 is in 1/3 vicinity of the evader

11.4 Conclusion

We have studied pursuit and evasion differential games on the edge graph of four dimensional cube. We have established that in the differential game of four pursuers and one evader pursuit can be completed. Here, one of the central results of the paper is the construction of strategies for the pursuers.

Next, we proved that in the differential game of three pursuers and one evader evasion is possible. Also, we have proposed a strategy for the evader that ensures evasion in the differential game.

Based on the two theorems proved for the pursuit and evasion differential games we can conclude that the optimal number of pursuers N in the game is $N = 4$.

For the future work, we suggest studying the problem about the optimal number of pursuers in n-dimensional cube as an open problem. The research questions are (1) what is the optimal number of pursuers in the game on the edge graph of n-dimensional cube? (2) what are the pursuers' strategies? (3) what is the evader's strategy?

In the present paper, we have studied the case $1/3 \leq \rho_i < 1, i = 1, 2, 3, 4$. What if some of ρ_i are less than 1/3? For example, for the future work, we can recommend to study the following problem: Is the evasion possible from some initial states if $0 < \rho_i < 1/3, i = 1, 2, 3, 4$?

References

1. Isaacs, R.: Differential Games. Wiley, New York, NY, USA (1965)
2. Petrosyan, L.A.: Differential Games of Pursuit. World Scientific, Singapore, London (1993)
3. Satimov, N.Y., Rikhsiev, B.B.: Methods of Solving of Evasion Problems in Mathematical Control Theory. Fan, Tashkent, Uzbekistan (2000)
4. Blagodatskikh, A.I., Petrov, N.N.: Simultaneous multiple capture of rigidly coordinated evaders. Dyn. Games Appl. **9**, 594–613 (2019). https://doi.org/10.1007/s13235-019-00300-8
5. Azamov, A.A., Kuchkarov, ASh., Holboyev, A.G.: The pursuit-evasion game on the 1-skeleton graph of the regular polyhedron. III. Mat. Teor. Igr Pril. **11**(4), 5–23 (2019)
6. Scott, W.L., Leonard, N.E.: Optimal evasive strategies for multiple interacting agents with motion constraints. Autom. J. IFAC **94**, 26–34 (2018)
7. Yan, R., Shi, Z., Zhong, Y.: Cooperative strategies for two-evader-one-pursuer reach-avoid differential games. Int. J. Syst. Sci. **52**(9), 1894–1912 (2021). https://doi.org/10.1080/00207721.2021.1872116
8. Sh, K.A., Ibragimov, G.I., Khakestari, M.: On a linear differential game of optimal approach of many pursuers with one evader. J. Dyn. Control Syst. **19**(1), 1–15 (2013). https://doi.org/10.1007/s10883-013-9161-z
9. Ibragimov, G.I., Ferrara, M., Ruziboev, M., Pansera, B.A.: Linear evasion differential game of one evader and several pursuers with integral constraints. Int. J. Game Theor. **50**, 729–750 (2021). https://doi.org/10.1007/s00182-021-00760-6
10. Ibragimov, G.I., Salleh, Y.: Simple motion evasion differential game of many pursuers and one evader with integral constraints on control functions of players. J. Appl. Math. 748096 (2012). https://doi.org/10.1155/2012/748096
11. Kumkov, S.S., Le Ménec, S., Patsko, V.P.: Zero-sum pursuit-evasion differential games with many objects: survey of publications. Dyn. Games Appl. **7**, 609–633 (2017). https://doi.org/10.1007/s13235-016-0209-z
12. Blagodatskikh A.I. : Multiple capture of rigidly coordinated evaders. Vestn. Udmurtsk. Univ. Mat. Mekh. Komp. Nauki **26**(1), 46–57 (2016)
13. Garcia, E., Casbeer, D.W., Von Moll, A., Pachter, M.: Multiple pursuer multiple evader differential games. IEEE Trans. Autom. Control **66**(5), 2345–2350 (2021). https://doi.org/10.1109/TAC.2020.3003840
14. Belousov, A.A.: O lineinykh differentsialnykh igrakh presledovaniya s integralnymi ogranicheniyami. In-t matematiki im. V. A. Steklova RAN, MGU im. M. V. Lomonosova, M., 321–322 (2008)
15. Chikrii, A.A., Belousov, A.A.: On linear differential games with convex integral constraints. Trudy Inst. Mat. i Mekh. UrO RAN **19**(4), 308–319 (2013)
16. Ibragimov, G.I.: A game of optimal pursuit of one object by several. J. Appl. Maths Mechs. **62**(2), 187–192 (1998) (Prikladnaya Matematika i Mekhanika **62**(2): 199–205, 1998)
17. Ibragimov, G.I., Ferrara, M., Kuchkarov, ASh., Pansera, B.A.: Simple motion evasion differential game of many pursuers and evaders with integral constraints. Dyn. Games Appl. **8**, 352–378 (2018). https://doi.org/10.1007/s13235-017-0226-6
18. Satimov, N.Y., Ibragimov, G.I.: One class of simultaneous pursuit games. Izv. Vyssh. Uchebn. Zaved. Mat. **5**, 46–55 (2012)
19. Ibragimov, G.I., Salimi, M., Amini, M.: Evasion from many pursuers in simple motion differential game with integral constraints. Euro. J. Oper. Res. **218**(2), 505–511 (2012)
20. Ibragimov, G., Alias, I.A., Tukhtasinov, M., Hasim, R.M.: A pursuit problem described by infinite system of differential equations with coordinate-wise integral constraints on control functions. Malaysian J. Math. Sci. **9**(1), 67–76 (2015)
21. Ferrara, M., Ibragimov, G.I., Salimi, M.: Pursuit-evasion game of many players with coordinate-wise integral constraints on a convex set in the plane. Atti della Accademia Peloritana dei Pericolanti-Classe di Scienze Fisiche, Matematiche e Naturali **95**(2), 1–6 (2017)
22. Azamov, A.A.: Lower bound for the advantage coefficient in the graph search problem. Differential Eqs. **44**(12), 1764–1767 (2008)

23. Fomin, F.V., Thilikos, D.M.: An annotated bibliography on guaranteed graph searching. Theoret. Comput. Sci. **399**, 236–245 (2008)
24. Ibragimov, G.I., Luckraz, Sh.: On a characterization of evasion strategies for pursuit-evasion games on graphs. J. Optim. Theor. Appl. **175**, 590–596 (2017). https://doi.org/10.1007/s10957-017-1155-7
25. Bonato, A., Golovach, P., Hahn, G., Kratochvil, J.: The capture time of a graph. Discr. Math. **309**(18), 5588–5595 (2009)
26. Azamov, A., Ibaydullaev, T.: A pursuit-evasion differential game with slow pursuers on the edge graph of simplexes I. Math. Game Theor. Appl. **12**(4), 7–23 (2020). https://doi.org/10.17076/mgta_2020_4_23
27. Andreae, T., Hartenstein, F., Wolter, A.: A two-person game on graphs where each player tries to encircle his opponent's men. Theoret. Comput. Sci. (Math. Games) **215**, 305–323 (1999)
28. Azamov, A.A., Kuchkarov, ASh., Holboyev, A.G.: The pursuit-evasion game on the 1-skeleton graph of the regular polyhedron. II. Mat. Teor. Igr Pril. **8**(4), 3–13 (2016)
29. Azamov, A.A., Kuchkarov, A.S., Holboyev, A.G.: The pursuit-evasion game on the 1-skeleton graph of the regular polyhedron. I. Mat. Teor. Igr Pril. **7**(3), 3–15 (2015)
30. Azamov, A.A., Ibaydullaev, T., Ibragimov, G.I., Alias, I.A.: Optimal number of pursuers in the differential games on the 1-skeleton of orthoplex. Symm. (Game Theor. Symm. Dyn. Process.) **13**(11), 2170 (2021). https://doi.org/10.3390/sym13112170

Part IV
Optimal Control and Applications

Chapter 12
Development of an Operation Support Device for Portable Underwater Vehicles

Yu Kimura, Norimitsu Sakagami, Takahiro Wada, and Koichi Koganezawa

Abstract This paper describes the development of a device for operational support of portable underwater vehicles during underwater tasks. In recent years, small, lightweight, and inexpensive underwater vehicles of many types have been released and used for monitoring, inspection, and educational purposes. However, the control of the vehicle position in three dimensions is difficult for untrained operators. To resolve that difficulty, we use haptic shared control for underwater vehicles. Haptic shared control realizes cooperative work of an operator and a control system through a single control device such as a steering wheel or joystick. For this work, we designed and developed a control device that enables three-dimensional position control of an underwater vehicle and presents forces for haptic guidance to an operator. Using this control device, we constructed the operation support device based on haptic shared control. Additionally, we conducted a preliminary experiment to evaluate the performance of the developed device. From the experimentally obtained results, we confirmed that the developed device is useful for manual control and haptic shared control for three-dimensional vehicle motion.

Keywords Haptic shared control · Underwater vehicles · Operation support

Y. Kimura · N. Sakagami (✉)
Tokai University, 3-20-1 Orido, Shizuoka 424-0907, Japan
e-mail: sakagami@tokai.ac.jp

T. Wada
Nara Institute of Science and Technology, 8916-5 Takayama-Cho, Ikoma 630-0101, Japan

K. Koganezawa
Tokai University, 4-1-1 Kitakaname, Hiratsuka 259-1292, Japan

D. Azimov (ed.), *Proceedings of the IUTAM Symposium on Optimal Guidance and Control for Autonomous Systems 2023*, IUTAM Bookseries 40,
https://doi.org/10.1007/978-3-031-39303-7_12

12.1 Introduction

The use of underwater vehicles such as autonomous underwater vehicles (AUVs) and remotely operated vehicles (ROVs) is spreading worldwide for purposes such as scientific research, resource development, infrastructure inspection, ship hull inspection, and education. Particularly noteworthy is that ROVs can be purchased easily from online shops. Many people increasingly have opportunities to use them. One attraction of underwater vehicles is that they can easily access difficult areas and allow the viewing of underwater environments that are difficult for humans to access directly.

Nevertheless, untrained users often show difficulty controlling the three-dimensional movement of ROVs because users generally cannot view the vehicle movement directly during underwater operation. In addition, a user on a small boat might have seasickness, which often makes it even more difficult for a user to operate an underwater vehicle. Such circumstances are expected to increase the mental workload of operators. Supporting the operations from an engineering perspective is important to reduce vehicle operator burdens.

Providing haptic information is one solution that can improve support of remotely controlled underwater vehicle operations. One report [1] has described a one degree-of-freedom (DOF) haptic driving system to control the surge motion of a simulated ROV. The joystick was used to present an operator with the presence of drag force acting on an underwater vehicle. Another study [2] developed a desktop haptic interface to control four DOF movements of an underwater vehicle. The developed interface presents haptic forces for external force cancellation and obstacle avoidance.

For one study [3], haptic guidance information was rendered to an operator for an underwater manipulation task of mating and demating a connector. Results of that study indicate that haptic cues are useful for underwater vehicle operations.

To reduce operators' mental workload, we use haptic shared control (HSC) to operate underwater vehicles. With haptic shared control, an operator and an automatic controller operate a controlled system using a single control device such as a steering wheel or flight joystick. The control device has an actuator(s) providing haptic guidance to an operator. The haptic guidance can be expected to reduce the mental workload of vehicle operators. Several studies have demonstrated the effectiveness of HSC application to controlled systems. For one earlier study [4], an HSC was applied to a suspended grab for deep-sea excavation. The experimentally obtained results show that it reduced the cognitive workload of the operators. In other earlier studies [5, 6], we applied an HSC to an underwater vehicle for visual identification tasks. The experimentally obtained results indicated that haptic cues helped the vehicle operators and reduced the mental workload. In those experiments, for simplicity, we evaluated only sway and yaw motions of the vehicle using a desktop joystick for two-axis control. Unfortunately, a two-axis joystick is inadequate for steering three-dimensional vehicle motion. Moreover, some risk exists that shaking of the human body caused by ship motion leads to erroneous input to a desktop

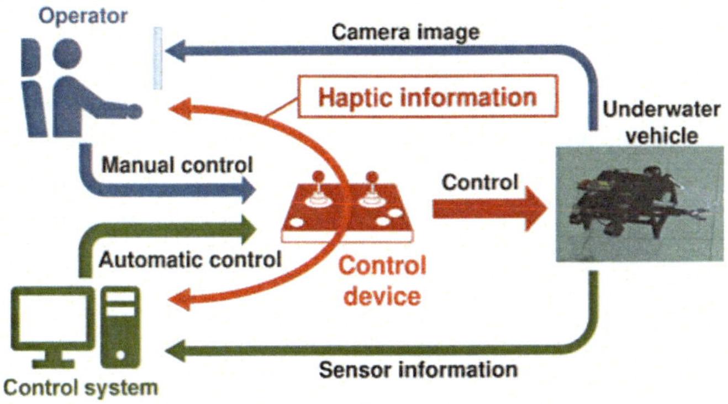

Fig. 12.1 Concept of haptic shared control for an underwater vehicle. An operator and automatic control system simultaneously use a single control device to control the underwater vehicle. The control device can present haptic guidance to an operator

joystick. In addition to these difficulties, because of the large mass of a desktop joystick, its response tends to be slow when providing haptic force.

For the work described herein, we designed and developed a small control device with two thumb sticks for HSC. Each stick has two DOF. The two sticks are useful to control three-dimensional vehicle motion. The thumb sticks can reduce the poor response depending on the mass because they are small and very light. Moreover, erroneous input caused by ship motion can be prevented easily because the developed control device can be placed on the lap and can be moved according to body and arm motion.

Figure 12.1 presents the HSC concept for an underwater vehicle. A human operator judges the situation and responds flexibly during a mission in an unknown environment. Simultaneously, the HSC executes automatic control to reduce the physical and mental workload on the human operator. Therefore, the HSC leverages the strengths of human and automation systems to compensate for their respective shortcomings.

12.2 Design and Development of the Control Device

This section presents the designed and developed control device that enables three-dimensional position control of an underwater vehicle and which presents forces for haptic guidance to an operator for HSC.

12.2.1 Control Device Concept

We designed and developed a compact, lightweight and laptop control device. The control device resembles a video game controller because video game controllers are devices with which many people are familiar. Video game controllers are held with both hands. Users, simply using their thumbs, can control their two small joysticks. Each thumb joystick usually has two degrees of freedom. Two thumb joysticks can be allocated for the surge, heave, yaw, and sway motions of underwater vehicles. Additionally, they present the benefit that it is difficult for incorrect input to enter the sticks because of shaking of the operator's body on a small boat because the entire control device can be moved according to the human body movement. However, if users use a desktop-style flight joystick when holding only the stick, incorrect input might occur if the user's body shakes. Moreover, the thumb sticks might reduce dynamic response lag because of their mass: they are extremely small and lightweight.

12.2.2 Designed and Developed Control Device

As portrayed on the left side of Fig. 12.2, we designed a control device with two thumb sticks to control vehicle motion. Two small motors are placed at the base of each stick to present haptic guidance to an operator. The two shafts of two motors are arranged as mutually perpendicular. The thumb sticks are driven vertically and horizontally by two motors. In addition, the control device is designed to contain two dual motor drivers and three microcontrollers. It has 150 mm height, 150 mm width, and 71 mm depth. These sticks in Fig. 12.2 are 30 mm long, but they can be replaced easily with sticks of different lengths. These sticks can be tilted up to 45°.

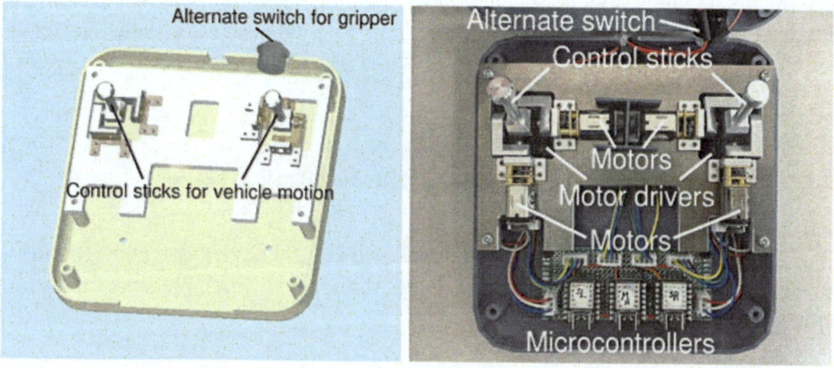

Fig. 12.2 Designed and developed control device of HSC for vehicles' operators. The axes and sticks are made of duralumin metal to improve the rotation accuracy. The housing is made using a 3D printer

Based on the design explained above, we developed the control device as shown on the right side of Fig. 12.2. We selected a gear motor (50:1 Micro Metal Gear-motor HPCB; Pololu Corp.) with a gear ratio of 1:50 and with an optical encoder to control the stick angle. Under this gear ratio, it is possible to back drive with the force from a thumb. An operator can manipulate the sticks according to or against automatic control during HSC. To implement feedback control of the stick angles, we chose microcontrollers (Seeeduino Xiao; Seeed Technology Co., Ltd.) and dual motor drivers (Sabertooth dual 5A motor driver; Dimension Engineering, LLC). One microcontroller and one motor driver are used to implement the angle control of each thumb stick. The maximum force of 20 N is produced at the tip of the sticks. Another microcontroller is used for data communication with a laptop computer for operation. The developed control device has approximate weight of 0.43 kg. An alternate switch for opening and closing a robotic gripper was attached to the top of the control device.

12.3 Haptic Shared Control for Underwater Vehicles

We constructed an operation support device based on HSC including the developed control device for an underwater vehicle. We used a small underwater vehicle with six thrusters for surge, sway, heave, and yaw motions. The vehicle has 350 mm height, 520 mm width, 500 mm depth, and approximately 13.3 kg weight. The vehicle has a robotic gripper (Newton Subsea Gripper; Blue Robotics Inc.) to grab underwater objects.

The vehicle has one waterproof housing containing a battery pack and another housing containing electrical components. The electronic components include a single board computer (Jetson Nano; NVIDIA Corp.), cameras, and speed controllers for the thrusters. The vehicle has an image processing function to detect underwater objects. The image processing used for this study was simple color recognition, but the recognition method can be changed by programming according to underwater objects. The camera image and sensor information of the vehicle are sent to a laptop computer on land through an umbilical cable. A laptop computer is used to present real-time images sent from the vehicle cameras to an operator. It is also used to communicate data between the single board computer inside the vehicle and a microcomputer inside the control device. Details of the hardware of the control device are described above. The microcomputers inside the control device implement feedback control of the thumb sticks to present a haptic force for an operator. Figure 12.3 portrays the control map between the control device and the vehicle motion. For this experiment, forces (f_x, f_y) exerted from the sticks are presented to an operator by implementing PD feedback control for the stick angles proportional to the target errors (e_x, e_y) from the image center as portrayed in Fig. 12.4. The controller works with an operator to control the yaw angle and depth (heave) of the underwater vehicle using the control device. The control law was implemented as described in an earlier report [5]. With feedback control disabled, the control device

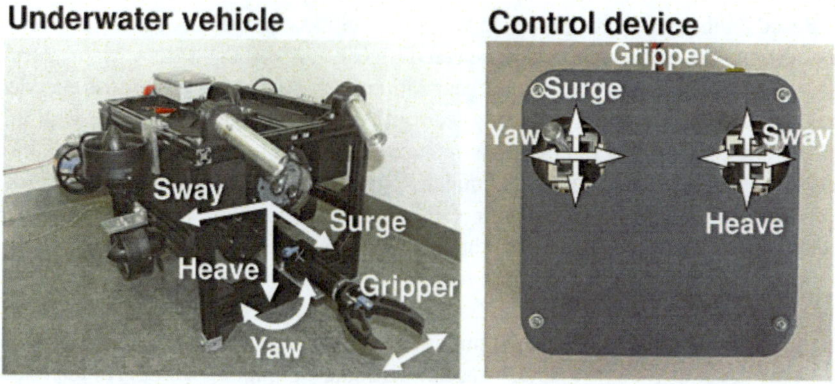

Fig. 12.3 Control map of the vehicle movement with the control device

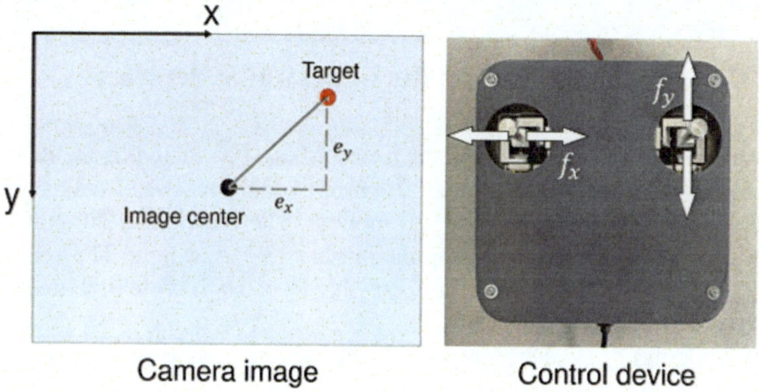

Fig. 12.4 Position errors and haptic force. HSC generates forces to an operator according to the target position on the image

only supports manual control. The opening–closing operation was performed by an operator using an alternate switch attached to the top of the control device.

12.4 Experiments

We conducted preliminary experiments to test the performance of the developed control device and HSC.

Fig. 12.5 Experiment setup

12.4.1 Experiment Setup

Figure 12.5 shows the experiment setup. As described above, we used the developed control device, an underwater vehicle, and a laptop computer for a grasping task. Two operators (operators A and B) used the control device to control the vehicle, and open or close its gripper while watching real-time images sent from the vehicle camera displayed on the laptop computer. The computer was placed next to the water tank. An underwater target (a pipe painted red) was placed 2 m in front of the vehicle. The target (pipe) was inserted into a vertical bar. It was able to be pulled out by the gripper.

The operators started operating the vehicle to grab and retrieve the target using the gripper. One operator (operator A) first tested the manual control and then the HSC. Another operator (operator B) first tested the HSC and then the manual control. During the experiment, the operators were unable to view the vehicle or target directly.

12.4.2 Manual Control

Using manual control, we investigated whether the developed control device is useful for remote operation. During manual control, automatic control was disabled. The operators used the control device to bring the vehicle closer to the target (pipe), and then grabbed it with the gripper. Figures 12.6 and 12.7 show time-series trajectories of the target on the vehicle camera image. When the target was not visible in the camera image, the trajectory was not recorded (lines in Figs. 12.6 and 12.7 are broken). For

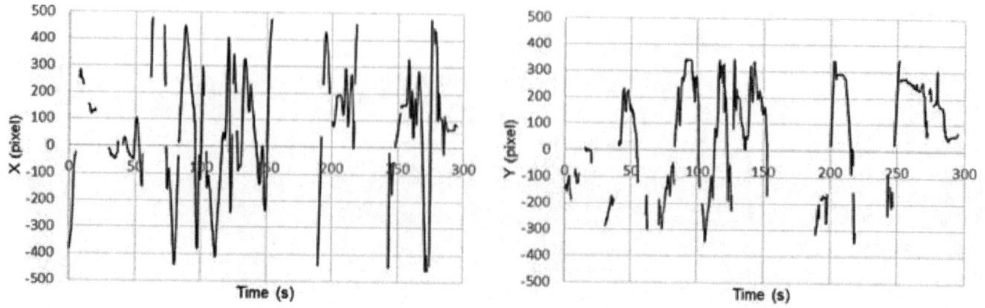

Fig. 12.6 Experimentally obtained results for manual control (operator A). The left panel shows the time-series trajectory of the horizontal position of the target with the origin at the center of the camera image. The right panel presents the time-series trajectory of the vertical position of the target

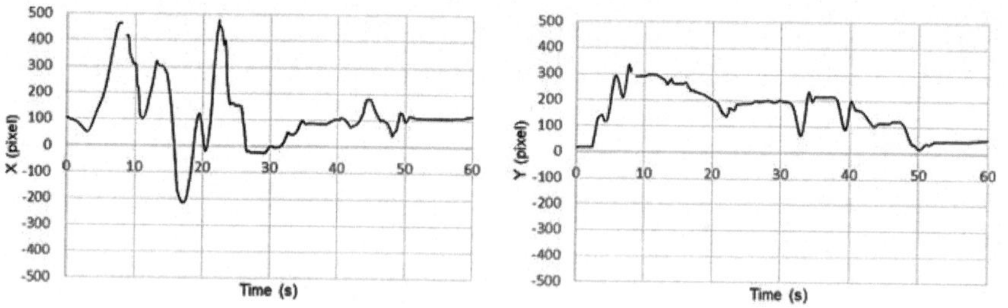

Fig. 12.7 Experiment results obtained using manual control (operator B). The left panel shows the time-series trajectory of the horizontal position of the target with the origin at the center of the camera image. The right panel presents the time-series trajectory of the vertical position of the target

operator A, it took about 300 s to grab and retrieve the pipe. For operator B, it took about 60 s to complete the same task.

These results show that although the two operators took different amounts of time to complete the same task, they were able to use the underwater vehicle and its gripper, and were able to grasp and retrieve the target using the developed control device.

12.4.3 Haptic Shared Control

Next, haptic shared control was tested using the same task. The automatic control and the operator controlled the yaw and heave motions of the vehicle using the same control device. The vehicle surge and gripper opening and closing movement were controlled only by the operator. Figures 12.8 and 12.9 portray time-series trajectories of the target on the camera image. For operator A, only about 100 s were needed to complete the task. The time to complete the task was reduced to about one-third

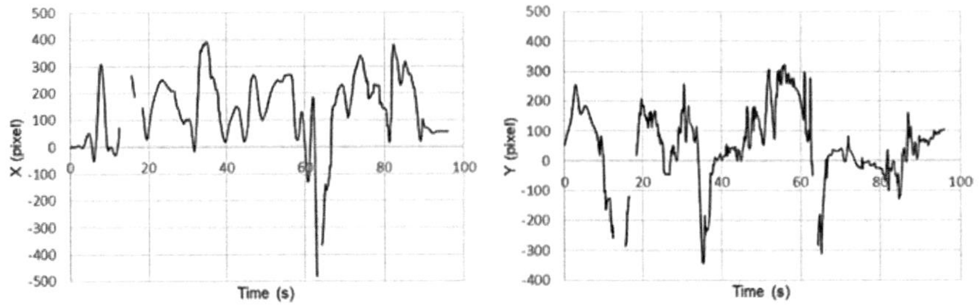

Fig. 12.8 Time-series trajectories of the target in the camera image during HSC (operator A)

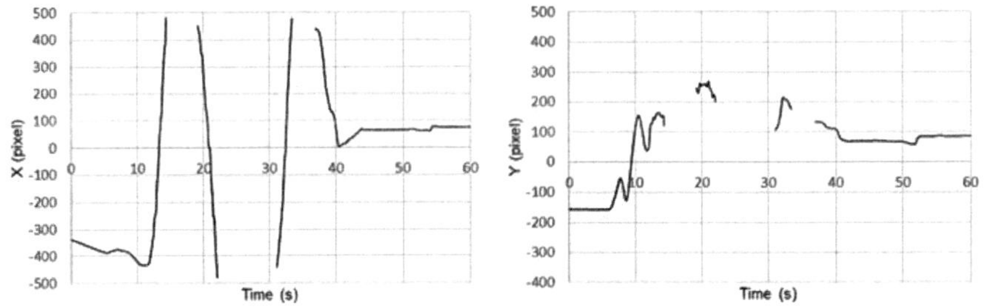

Fig. 12.9 Time-series trajectories of the target in the camera image during HSC (operator B)

compared to that when using manual control. For operator B, only about 60 s we needed to complete the same task. This result was almost identical to those for the time taken for manual control (Fig. 12.7).

12.5 Discussion and Future Work

For this work, after constructing an operation support device based on HSC, we conducted preliminary experiments. The experimentally obtained results show that the control device is useful for manual control of an underwater vehicle. For one operator (operator A), the HSC reduced the time to complete the grasping task. However, the same trend was not observed for another operator (operator B). This operator took only a short time to complete the task, even when using manual control. For that reason, it is possible that the HSC was ineffective. The main purpose of this work was to design and develop a control device for implementing HSC. We achieved this purpose. As future work, we plan to conduct subject experiments to evaluate operators' mental workload and attention allocation to assess the effectiveness of HSC for operating underwater vehicles.

12.6 Conclusion

As described herein, we designed and developed a control device for human-portable underwater vehicles with a gripper. Using the developed control device, we investigated the potential use of HSC for underwater grasping tasks. As described in this paper, we obtained the results explained below.

We designed and developed the control device with two thumb sticks to control three-dimensional vehicle motion. We conducted a preliminary experiment using an underwater vehicle to evaluate the effectiveness of the developed device. The developed device is useful for manual control. Moreover, these sticks can present forces to an operator and can also be used for HSC.

As future work, we will conduct further experiments to investigate the operator's mental workload and to evaluate the potential utility of HSC.

References

1. Duy Le, K., Nguyen, H. D., Ranthumugala, D., Forrest, A.: Haptic driving system for surge motion control of underwater remotely operated vehicles. In: Proceedings of 2014 International Conference on Modelling, Identification and Control, Melbourne, pp. 207–212. VIC, Australia (2014)
2. Choi, J., Lee, Y., Choi, H.-T., Kang, J.-J., Ryu, J.-H.: A preliminary study on development of haptic interface for underwater vehicles. In: Proceedings of 2015 IEEE Underwater Technology, pp. 1–4. Chennai, India (2015)
3. Stewart, A., Ryden, F., Cox, R.: An interactive interface for multi-pilot ROV intervention. In: Proceedings of OCEANS 2016, pp. 1–6. Shanghai, China (2016)
4. Kuiper, K. J., Frumau, J.C., van der Helm, F.C., Abbink, D.A.: Haptic support for bi-manual control of a suspended grab for deep-sea excavation. In: IEEE International Conference on Systems, Man, and Cybernetics, pp. 1822–1827 (2013)
5. Sakagami, N., Suka, M., Kimura, Y., Sato, E., Wada, T.: Haptic shared control applied for ROV operation support in flowing water. Artif. Life Robot. **27**(4), 867–875 (2021)
6. Sato, E., Liu, H., Orita, Y., Sakagami, N., Wada, T.: Cooperative path-following control of a remotely operated underwater vehicle for human visual inspection task. Front. Control Eng. **3**, 1056937 (2022). https://doi.org/10.3389/fcteg.2022.1056937

Chapter 13
Optimal Control Problems for the Whitham Type Nonlinear Differential Equations with Impulse Effects

Tursun K. Yuldashev, Najmiddin N. Qodirov, Mansur P. Eshov, and Gulnora K. Abdurakhmanova

Abstract In the article the questions of solving optimal control problem for the nonlinear Whitham-type partial differential equations with impulsive effects are studied. The modified method of characteristics allows partial differential equations of the first order to be represented as ordinary differential equations that describe the change of an unknown function along the line of extended characteristics. The unique solvability of the initial value problems for the fixed values of control function is proved by the method of successive approximations and contraction mappings. The quality functional has a quadrate form. The necessary optimality conditions for nonlinear control are formulated. Method of maximum principle is used. The definition of the unknown control function is reduced to solving the nonlinear functional equations.

Keywords Optimal control problem · Whitham-type equations · Necessary conditions of optimality · Method of successive approximations · Unique solvability

13.1 Formulation of the Problem

A large number of works have been devoted to the study of linear and nonlinear waves (see [1–6] for an overview). In particulary, in [1] by Castro, Palacios and Zuazua an alternating descent method for the optimal control of the inviscid Burgers equation

T. K. Yuldashev (✉) · N. N. Qodirov · M. P. Eshov · G. K. Abdurakhmanova
Tashkent State University of Economics, Karimov street 49, 100066 Tashkent, Uzbekistan
e-mail: tursun.k.yuldashev@gmail.com

N. N. Qodirov
e-mail: qodirov@tsue.uz

M. P. Eshov
e-mail: m.eshov@tsue.uz

G. K. Abdurakhmanova
e-mail: g.abdurakhmanova@tsue.uz

D. Azimov (ed.), *Proceedings of the IUTAM Symposium on Optimal Guidance and Control for Autonomous Systems 2023*, IUTAM Bookseries 40,
https://doi.org/10.1007/978-3-031-39303-7_25

is considered in the presence of shocks. It is computed numerical approximations of minimizers for optimal control problems governed by scalar conservation laws.

Partial differential equations of the first order are locally solved by methods of the theory of ordinary differential equations by reducing them to a characteristic system. The application of the method of characteristics to the solution of partial differential equations of the first order makes it possible to reduce the study of wave evolution [7]. In [8, 9], methods for integrating of nonlinear partial differential equations of the first order were developed. Further, many papers appeared devoted to the study of questions of the unique solvability of the Cauchy problem for different types of partial differential equations of first order (see, for example, [10–14]).

The theory of optimal control for systems with distributed parameters is widely used in solving problems of aerodynamics, chemical reactions, diffusion, filtration, combustion, heating, etc. (see, [15–19]). Various analytical and approximate methods for solving problems of optimal control systems with distributed parameters are being developed and effectively used (see, for example, [20–22]). Interesting results are obtained in the works [23–28]).

In this paper we consider the questions of unique solvability and determination of the control function in the nonlinear initial value problem for a Whitham-type partial differential equation with nonlinear impulse conditions. So, in the domain $\Omega \equiv [0; T] \times \mathrm{R}$ for $t \neq t_i, i = 1, 2, \ldots, k$ we study the following quasilinear equation

$$\frac{\partial u(t,x)}{\partial t} + u(t,x)\frac{\partial u(t,x)}{\partial x} = F\left(t, x, u(t,x), \alpha(t)\right) \tag{13.1}$$

with nonlinear initial value condition

$$u(t,x)_{|t=0} = \varphi\left(x, \int\limits_0^T K(\xi)u(\xi,x)d\xi\right), \quad x \in \mathrm{R} \tag{13.2}$$

and nonlinear impulsive condition

$$u\left(t_i^+, x\right) - u\left(t_i^-, x\right) = G_i\left(u\left(t_i, x\right)\right), \quad i = 1, 2, \ldots, k, \tag{13.3}$$

where $u(t,x)$ is unknown state function, $\alpha(t)$ is unknown control function, $t \neq t_i, i = 1, 2, \ldots, k$, $0 = t_0 < t_1 < \cdots < t_k < t_{k+1} = T < \infty$, $\mathrm{R} \equiv (-\infty, \infty)$, $F(t,x,u,\alpha) \in C^{0,1,1,1}(\Omega \times \mathrm{R} \times \Upsilon)$, $\Upsilon \equiv [0, M^*]$, $0 < M^* < \infty$, $\varphi(x,u) \in C^1(\mathrm{R}^2)$, $\int_0^T |K(\xi)|\, d\xi < \infty$, $u\left(t_i^+, x\right) = \lim_{\nu \to 0^+} u\left(t_i + \nu, x\right)$, $u\left(t_i^-, x\right) = \lim_{\nu \to 0^-} u\left(t_i - \nu, x\right)$ are right-hand side and left-hand side limits of function $u(t,x)$ at the point $t = t_i$, respectively. Here we note that the points $t = t_i$ $(i = 1, 2, \ldots, k,)$ are given and constants. In addition, all functions involved in this work are scalar; vector functions are not used.

In most literature, the Burgers equation is called the equation (see, for examples, [6, Formula 4.1, Page 106], [7, Formula 14.9, Page 82])

$$\frac{\partial u(t,x)}{\partial t} + u(t,x)\frac{\partial u(t,x)}{\partial x} = \mu \frac{\partial^2 u(t,x)}{\partial x^2}.$$

When $\mu = 0$ this equation is called the Hopf equation, which describes the Riemann wave (see [7, Formula 4.6, Page 29]). Equation (13.1) has a nonlinear right hand side. In this form, it is called a nonlinear Whitham-type equation (see [9]). One of the features of this Eq. (13.1) is that it cannot be integrated in the usual characteristic. Therefore, in this paper, an extended characteristic is introduced.

The differential equation (13.1) also describes the process of regulating the behavior of students within the discipline and moral ethics established at the university for many decades.

We use some Banach spaces: the space $C(\Omega, \mathrm{R})$ consists continuous function $u(t,x)$ with the norm

$$\|u\|_C = \sup_{(t,x)\in\Omega} |u(t,x)|;$$

we also need in using the linear space

$$PC(\Omega, \mathrm{R}) = \left\{u : \Omega \to \mathrm{R};\, u(t,x) \in C\left(\Omega_{i,i+1}, \mathrm{R}\right), \quad i = 1, \ldots, k\right\}$$

with the following norm

$$\|u\|_{PC} = \max\left\{\|u\|_{C(\Omega_{i,i+1})}, \quad i = 1, 2, \ldots, k\right\},$$

where $\Omega_{i,i+1} = (t_i, t_{i+1}] \times \mathrm{R}$, $u\left(t_i^+, x\right)$ and $u\left(t_i^-, x\right)$ $(i = 0, 1, \ldots, k)$ exist and are bounded; $u\left(t_i^-, x\right) = u(t_i, x)$.

To determine the control function $\alpha(t)$ in the initial value problem (13.1)–(13.3), we use the following quadratic cost function

$$J[u(t,x)] = \int_0^T \gamma(t)\alpha^2(t)\mathrm{d}t, \tag{13.4}$$

where $\gamma(t) \in C([0; T], \mathrm{R})$.

Problem Statement. Find a state functions $u(t,x) \in PC(\Omega, \mathrm{R})$ and control function $\alpha(t) \in \{\alpha : |\alpha(t)| \le M^*, t \in \Omega_T\}$, delivering a minimum to functional (13.4) and the state function $u(t,x)$ for all $(t,x) \in \Omega, t \ne t_i, i = 1, 2, \ldots, k$ satisfies the differential equation (13.1), initial value condition (13.2), for $(t,x) \in \Omega, t = t_i, i = 1, 2, \ldots, k$ satisfies the nonlinear impulsive condition (13.3).

13.2 Reducing the Problem (13.1)–(13.3) to a Functional-Integral Equation

Since condition (13.2) is specified at the initial point of the interval $[0, T]$, then problem (13.1)–(13.3) will be called the nonlinear initial value problem with impulse effects or, in short, initial value problem. Let the function $u(t, x) \in PC(\Omega, \mathrm{R})$ be a solution of the initial value problem (13.1)–(13.3). We write the domain Ω as $\Omega = \Omega_{0,1} \cup \Omega_{1,2} \cup \cdots \cup \Omega_{k,k+1}$, where $\Omega_{i,i+1} = \left(t_i, t_{i+1}\right] \times \mathrm{R}$. On the first domain $\Omega_{0,1}$ the Eq. (13.1) we rewrite as

$$D_u[u] = F\left(t, x, u(t, x), \alpha(t)\right), \tag{13.5}$$

where $D_u = \left(\frac{\partial}{\partial t} + u(t, x)\frac{\partial}{\partial x}\right)$ is Whitham operator.

Now we introduce the notation, which is called the extended characteristic:

$$p(t, s, x) = x - \int_s^t u(\theta, x)\mathrm{d}\theta, \quad p(t, t, x) = x.$$

We introduce a function of three dimensional argument $w(t, s, x) = u\left(s, p(t, s, x)\right)$, such that for $t = s$ it takes the form $w(t, t, x) = u\left(t, p(t, t, x)\right) = u(t, x)$. Differentiate the function $w(t, s, x)$ with respect to the new argument s

$$w_s(t, s, x) = u_s\left(s, p(t, s, x)\right) + u_p\left(s, p(t, s, x)\right) \cdot p_s(t, s, x).$$

Then, taking into account the last relation, we rewrite Eq. (13.5) in the following extended form

$$\frac{\partial}{\partial s} w(t, s, x) = F\left(s, p(t, s, x), w(t, s, x), \alpha(s)\right). \tag{13.6}$$

Integrating equations (13.6) along the extended characteristic, we obtain

$$\int_0^{t_1} F\left(s, p(t, s, x), w\left(t, s, x\right), \alpha(s)\right) \mathrm{d}s$$
$$= w(t, t_1^-, x) - w(t, 0^+, x), \quad t \in (0, t_1], \tag{13.7}$$

$$\int_{t_1}^{t_2} F\left(s, p(t, s, x), w\left(t, s, x\right), \alpha(s)\right) \mathrm{d}s$$
$$= w(t, t_2^-, x) - w(t, t_1^+, x), \quad t \in (t_1, t_2], \tag{13.8}$$

$$\vdots$$

$$\int_{t_k}^{t_{k+1}} F\left(s, p(t,s,x), w\left(t,s,x\right), \alpha(s)\right) \mathrm{d}s$$
$$= w(t, t_{k+1}^-, x) - w(t, t_k^+, x), \quad t \in \left(t_k, t_{k+1}\right], \quad t_{k+1} = T. \tag{13.9}$$

Taking $w(t, 0^+, x) = w(t, 0, x)$, $w(t, t_{p+1}^-, x) = w(t, s, x)$ into account, $(0, T]$ from the integral relations (13.7)–(13.9) we have on the interval:

$$\int_0^s F\left(\varsigma, p(t,\varsigma,x), w\left(t,\varsigma,x\right), \alpha(\varsigma)\right) \mathrm{d}\varsigma = \left[w\left(t, t_1, x\right) - w\left(t, 0^+, x\right)\right]$$
$$+ \left[w\left(t, t_2, x\right) - w\left(t, t_1^+, x\right)\right] + \cdots + \left[w(t,s,x) - w\left(t, t_p^+, x\right)\right]$$
$$= -w(t,0,x) - \left[w\left(t, t_1^+, x\right) - w\left(t, t_1, x\right)\right] - \left[w\left(t, t_2^+, x\right) - w\left(t, t_2, x\right)\right]$$
$$- \cdots - \left[w\left(t, t_k^+, x\right) - w\left(t, t_k, x\right)\right] + w(t,s,x). \tag{13.10}$$

Taking into account the impulsive condition (13.3), the last equality (13.10) we rewrite as

$$w(t,s,x) = w(t,0,x) + \int_0^s F\left(\varsigma, p(t,\varsigma,x), w\left(t,\varsigma,x\right), \alpha(\varsigma)\right) \mathrm{d}\varsigma$$
$$+ \sum_{0<t_i<s} G_i\left(w\left(t, t_i, x\right)\right), \tag{13.11}$$

where $w(t, 0, x)$ is arbitrary constant along the extended characteristic, which to be determined. The initial value condition (13.2) for Eq. (13.11) takes the form

$$w(t,0,x) = \varphi\left(p(t,0,x), \int_0^T K(\xi) w(t,\xi,x) \mathrm{d}\xi\right).$$

Then, taking into account this initial value condition (13.2), from (13.11) we obtain that

$$
\begin{aligned}
w(t,s,x) = \varphi\left(p(t,0,x),\int\limits_0^T K(\xi)w(t,\xi,x)\mathrm{d}\xi\right) \\
+ \int\limits_0^s F\left(\varsigma, p(t,\varsigma,x), w\left(t,\varsigma,x\right), \alpha(\varsigma)\right) d\varsigma \\
+ \sum_{0<t_i<s} G_i\left(w\left(t,t_i,x\right)\right).
\end{aligned} \tag{13.12}
$$

For $t = s$, from (13.12) we arrive at the nonlinear functional-integral equation

$$
\begin{aligned}
u(t,x) = \Theta(t,x;u) \equiv \varphi\left(p(t,0,x),\int\limits_0^T K(\xi)u(\xi,p(t,\xi,x))\mathrm{d}\xi\right) \\
+ \int\limits_0^t F\left(s, p(t,s,x), u\left(s, p(t,s,x)\right), \alpha(s)\right) \mathrm{d}s \\
+ \sum_{0<t_i<t} G_i\left(u\left(t_i, p(t,t_i,x)\right)\right),
\end{aligned} \tag{13.13}
$$

where $p(t,s,x)$ is defined from the integral equation

$$
p(t,s,x) = x - \int\limits_s^t u\left(\theta, p(t,\theta,x)\right)\mathrm{d}\theta, \quad p(t,t,x) = x, \tag{13.14}
$$

$x \in \mathrm{R}$ plays the role of a parameter.

13.3 Solvability of the Functional-Integral Equation (13.13)

For fixed values of control function $\alpha(t)$, we study the solvability of the functional-integral equation (13.13).

Theorem 1 *Let the positive quantities* Δ_φ, $\Delta_f(t)$, Δ_{G_i}, χ_1, $\chi_{21}(t)$, $\chi_{22}(t)$, χ_{3i}, $i = 1, 2, \ldots, k$ *be existed and for them following conditions be satisfied:*

1. $0 < \sup_{x\in\mathrm{R}} |\varphi(x,0)| \le \Delta_\varphi < \infty$;
2. $\sup_{x\in\mathrm{R}} |F(t,x,0,\alpha)| \le \Delta_f(t)$, $0 < \Delta_f(t) \in C[0;T]$;

3. $0 < |G_i(0)| \leq \Delta_{G_i} < \infty, i = 1, 2, \ldots, k;$

4. $|\varphi(x_1, u_1) - \varphi(x_2, u_2)| \leq \chi_1 (|x_1 - x_2| + |u_1 - u_2|)\,,\ 0 < \chi_1 = \text{const};$

5. $|F(t, x_1, u_1, \alpha) - F(t, x_2, u_2, \alpha)| \leq \chi_{21}(t)\,|x_1 - x_2| + \chi_{22}(t)\,|u_1 - u_2|\,;$

6. $|G_i(u_1) - G_i(u_2)| \leq \chi_{3i}\,|u_1 - u_2|\,, 0 < \chi_{3i} = \text{const}, i = 1, 2, \ldots, k;$

7. $\rho = \max_{t\in[0;T]} \int_0^t H(t, s)ds + \sum_{i=1}^k \chi_{3i} < 1$, *where*

$$H(t, s) = \chi_1 (1 + |K(s)|) + \chi_{21}(s)(t - s) + \chi_{22}(s).$$

Then, for fixed values of $\alpha(t)$, the functional-integral equation (13.13) with (13.14) has a unique solution in the domain Ω. This solution can be obtained by the following successive approximations:

$$u_0(t, x) = 0, \quad u_{n+1}(t, x) \equiv \Theta(t, x; u_n, p_n), \quad n = 0, 1, 2, \ldots, \tag{13.15}$$

where $p_n(s, t, x)$ is defined from the following iteration

$$p_0(t, t, x) = x, \quad p_n(t, s, x) = x - \int_s^t u_{n-1}(\theta, p_{n-1}(t, \theta, x))\, d\theta.$$

Proof By virtue of the conditions of the theorem, we obtain that the following estimate holds for the first difference of approximation (13.15):

$$|u_1(t, x) - u_0(t, x)| \leq \sup_{x\in R} |\varphi(x, 0)| + \sum_{0<t_i<T} |G_i(0)|$$
$$+ \max_{t\in[0;T]} \int_0^t \Delta_f(s)\mathrm{d}s \leq \Delta_\varphi + \sum_{i=1}^k \Delta_{G_i} + \Delta_1 < \infty, \tag{13.16}$$

where

$$\Delta_1 - \max_{t\in[0;T]} \int_0^t \Lambda_f(s)\mathrm{d}s < \infty.$$

Taking into account estimate (13.16) and the conditions of the theorem, we obtain that for an arbitrary difference of approximation (13.15) the following estimate holds:

$$
\begin{aligned}
|u_{n+1}(t,x)-u_n(t,x)| \le & \left|\varphi\left(p_{n+1}(t,0,x), \int\limits_0^T K(\xi)u_n(\xi, p_n(t,\xi,x))\mathrm{d}\xi\right)\right.\\
& \left.-\varphi\left(p_n(t,0,x), \int\limits_0^T K(\xi)u_{n-1}(\xi, p_{n-1}(t,\xi,x))\mathrm{d}\xi\right)\right|\\
& + \int\limits_0^t |F\left(s, p_{n+1}(t,s,x), u_n\left(s, p_n(t,s,x)\right), \alpha(s)\right)\\
& - F\left(s, p_n(t,s,x), u_{n-1}\left(s, p_{n-1}(t,s,x)\right), \alpha(s)\right)|\,\mathrm{d}s\\
& + \sum_{0<t_i<t} |G_i\left(u_n\left(t_i, p_n(t,t_i,x)\right)\right)\\
& - G_i\left(u_{n-1}\left(t_i, p_{n-1}(t,t_i,x)\right)\right)|\\
\le & \,\chi_1\Big[\int\limits_0^t |u_n(s,x)-u_{n-1}(s,x)|\,\mathrm{d}s\\
& + \int\limits_0^T |K(s)|\cdot|u_n(s,x)-u_{n-1}(s,x)|\,\mathrm{d}s\Big]\\
& + \int\limits_0^t \Big[\chi_{21}(s)\int\limits_s^t |u_n(\theta,x)-u_{n-1}(\theta,x)|\,\mathrm{d}\theta\\
& + \chi_{22}(s)\,|u_n(s,x)-u_{n-1}(s,x)|\Big]\mathrm{d}s\\
& + \sum_{0<t_i<t} \chi_{3i}\,|u_n\left(t_i,x\right)-u_{n-1}\left(t_i,x\right)|\\
\le & \max_{t\in[t_i;t_{i+1}]} \int\limits_0^t H(t,s)\,|u_n(s,x)-u_{n-1}(s,x)|\,\mathrm{d}s\\
& + \sum_{i=1}^k \chi_{3i} \max_{t\in[t_i;t_{i+1}]} |u_n\left(t,x)\right)-u_{n-1}\left(t,x\right)|, \qquad (13.17)
\end{aligned}
$$

where

$$
H(t,s)=\chi_1\left(1+|K(s)|\right)+\chi_{21}(s)(t-s)+\chi_{22}(s).
$$

In estimate (13.17), we pass to the norm in the space $PC\left(\Omega, \mathrm{R}\right)$ and arrive at the estimate

$$
\|u_{n+1}(t,x)-u_n(t,x)\|_{PC} \le \rho\cdot\|u_n(t,x)-u_{n-1}(t,x)\|_{PC}\,, \qquad (13.18)
$$

where

$$\rho = \max_{t\in[t_i;t_{i+1}]} \int_0^t H(t,s)\mathrm{d}s + \sum_{i=1}^{k} \chi_{3i}.$$

Since $\rho < 1$, it follows from estimate (13.18) that the sequence of functions $\{u_n(t,x)\}_{n=1}^{\infty}$, defined by formula (13.15), converges absolutely and uniformly in the domain Ω. In addition, it follows from the existence of a unique fixed point of the operator $\Theta(t,x;u)$ on the right-hand side of (13.13) that the functional-integral equation (13.13) has a unique solution in the domain Ω. The theorem has been proven.

Corollary *Let all the conditions of Theorem be satisfied. Then, for fixed values of the control function $\alpha(t)$, the initial value problem (13.1)–(13.3) with impulse effects has a unique solution in the domain Ω.*

13.4 Determination of the Control Function

Let $\alpha^*(t)$ be optimal control function:

$$\Delta J\left[\alpha^*(t)\right] = J\left[\alpha^*(t) + \Delta\alpha^*(t)\right] - J\left[\alpha^*(t)\right] \geq 0,$$

where $\alpha^*(t) + \Delta\alpha^*(t) \in C(\Upsilon)$. We apply the maximum principle to our problem to find the necessary conditions for optimality. To build the Pontryagin's function we use the Eq. (13.6):

$$H\left(\alpha(s), w(t,s,x)\right) = \psi(s)F\left(s, p(t,s,x), w(t,s,x), \alpha(s)\right) - \gamma(s)\alpha^2(s), \tag{13.19}$$

where for the new unknown function $\psi(t)$ by differentiation (13.19) with respect to the function $w(t,s,x)$ we will obtain the following equation

$$\dot{\psi}(s) = -\psi(s)F_w\left(s, p(t,s,x), w(t,s,x), \alpha(s)\right). \tag{13.20}$$

Similarly, by differentiation (13.19) with respect to the control function $\alpha(t)$ we obtain the following equation

$$\psi(s)F_\alpha\left(s, p(t,s,x), w(t,s,x), \alpha(s)\right) - 2\gamma(s)\alpha(s) = 0. \tag{13.21}$$

By differentiation the Eq. (13.21) with respect to the control function $\alpha(t)$ we obtain the necessary condition for optimality

$$\psi(s)F_{\alpha\alpha}\left(s, p(t,s,x), w(t,s,x), \alpha(s)\right) - 2\gamma(s) \leq 0. \tag{13.22}$$

In (13.21) we express $\psi(s)$ and substitute it into (13.22). Then we have

$$F_\alpha\left(s, p(t,s,x), w(t,s,x), \alpha(s)\right) \cdot \left(\frac{\alpha(s)}{F_\alpha\left(s, p(t,s,x), w(t,s,x), \alpha(s)\right)}\right)_\alpha > 0.$$

Equation (13.21) we rewrite as

$$\alpha(s) = \frac{\psi(s)}{2\gamma(s)} F_\alpha\left(s, p(t,s,x), w(t,s,x), \alpha(s)\right). \tag{13.23}$$

Solving differential equation (13.20) with end-point condition $\psi(T) = \psi_T$, we obtain

$$\psi(s) = \psi_T \cdot \exp\left\{\int_s^T F_w\left(\theta, p(t,\theta,x), w(t,\theta,x), \alpha(\theta)\right) \mathrm{d}s\right\}. \tag{13.24}$$

Substitute the presentation (13.24) into (13.23), we have

$$\alpha(s) = \frac{\psi_T}{2\gamma(s)} F_\alpha\left(s, p(t,s,x), w(t,s,x), \alpha(s)\right)$$
$$\times \exp\left\{\int_s^T F_w\left(\theta, p(t,\theta,x), w(t,\theta,x), \alpha(\theta)\right) \mathrm{d}\theta\right\}. \tag{13.25}$$

Thus, in general case, the problem of optimal control is reducing to solve the Eqs. (13.13) and (13.20) together as one system of two equations. It can be solved by the method of successive approximations. But, if we consider the concrete form of function, for example,

$$F\left(s, p(t,s,x), w(t,s,x), \alpha(s)\right) = 2s - 3p(t,s,x) + \beta(s)w(t,s,x) + \delta(s)\alpha(s), \tag{13.26}$$

then the Eq. (13.20) takes the form $\dot{\psi}(s) = -\psi(s)\beta(s)$ and the Eq. (13.25) takes the form

$$\alpha(s) = \frac{\psi_T}{2\gamma(s)} \delta(s) \exp\left\{\int_s^T \beta(\theta) \mathrm{d}\theta\right\}, \tag{13.27}$$

where $\delta(s)$ and $\beta(s)$ are given continuous on $[0, T]$ functions.

For $s = t$ the obtained value on control function (13.27) we substitute into Eq. (13.13):

$$u(t,x) = \varphi\left(p(t,0,x), \int_0^T K(\xi)u(\xi, p(t,\xi,x))d\xi\right)$$
$$+ \int_0^t \Big[2s - 3p(t,s,x) + \beta(s)u\,(s, p(t,s,x))$$
$$+ \frac{\psi_T}{2\gamma(s)}\delta^2(s)\exp\left\{\int_s^T \beta(\theta)d\theta\right\}\Big]ds + \sum_{0<t_i<t} G_i\,(u\,(t_i, p(t,t_i,x))) \tag{13.28}$$

with (13.14). Equation (13.28) consists only one unknown function $u(t,x)$ and it can be easily solved by the method of successive approximations. In this case the necessary condition for optimality (13.22) takes form $\gamma(t) > 0$.

13.5 Conclusion

In this paper the questions of unique solvability and determination of the control function $\alpha(t)$ in the initial value problem (13.1)–(13.3) for a Whitham type partial differential equation with impulse effects are studied. The modified method of characteristics allows partial differential equations of the first order to be represented as ordinary differential equations that describe the change of an unknown function along the line of characteristics. The nonlinear functional-integral equation (13.13) is obtained. The unique solvability of the initial value problem (13.1)–(13.3) is proved by the method of successive approximations and contraction mappings. The determination of the unknown control function $\alpha(t)$ is reduced to solving the Eq. (13.25). In the particular case of (13.26), solving of the optimal control problem is simple. The process of solving state function $u(t,x)$ reduced to solve functional-integral equation (13.28) by the method of successive approximations.

References

1. Castro, C., Palacios, F., Zuazua, E.: An alternating descent method for the optimal control of the inviscid Burgers equation in the presence of shocks. Math. Models Methods Appl. Sci. **18**(3), 369–416 (2008). https://doi.org/10.1142/S0218202508002723
2. Il'yichev, A.T.: Solitary Waves in Hydromechanics Models, p. 256. Fizmatlit, Moscow (2003). (in Russian)
3. Lyapidevsky, V.Yu., Teshukov, V.M.: Mathematical Models of Propagation of Long Waves in an Inhomogeneous Liquid. Publishing House of the Siberian Branch of the Russian Academy of Sciences, Novosibirsk (2000) (in Russian)
4. Kulikovskii, A., Sveshnikova, E.: Nonlinear Waves in Elastic Media. CRC Press, Boca Raton, Fl (1995)

5. Smoller, J.: Shock Waves and Reaction-Diffusion Equations. Grundlehren der mathem. Wissenschaften, p. 258. Springer, New York (1983)
6. Whitham, J.B.: Linear and Non-linear Waves. A Wiley-Interscience Publication, Gerald Beresford (1927)
7. Goritskiy, AYu., Kruzhkov, S.N., Chechkin, G.A.: Partial Differential Equations of the First Order, p. 95. Mekhmat MGU, Moscow (1999). (in Russian)
8. Imanaliev, M.I., Ved, Yu.A.: On a partial differential equation of the first order with an integral coefficient. Differ. Equ. **23**(3), 325–335 (1989)
9. Imanaliev, M.I., Alekseenko, S.N.: On the theory of systems of nonlinear integro-differential partial differential equations of the Whitham type. Doklady Math. **46**(1), 169–173 (1993)
10. Alekseenko, S.N., Dontsova, M.V.: Study of the solvability of a system of equations describing the distribution of electrons in the electric field of a sprite. Matematicheskiy vestnik pedvuzov i universitetov Volgo-Vyatskogo regiona **14**, 34–41 (2012). (in Russian)
11. Alekseenko, S.N., Dontsova, M.V., Pelinovsky, D.E.: Global solutions to the shallow water system with a method of an additional argument. Appl. Anal. **96**(9), 1444–1465 (2017). https://doi.org/10.1080/00036811.2016.1208817
12. Dontsova, M.V.: Nonlocal solvability conditions for Cauchy problem for a system of first order partial differential equations with special right-hand sides. Ufa Math. J. **6**(4), 68–80 (2014)
13. Imanaliev, M.I., Alekseenko, S.N.: To the problem of the existence of a smooth bounded solution for a system of two nonlinear partial differential equations of the first order. Doklady Math. **64**(1), 10–15 (2001)
14. Yuldashev, T.K., Shabadikov, K.Kh.: Initial-value problem for a higher-order quasilinear partial differential equation. J. Math. Sci. **254**(6), 811–822 (2021). https://doi.org/10.1007/s10958-021-05343-0
15. Evtushenko, Yu.G.: Methods for Solving Extremal Problems and Their Application in Optimization Systems. Nauka, Moscow (1982). (in Russian)
16. Egorov, A.I.: Optimal Control of Thermal and Diffusion Processes. Nauka, Moscow (1978). (in Russian)
17. Girsanov, I.V.: Lectures on the Mathematical Theory of Extremum Problems. Springer, New York (1972)
18. Lions, J.L.: Optimal Control of Systems Governed by Partial Differential Equations. Springer, New York (1971)
19. Lur'ye, K.A.: Optimal Control in the Problems of Mathematical Physics. Nauka, Moscow (1975). (in Russian)
20. Kerimbekov, A.K.: On solvability of the nonlinear optimal control problem for processes described by the semilinear parabolic equations. In: Proceedings World Congress on Engineering, London, UK, vol. 1, pp. 270–275 (2011)
21. Miller, B.M., Rubinovich, E.Ya.: Discontinuous solutions in the optimal control problems and their representation by singular space-time transformations. Autom. Rem. Control **74**(12), 1969–2006 (2013). https://doi.org/10.1134/S0005117913120047
22. Yuldashev, T.K.: Nonlinear optimal control of thermal processes in a nonlinear inverse problem. Lobachevskii J. Math. **41**(1), 124–136 (2020). https://doi.org/10.1134/S1995080220010163
23. Azimov, D.M.: Active legs of extremal trajectories in linear central field. Autom. Rem. Control **66**(10), 1533–1551 (2005)
24. Azimov, D.M.: Active rocket trajectory arcs: a review. Autom. Rem. Control **66**(11), 1715–1732 (2005)
25. Azimov, D.M.: Two classes of extreme motion trajectories of a point with varying specific impulse. Autom. Rem. Control **68**(6), 935–947 (2007)
26. Ruzmatov, M.I., Azimov, D.M., Korshunova, N.A.: A new class of intermediate thrust arcs in the restricted three-body problem. Adv. Space Res. **71**(1), 369–374 (2023)

27. Nikiforov, V.O., Paramonov, A.V., Gerasimov, D.N.: Adaptive control algorithms in MIMO linear systems with control delay. Autom. Rem. Control **81**(6), 1091–1106 (2020). https://doi.org/10.1134/S0005117920060107
28. Paraev, Y.I., Poluektova, K.O.: Optimal control of a single-sector economy under random variations of fixed capital and labor resources. Autom. Rem. Control **81** (4), 704–712 (2020). https://doi.org/10.1134/S0005117920040116

Chapter 14
Hamiltonian Optimal Control of Distributed Lagrangian Systems

Nhan Nguyen

Abstract This paper presents a distributed Hamiltonian optimal control method method for a class of distributed Lagrangian infinite-dimensional systems coupled to nonlinear finite-dimensional systems. A virtual control concept in conjunction with an optimal control allocation strategy is employed using the Hamilton's principle to provide a distributed Hamiltonian control solution that spans an infinite-dimensional eigenfunction space. The distributed optimal control theory is formulated using a semi-group abstraction resulting in an integro-differential Riccati equation. An aircraft flight control application is presented to illustrate the theory.

Keywords Hamiltonian control · Distributed-parameter optimal control · Lagrangian infinite-dimensional systems

14.1 Introduction

Distributed systems are dynamical systems which have both temporal and spatial distributions, and are modeled by partial differential equations (PDEs). Control theories for distributed systems have been studied by numerous researchers. However, they are generally not universally employed in practical control system design and analysis. This is due in part to the mathematical complexity of the underlying PDEs. Oftentimes, the tasks of solving PDEs can be so overwhelming that they tend to detract the control designer or analyst from the real objective of attaining a practical control system design. Distributed control systems, however, enjoy a considerable attention within the mathematical research community [1–6]. Applied work on distributed control for practical control engineering applications is generally quite limited and is typically restricted to simple applications [7–9]. Some published work on distributed control for complex practical applications to flexible aircraft and flow control can be found in references [10–12].

N. Nguyen (✉)
NASA Ames Research Center, Moffett Field, Mountain View, CA 95050, USA
e-mail: nhan.t.nguyen@nasa.gov

D. Azimov (ed.), *Proceedings of the IUTAM Symposium on Optimal Guidance and Control for Autonomous Systems 2023*, IUTAM Bookseries 40,
https://doi.org/10.1007/978-3-031-39303-7_13

In terms of practical flight control applications, very few distributed control theories are employed for realistic aircraft flight control problems. The author has developed a Hamiltonian distributed control theory in recent years in an attempt to bridge the gap between theory and practice for flight control of flexible aircraft [13–16]. The distributed control theory has been applied to realistic aircraft applications.

The dynamical equations of many distributed systems can be described by the Euler-Lagrange equation which is formalized by the existence of a positive-definite energy notion. The author develops a generalized potential energy density formulation for an arbitrary order PDE system that covers a wide range of physical problems [13]. This formulation thus provides a unified control theory for distributed infinite-dimensional elastic systems and lump-parameter finite-dimensional systems. A distributed control is developed based on the Hamilton's principle using the Hamiltonian function [13]. A model-reference adaptive control theory is developed for the proposed class of distributed systems in the presence of uncertainties. This theory is demonstrated by an aircraft wing flutter suppression control to stabilize the wing structure operating beyond a flutter speed.

A follow-on study [14] presents the development of a distributed optimal control theory for minimizing a performance objective. The distributed optimal control theory is formulated using a semi-group abstraction which leads to an integro-differential Riccati equation in terms of the semi-group operators. A concrete realization of the semi-group abstraction is constructed. The study includes a realistic application of gust load alleviation flight control for an elastic aircraft wing using distributed sensing to measure the distributed state information.

The work on optimal control for Lagrangian infinite-dimensional systems is further extended to address state and disturbance estimation by a distributed Kalman filter operator [15]. The Kalman filter development for the distributed Lagrangian systems leads to a Riccati equation defined by a semi-group abstraction. A disturbance rejection optimal control can be synthesized using the disturbance estimates computed by the Kalman filter. A gust load alleviation control application for a flexible aircraft wing demonstrates the proposed theory.

In a recent work, distributed optimal control has been developed for a class of Lagrangian infinite-dimensional systems coupled to finite-dimensional systems [16]. This class of systems represents a wide range of practical systems such as those in aerospace applications for aircraft and space launch vehicles which exhibit a high degree of structural interactions. The work is applied to a flexible aircraft flight control application.

One distinguishing feature of the present work [13–16] is the use of distributed sensors for measurements of the distributed state information. Many natural and bio-inspired systems can sense environmental inputs through sensor systems distributed throughout those systems. For example, avian wings are known to be able to sense air flow over the wings which allows birds to adjust their flight path by actuating their feather systems to change the wing cambers or curvatures [17]. In recent years, advancements in distributed sensor technologies which allow measurements of both temporal and spatial information of physical signals have made steady progress.

Distributed fiber optic sensors offer new possibilities in control system designs that could bring distributed control research into a practical reality [18].

In this paper, we introduce a virtual control concept in conjunction with an optimal control allocation strategy to provide a distributed Hamiltonian control solution that spans an infinite-dimensional eigenfunction space. This new control concept removes a limitation in the previous theory [13–16] which employs an approximation of a finite-dimensional eigenfunction space whereby the number of eigenfunctions are required to be no greater than the number of distributed control. Moreover, we also introduce a notion of distributed feedback gain operator for the finite-dimensional system. An aircraft flight control application is presented to demonstrate the theory.

14.2 Distributed Lagrangrian System

Consider a finite-dimensional or lumped-parameter (P) system coupled to an infinite-dimensional or distributed-parameter Lagrangian system (S)

$$\dot{z} = Az + \int_{\Omega} r\left(x, w, \frac{\partial w}{\partial x}, \frac{\partial w}{\partial t}\right) \mathrm{d}x + B_z u_z + g_z(t) \tag{14.1}$$

$$\begin{aligned} m(x) \frac{\partial^2 w}{\partial t^2} - f\left(x, w, \frac{\partial w}{\partial x}, \ldots, \frac{\partial^m w}{\partial x^m}\right) \\ = q_c\left(x, w, \frac{\partial w}{\partial x}, \frac{\partial w}{\partial t}\right) + b(x) u_w + g_w(x, t) \end{aligned} \tag{14.2}$$

where $z(t) : [0, \infty) \to \mathbb{R}^q$ is a state vector of (P), $w(x, t) : \Omega \times [0, \infty) \to \mathbb{R}^n$ is a distributed state vector of (S) defined over an open bounded domain $x \in \Omega \subset \mathbb{R}$ with suitable boundary conditions on the boundary $\partial\Omega$ for all $t \in [0, \infty)$, $u_z(t) : [0, \infty) \to \mathbb{R}^q$ is a control vector for (P), and $u_w(t) : [0, \infty) \to \mathbb{R}^p$ is a control vector for (S). The distributed state vector $w(x, t) \in C_c^\infty$ is assumed to be continuous and infinitely differentiable. The initial values are specified by $z(0) = z_0$ and $w(x, 0) = w_0(x)$ and $\frac{\partial w(x,0)}{\partial t} = w_{t_0}(x)$.

The function $m(x) > 0 \in \mathbb{R}^{n \times n}$ is a positive-valued function distributed over x. The function $q_c(\ldots) \in \mathbb{R}^n$ is a non-homogeneous term that depends on $w(x, t)$ as well as the spatial and temporal partial derivatives of the solution $w(x, t)$. The functions $g_z(t)$ and $g_w(x, t)$ represent bounded exogenous disturbances and are assumed to be known. We assume the pair (A, B_z) is controllable.

Without loss of generality, let

$$r = r_1(x) w + r_2(x) \frac{\partial w}{\partial x} + r_3(x) \frac{\partial w}{\partial t} \tag{14.3}$$

$$q_c = q_1(x)\, w + q_2(x)\, \frac{\partial w}{\partial x} + q_3(x)\, \frac{\partial w}{\partial t} \tag{14.4}$$

The control objective is to minimize a performance objective $p(t) : [0, \infty) \to \mathbb{R}^l$, $l \le n$ that exists inside the bounded domain Ω or on the boundary $\partial\Omega$. This performance objective is defined as

$$p(x,t) = \begin{bmatrix} P_z z(t) \\ P_w w(x,t) \end{bmatrix} \tag{14.5}$$

where $P_z \in \mathbb{R}^{k \times q}$ is a matrix and $P_w \in \mathbb{R}^{(l-k)\times n}$ is an differential or integral operator [14–16].

The development of the distributed control utilizes calculus of variations. In the previous work [13], it is shown that there exists a positive-valued scalar function $\nu\left(x, w, \frac{\partial w}{\partial x}, \ldots, \frac{\partial^m w}{\partial x^m}\right) > 0 \in \mathbb{R}$ such that the virtual work done by $f\left(x, w, \frac{\partial w}{\partial x}, \ldots, \frac{\partial^m w}{\partial x^m}\right)$ is related to $\nu\left(x, w, \frac{\partial w}{\partial x}, \ldots, \frac{\partial^m w}{\partial x^m}\right)$ by

$$\delta w \cdot f = -\delta w \cdot \sum_{i=0}^{m} (-1)^i \frac{\partial^i}{\partial x^i} \left(\frac{\partial \nu}{\partial \left(\frac{\partial^i w}{\partial x^i}\right)} \right) \tag{14.6}$$

where the $\cdot$ symbol denotes a scalar product operation of two vectors, and

$$\delta\phi_{ij}(x) = \left(\frac{\partial^{j-i-1} \delta w}{\partial x^{j-i-1}} \right) \cdot \frac{\partial^i}{\partial x^i} \left(\frac{\partial \nu}{\partial \left(\frac{\partial^j w}{\partial x^j}\right)} \right) \tag{14.7}$$

$\nu(\ldots)$ is called a potential energy density and has compact support in Ω with vanishing boundary values on $\partial\Omega$ [19] for all $0 \le i \le m-1$ and $1 \le j \le m$.

In the previous work, a generalized quadratic potential energy density of arbitrary order $2m$ has been introduced as [13]

$$\nu = \sum_{i=0}^{m} \frac{1}{2} \frac{\partial^i w}{\partial x^i} \cdot k_i(x) \frac{\partial^i w}{\partial x^i} \tag{14.8}$$

where $k_i(x) \ge 0 \in \mathbb{R}^{n\times n}$, $i = 0, 1, \ldots, m$, can be positive semi-definite but $\nu(\ldots)$ is a positive-definite function. Thus, the distributed system (S) is $2m$th-order in space and second-order in time.

Further, a linear operator associated with the generalized potential energy density has been defined [13]. Let $\mathcal{H}$ be a Hilbert space with the an inner product definition

$$\langle \xi, \upsilon \rangle = \int_{\Omega} \upsilon \cdot m(x)\, \xi \mathrm{d}x \tag{14.9}$$

on a Lebesgue square integrable inner product space $\mathscr{L}^2(\Omega)$where $\xi(x) \in \mathscr{H}(\Omega)$ and $\upsilon(x) \in \mathscr{H}(\Omega)$. Then, a linear operator $L : \mathscr{H} \to \mathscr{H}$ operated on $\mathscr{H}(\Omega)$ is defined by [13]

$$L = m^{-1}(x) \sum_{i=0}^{m} (-1)^i \frac{\partial^i}{\partial x^i} \cdot \left(k_i(x) \frac{\partial^i}{\partial x^i} \right) \tag{14.10}$$

Let $L^* : \mathscr{H} \to \mathscr{H}$ be an linear operator on $\mathscr{H}(\Omega)$. It is shown that L is a self-adjoint operator [13] such that

$$\langle \upsilon, L\xi \rangle = \langle \xi, L^* \upsilon \rangle \tag{14.11}$$

Therefore, L is a positive-definite operator and is related to the generalized potential energy density as

$$\langle w, Lw \rangle = 2 \int_{\Omega} \nu \mathrm{d}x > 0 \tag{14.12}$$

Thus, (S) in Eq. (14.2) can be expressed in a linear operator form as [13]

$$m(x) \frac{\partial^2 w}{\partial t^2} + m(x) Lw = q_c \left(x, w, \frac{\partial w}{\partial x}, \frac{\partial w}{\partial t} \right) + b(x) u_w + g_w(x, t) \tag{14.13}$$

Example 1 Consider the motion of a rectangular space structure where $(x, y) \in \Omega \subset \mathbb{R}^2$. The operator L is given by

$$\begin{aligned} m(x) Lw &= (-1)^2 \begin{bmatrix} \frac{\partial^2}{\partial x^2} & \frac{\partial^2}{\partial y^2} \end{bmatrix} \left(\begin{bmatrix} k_{11}(x, y) & k_{12}(x, y) \\ k_{12}(x.y) & k_{22}(x, y) \end{bmatrix} \begin{bmatrix} \frac{\partial^2 w}{\partial x^2} \\ \frac{\partial^2 w}{\partial y^2} \end{bmatrix} \right) \\ &= \frac{\partial^2}{\partial x^2} \left(k_{11}(x, y) \frac{\partial^2 w}{\partial x^2} + k_{12}(x, y) \frac{\partial^2 w}{\partial y^2} \right) \\ &\quad + \frac{\partial^2}{\partial y^2} \left(k_{12}(x) \frac{\partial^2 w}{\partial x^2} + k_{22}(x) \frac{\partial^2 w}{\partial y^2} \right) \end{aligned} \tag{14.14}$$

The general homogeneous solution of Eq. (14.13) with $q(\ldots) = 0$ can be expressed in the form

$$w(x, t) = \sum_{i=1}^{\infty} \phi_i(x) \theta_i(t) = \phi(x) \theta(t) \tag{14.15}$$

where $\phi_i(x) : \Omega \to \mathbb{R}^n$ is a solution of the eigenvalue problem $(L - \lambda)\phi = 0$ and $\theta_i(t) : [0, \infty) \to \mathbb{R}$. It follows that

$$\langle \phi, L\phi \rangle = \lambda \langle \phi, \phi \rangle \tag{14.16}$$

where $\{\phi_i(x)\}_{i=1}^{\infty}$ form a set of orthogonal basis functions which span an infinite-dimensional solution space. Due to $q(\ldots)$ as a function of $w(x,t)$ and $\frac{\partial w}{\partial x}$, $\phi(x)$ represents the solution of Eq. (14.13) in an integral sense.

Consider an integral solution of (S) with $\upsilon(x) \triangleq D^m\phi(x) \in \mathscr{W}^{m,2} = \mathscr{H}^m(\Omega)$, where $\mathscr{W}^{m,2}(\Omega)$ is a Sobolev space with the $\mathscr{L}^2$ Lebesque measure, defined as the weak derivative of $\phi(x) \in H^m(\Omega)$ of order m [19] such that

$$\int_{\Omega} \phi \frac{\partial^m \varphi}{\partial x^m} \mathrm{d}x = (-1)^m \int_{\Omega} \upsilon\varphi \mathrm{d}x \tag{14.17}$$

for all test functions $\varphi(x) \in C_c^{\infty}(\Omega)$ with compact support in Ω.

Lemma 1 *Let $\phi(x)$ be a steady-state solution of (S) that solves the steady-state equation*

$$m(x) L\phi = q_c\left(x, \phi, \frac{\partial \phi}{\partial x}\right) = q_0(x) + \sum_{i=0}^{m} q_{i+1}(x) \frac{\partial^i \phi}{\partial x^i} \tag{14.18}$$

where $q(\ldots)$ only depends on $\phi(x)$ and x

Let $\mathscr{B}(\phi,\varphi) : \mathscr{H} \times \mathscr{H} \to \mathbb{R}$ be a bilinear mapping defined by

$$\mathscr{B}(\phi,\varphi) = \left\langle \varphi, m^{-1}(x) q_0(x)\right\rangle = \langle \varphi, L\phi\rangle - \left\langle \varphi, \sum_{i=0}^{m} m^{-1}(x) q_{i+1}(x) \frac{\partial^i \phi}{\partial x^i}\right\rangle \tag{14.19}$$

Then, (S) admits a unique weak solution $\phi(x)$ if

$$\mathscr{B}(\phi,\phi) \geq \kappa \|\phi\|^2 \tag{14.20}$$

for some $\kappa > 0$, which implies

$$\left\langle \phi, \sum_{i=0}^{m} m^{-1}(x) q_{i+1}(x) \frac{\partial^i \phi}{\partial x^i}\right\rangle < 2\int_{\Omega} \nu(\phi)\, dx \tag{14.21}$$

Proof Since $\phi(x)$ and $\varphi(x)$ are infinitely differentiable, the bilinear mapping $\mathscr{B}(\phi,\varphi)$ is a continuous bounded functional. This implies $|\mathscr{B}(\phi,\varphi)| \leq \eta \|\phi\| \|\varphi\|$. If $\mathscr{B}(\xi,\varphi)$ satisfies In Eq. (14.20), then $\mathscr{B}(\xi,\varphi)$ is coercive [19]. Thus, (S) possesses finite energy. According to the Lax-Milgram theorem [19], Eq. (14.18) admits an integral solution if

$$\mathscr{B}(\phi,\phi) = \langle \phi, L\phi\rangle - \left\langle \phi, \sum_{i=0}^{m} m^{-1}(x) q_{i+1}(x) \frac{\partial^i \phi}{\partial x^i}\right\rangle \geq \kappa \|\phi\|^2 > 0 \tag{14.22}$$

Since $\langle\phi, L\phi\rangle = 2\int_{\Omega} \nu(\phi)\,\mathrm{d}x$, it follows that

$$\left\langle \phi, \sum_{i=0}^{l} m^{-1}(x)\, q_{i+1}(x) \frac{\partial^i \phi}{\partial x^i} \right\rangle < 2 \int_{\Omega} \nu(\xi)\,\mathrm{d}x \tag{14.23}$$

Example 2 Consider the steady-state equation

$$m(x)\, Lw = q_0(x) + q_1(x)\, w \tag{14.24}$$

with $q_1(x) > 0$. Let $w(x) = \phi(x)\,\theta$. Then,

$$\left(\langle\phi, L\phi\rangle - \left\langle\phi, m^{-1}(x)\, q_1(x)\,\phi\right\rangle\right)\theta = \left\langle\phi, m^{-1}(x)\, q_0(x)\right\rangle \tag{14.25}$$

Clearly, if $\left\langle\phi, m^{-1}(x)\, q_1(x)\,\phi\right\rangle = 2\int_{\Omega} \nu(\xi)\,\mathrm{d}x = \langle\phi, L\phi\rangle$, the solution θ is unbounded. If $\left\langle\phi, m^{-1}(x)\, q_1(x)\,\phi\right\rangle > 2\int_{\Omega} \nu(\xi)\,\mathrm{d}x$, then fundamentally the notion of positive-valued potential energy ceases to exist.

14.3 Hamiltonian Control

The Hamiltonian function which represents the storage energy of distributed system (S) is defined as [13]

$$H = \frac{1}{2}\left\langle \frac{\partial w}{\partial t}, \frac{\partial w}{\partial t} \right\rangle + \frac{1}{2}\langle w, Lw\rangle \tag{14.26}$$

A stabilizing control is a control policy that renders that the system dissipative for which $\dot{H} \le 0$. The time derivative of the Hamiltonian function for zero disturbance is evaluated as

$$\dot{H} = \left\langle \frac{\partial w}{\partial t}, \frac{\partial^2 w}{\partial t} + Lw \right\rangle = \left\langle \frac{\partial w}{\partial t}, m^{-1}(x)\left(q_c + b u_w^*\right) \right\rangle \tag{14.27}$$

where $u_w^*(t)$ is a stabilizing control without the disturbance input, i.e., $g_w(x,t) = 0$.

Equation (14.27) is the result of the passivity theorem for distributed systems. The right hand side of Eq. (14.27) represents the rate of excess energy which must be decreased for the system to be dissipative.

Let $u_v^*(x,t) = b(x)\, u_w^*(t)$ be a virtual control signal. A stabilizing control for (S) is designed as

$$u_v^* = -q_c - m(x)\, c \frac{\partial w}{\partial t} + \mu m(x)\, Lw \tag{14.28}$$

The distributed control $u_w^*(t)$ can be obtained by a control allocation strategy by minimizing the cost function

$$\min_{u_w^*} J = \left\| W_w^{\frac{1}{2}} \left\langle \frac{\partial w}{\partial t}, m^{-1}(x)\left(b u_w^* - u_v^*\right)\right\rangle \right\|^2 \tag{14.29}$$

where $W_w > 0$. Utilizing Eq. (14.15), the optimal control allocation is obtained as

$$u_w^* = B^\dagger \left\langle \phi, m^{-1}(x)\, u_v^* \right\rangle \tag{14.30}$$

where $B^\dagger = \left(B^\top W_w B\right)^{-1} B^\top W_w$ and $B = \left\langle \phi, m^{-1}(x)\, b \right\rangle$.

$\dot{H}(t)$ becomes

$$\dot{H} = \left\langle \frac{\partial w}{\partial t}, m^{-1}(x)\left[q_c + bB^\dagger \left\langle \phi, m^{-1}(x)\left(-q_c - m(x)\, c\frac{\partial w}{\partial t} + \mu m(x)\, Lw\right)\right\rangle\right]\right\rangle \tag{14.31}$$

We choose c such that

$$\left\langle \phi, m^{-1}(x)\, bB^\dagger \left\langle \phi, c\frac{\partial w}{\partial t}\right\rangle\right\rangle > \left\langle \phi, m^{-1}(x)\left[q_3\frac{\partial w}{\partial t} - bB^\dagger \left\langle \phi, m^{-1}(x)\, q_3\frac{\partial w}{\partial t}\right\rangle\right]\right\rangle \tag{14.32}$$

Applying the Lax-Milgram theorem [19], the solution exists if

$$\begin{aligned}\mathscr{B}(\phi,\phi) = \langle \phi, L\phi\rangle - \Bigg\langle \phi, m^{-1}(x)\Bigg[q_1\phi + q_2\frac{\partial \phi}{\partial x} + bB^\dagger \left\langle \phi, m^{-1}(x)\left(-q_1\phi - q_2\frac{\partial \phi}{\partial x}\right)\right\rangle \\ + bB^\dagger \langle \phi, \mu L\phi\rangle \Bigg]\Bigg\rangle \geq \kappa \|\phi\|^2 > 0\end{aligned} \tag{14.33}$$

This yields

$$\begin{aligned}\left\langle \phi, m^{-1}(x)\, bB^\dagger \langle \phi, \mu L\phi\rangle\right\rangle < \langle \phi, L\phi\rangle - \Bigg\langle \phi, m^{-1}(x)\Bigg[q_1\phi + q_2\frac{\partial \phi}{\partial x} \\ - bB^\dagger \left\langle \phi, m^{-1}(x)\left(q_1\phi + q_2\frac{\partial \phi}{\partial x}\right)\right\rangle\Bigg]\Bigg\rangle\end{aligned} \tag{14.34}$$

The stabilizing control for (P) without the disturbance, i.e., $g_z(t) = 0$, is designed as

$$u_z^* = -K_z z - \mathscr{K}_w(w) - \mathscr{K}_{w_t}\left(\frac{\partial w}{\partial t}\right) \tag{14.35}$$

The distributed feedback gain operators $\mathscr{K}_w()$ and $\mathscr{K}_{w_t}()$ are obtained by minimizing a weighted least-squares cost function

$$\min_{\mathscr{K}_w, \mathscr{K}_{w_t}} J = \left\| W_z^{\frac{1}{2}} \left[\int_\Omega r\,dx - B_z \mathscr{K}_w(w) - B_z \mathscr{K}_{w_t}\left(\frac{\partial w}{\partial t}\right)\right]\right\|^2 \tag{14.36}$$

where $W_z > 0$.

The weighted least-squares solution yields

$$\mathcal{K}_w\,() = B_z^{\dagger} \int\limits_{\Omega} \left(r_1 + r_2 \frac{\partial}{\partial x} \right) ()\, \mathrm{d}x \tag{14.37}$$

$$\mathcal{K}_{w_t}\,() = B_z^{\dagger} \int\limits_{\Omega} r_3\, ()\, \mathrm{d}x \tag{14.38}$$

where $B_z^{\dagger} = \left(B_z^{\top} W_z B_z \right)^{-1} B_z^{\top} W_z$.

The feedback gain K_z is a stabilizing gain if there exists a positive definite matrix $P > 0$ such that

$$A_c^{\top} P + P A_c = -Q \tag{14.39}$$

where $Q > 0$ and

$$A_c = A - B_z K_z \tag{14.40}$$

The closed-loop system (P + S) is expressed as

$$\dot{z} = A_c z \tag{14.41}$$

$$m\,(x) \frac{\partial^2 w}{\partial t^2} + m\,(x)\, \mathcal{G} \left(\frac{\partial w}{\partial t} \right) + m\,(x)\, \mathcal{F}\,(w) = 0 \tag{14.42}$$

where $\mathcal{G}\,()$ and $\mathcal{F}\,()$ are operators defined as

$$\mathcal{G}\,() = m^{-1}\,(x) \left\{ b\,(x)\, B^{\dagger} \int\limits_{\Omega} \phi \cdot [q_3 + m\,(x)\, c]\, ()\, \mathrm{d}x - q_3\, () \right\} \tag{14.43}$$

$$\mathcal{F}\,() = L + m^{-1}\,(x) \left\{ b\,(x)\, B^{\dagger} \int\limits_{\Omega} \phi \cdot \left[q_1 + q_2 \frac{\partial}{\partial x} - \mu m\,(x)\, L \right] ()\, \mathrm{d}x - q_1 - q_2 \frac{\partial}{\partial x} \right\} \tag{14.44}$$

From Eqs. (14.32) and (14.33), we see that $\mathcal{G}\,() > 0$ and $\mathcal{F}\,() > 0$.

Assertion: The closed-loop system is asymptotically stable.

Proof Choose a Lyapunov candidate function

$$V = \frac{1}{2} \left\langle \frac{\partial w}{\partial t}, \frac{\partial w}{\partial t} \right\rangle + \frac{1}{2} \langle w, \mathcal{F}\,(w) \rangle + z^{\top} P z \tag{14.45}$$

Then, $\dot{V}(t)$ is evaluated as

$$\begin{aligned}\dot{V} &= \left\langle \frac{\partial w}{\partial t}, \frac{\partial^2 w}{\partial t^2} + \mathscr{F}(w) \right\rangle + z^\top \left(A_c^\top P + P A_c\right) z \\ &= -\left\langle \frac{\partial w}{\partial t}, \mathscr{G}\left(\frac{\partial w}{\partial t}\right) \right\rangle - z^\top Q z \\ &\leq -\alpha \left\| \frac{\partial w}{\partial t} \right\|^2 - \lambda_{\min}(Q) \|z\|^2 < 0 \end{aligned} \tag{14.46}$$

where $\alpha \leq \frac{\left\langle \frac{\partial w}{\partial t}, \mathscr{G}\left(\frac{\partial w}{\partial t}\right)\right\rangle}{\left\|\frac{\partial w}{\partial t}\right\|^2}$. Therefore, the closed-loop is asymptotically stable.

14.4 Distributed Optimal Control

Let $u_w(t) = u_w^*(t) + \tilde{u}_w(t)$ and $u_z(t) = u_z^*(t) + \tilde{u}_z(t)$. The closed-loop system (P + S) can be expressed abstractly in a semi-group form as [16]

$$\frac{\partial \mathscr{X}}{\partial t} = \mathscr{A}\mathscr{X} + \mathscr{B}u + \mathscr{V} \tag{14.47}$$

where

$$\mathscr{X}(x,t) = \left[z^\top(t) \; w^\top(x,t) \; \frac{\partial w^\top(x,t)}{\partial t} \right]^\top : (\Omega \oplus \Delta) \times [0,\infty) \to \mathbb{R}^{m+2n}$$

is an abstract state vector with Δ a domain defined by the inner product definition

$$\langle z, z \rangle = \int_\Delta z \cdot z \mathrm{d}x \triangleq z^\top z \tag{14.48}$$

$u(t) = \left[\tilde{u}_z^\top(t) \; \tilde{u}_w^\top(t)\right]^\top : [0,\infty) \to \mathbb{R}^{p+q}$ is the incremental control vector, $\mathscr{A} = \begin{bmatrix} A_c & 0 & 0 \\ 0 & 0 & I \\ 0 & -\mathscr{F}() & -\mathscr{G}() \end{bmatrix} \in C_c^\infty(\Omega)$ is a semi-group operator with compact support and continuous derivatives, $\mathscr{B} = \begin{bmatrix} B_z & 0 \\ 0 & 0 \\ 0 & m^{-1}(x) b(x) \end{bmatrix} \in \mathscr{C}^0(\Omega)$ is a bounded function, and $\mathscr{V}(x,t) = \begin{bmatrix} g_z(t) \\ 0 \\ m^{-1}(x) g_w(x,t) \end{bmatrix} \in \mathscr{C}^0(\Omega) \times [0,\infty)$ represents a bounded disturbance.

Let $\mathscr{M}$ be an operator in $\mathscr{C}^0(\Omega)$. We define the following inner product with the norm definition:

$$\begin{aligned}\langle \mathscr{X}, \mathscr{X} \rangle = \|\mathscr{X}\|^2 &= \int\limits_{\Omega \oplus \Delta} \mathscr{X} \cdot \mathscr{M}\mathscr{X} \mathrm{d}x \\ &= \int\limits_{\Delta} z \cdot z \mathrm{d}x + \int\limits_{\Omega} w \cdot w \mathrm{d}x + \int\limits_{\Omega} \frac{\partial w}{\partial t} \cdot m(x) \frac{\partial w}{\partial t} \mathrm{d}x \end{aligned} \tag{14.49}$$

The objective is to minimize the performance objective signal $p(t)$ with the linear-quadratic regulator (LQR) cost functional

$$J = \frac{1}{2} \int\limits_0^\infty \left(\left\langle \mathscr{M}^{-1} \mathscr{Q} \chi, \chi \right\rangle + u^\top R u \right) dt \tag{14.50}$$

where $\mathscr{Q} = \begin{bmatrix} P_z^\top Q_z P_z & 0 & 0 \\ 0 & P_w^\top Q_w P_w & 0 \\ 0 & 0 & 0 \end{bmatrix} \in \mathscr{C}_c^\infty(\Omega)$ is a semi-group operator with $Q_z = Q_z^\top > 0 \in \mathbb{R}^{k \times k}$, $Q_w = Q_w^\top > 0 \in \mathbb{R}^{(l-k)\times(l-k)}$, $R = \operatorname{diag}(R_z, R_w) \in \mathbb{R}^{(p+q)\times(p+q)}$, $R_z = R_z^\top > 0 \in \mathbb{R}^{q \times q}$, and $R_w = R_w^\top > 0 \in \mathbb{R}^{p \times p}$ are the weighting matrices.

The optimal control that minimizes the LQR cost functional in Eq. (14.50) has been developed [16] and is given by

$$u = -R^{-1} \langle \mathscr{B}, \Psi W \langle \Psi, \chi \rangle + \Psi S \langle \Psi, \mathscr{V} \rangle \rangle \tag{14.51}$$

where $\Psi(x) = \begin{bmatrix} I & 0 & 0 \\ 0 & \phi(x) & 0 \\ 0 & 0 & \phi(x) \end{bmatrix} : \Omega \oplus \Delta \in \mathbb{R}^{(m+2n)\times(m+2nN)}$, $N \to \infty$, and W and S are the steady-state solutions of an integro-differential Riccati equation and its auxiliary equation defined as

$$\begin{aligned}&\left\langle \mathscr{M}^{-1} \mathscr{Q} \Psi + \Psi W \langle \Psi, \mathscr{A} \Psi \rangle \right. \\ &\quad \left. - \Psi W \left\langle \Psi, \mathscr{B} R^{-1} \langle \mathscr{B}, \Psi W \langle \Psi, \Psi \rangle \rangle \right\rangle + \mathscr{A}^* \Psi W \langle \Psi, \Psi \rangle, \Psi \right\rangle = 0 \end{aligned} \tag{14.52}$$

$$\begin{aligned}&\left\langle -\Psi W \left\langle \Psi, \mathscr{B} R^{-1} \langle \mathscr{B}, \Psi S \langle \Psi, \mathscr{V} \rangle \rangle \right\rangle + \Psi W \langle \Psi, \mathscr{V} \rangle \right. \\ &\quad \left. + \Psi \dot{S} \langle \Psi, \mathscr{V} \rangle + \Psi S \left\langle \Psi, \frac{\partial \mathscr{V}}{\partial t} \right\rangle + \mathscr{A}^* \Psi S \langle \Psi, \mathscr{V} \rangle, \Psi \right\rangle = 0 \end{aligned} \tag{14.53}$$

14.5 Aircraft Flight Control Application

Consider the lateral-directional dynamics of an aircraft coupled to the structural dynamics of the flexible wing [16]. The aircraft motion is described by

$$m V \dot{\beta} = Y_\beta \beta + Y_p p + (Y_r - mV) r + m g \phi + Y_{\delta_r} \delta_r + Y_g \tag{14.54}$$

$$\begin{aligned} I_{xx} \dot{p} + I_{xz} \dot{r} &= L_\beta + L_p + L_r \\ &+ \int_{-L}^{L} \left[l_w (x) w + l_{w_x} (x) \frac{\partial w}{\partial x} + l_{w_t} (x) \frac{\partial w}{\partial t} \right] x \cos \Lambda dx \\ &+ L_{\delta_r} \delta_r + L_\delta \delta_a + L_g \end{aligned} \tag{14.55}$$

$$I_{xz} \dot{p} + I_{zz} \dot{r} = N_\beta + N_p + N_r + N_{\delta_r} \delta_r + N_\delta \delta_a + N_g \tag{14.56}$$

$$\dot{\phi} = p \tag{14.57}$$

where $\beta (t)$ is the aircraft angle of sideslip, $\phi (t)$ is the aircraft bank angle, $p (t)$ is the aircraft roll rate, $r (t)$ is the aircraft yaw rate, $w (x, t) = \left[v (t) \; \theta (t) \right]^\top$ is a wing elastic state vector due to the wing vertical bending displacement $v (x, t)$ and wing torsional displacement $\theta (x, t)$ in anti-symmetric motion, $\delta_a (t)$ are the aileron control surface deflections, $\delta_r (t)$ is the rudder control surface deflection, $Y_g (t)$ is aerodynamic side force due to atmospheric gust, $L_g (t)$ is the aerodynamic rolling moment due to gust, and $N_g (t)$ is the aerodynamic yawing moment due to gust.

The bending-torsion motion of an aircraft wing is described by the aeroelastic equations of motion

$$\begin{aligned} \rho A \frac{\partial^2 v}{\partial t^2} + \rho A e_{cg} \frac{\partial^2 \theta}{\partial t^2} + \frac{\partial^2}{\partial x^2} \left(E I \frac{\partial^2 v}{\partial x^2} \right) &= l_{v_x} \frac{\partial v}{\partial x} + l_{v_t} \frac{\partial v}{\partial t} + l_{\theta_t} \frac{\partial \theta}{\partial t} \\ &+ l_\delta \delta + l_g (x, t) + v_n (x, t) \end{aligned} \tag{14.58}$$

$$\begin{aligned} \rho I_{xx} \frac{\partial^2 \theta}{\partial t^2} + \rho A e_{cg} \frac{\partial^2 v}{\partial t^2} - \frac{\partial}{\partial x} \left(G J \frac{\partial \theta}{\partial x} \right) &= m_{v_x} \frac{\partial v}{\partial x} + m_{v_t} \frac{\partial v}{\partial t} + m_{\theta_t} \frac{\partial \theta}{\partial t} \\ &+ m_\delta \delta + m_g (x, t) + \theta_n (x, t) \end{aligned} \tag{14.59}$$

where x denotes the coordinate along the wing elastic axis, $EI (x)$ is the bending stiffness, $GJ (x)$ is the torsional stiffness, $l_g (x, t)$ is aerodynamic lift due to gust, $m_g (x, t)$ is the aerodynamic pitching moment due to gust, $\delta (t)$ is a vector of flight control surface deflections on the aircraft wing, and $v_n (x, t)$ and $\theta_n (x, t)$ are the process noise due to atmospheric turbulence.

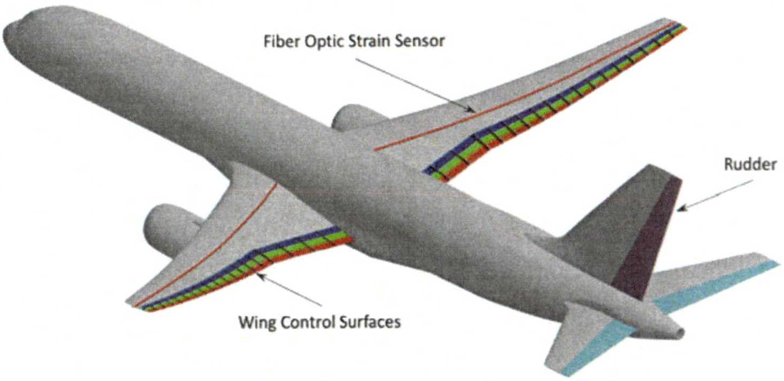

Fig. 14.1 Flexible wing aircraft with distributed fiber-optic strain sensors

The gust is modeled as a two-component sinusoidal vertical and lateral gust with a 1° equivalent angle of attack and a 1° equivalent angle of sideslip at a gust frequency of 30 rad/s. The vertical gust generates a symmetric gust load on the aircraft wing and the lateral gust generates an asymmetric gust load on the vertical tail. The gust load is defined as a differential operator

$$p(x,t) = \begin{bmatrix} P_z & 0 & 0 \\ 0 & -EI\frac{\partial^2}{\partial x^2}\delta(x-x_v) & 0 \\ 0 & 0 & GJ\frac{\partial}{\partial x}\delta(x-x_\theta) \end{bmatrix} \begin{bmatrix} z(t) \\ v(x,t) \\ \theta(x,t) \end{bmatrix} \tag{14.60}$$

where $z(t) = \left[\beta(t)\ p(t)\ r(t)\ \phi(t)\right]^\top$ is the aircraft state vector and $\delta(x - x_0)$ is the Dirac delta function.

The control input vector $u_w(t) = \left[\delta_1(t) \ldots \delta_p(t)\right]^\top \in \mathbb{R}^p$ comprises p distributed control surface deflections that span the wing. Figure 14.1 illustrates several distributed control surfaces of a flexible wing aircraft. The wing elastic motion is measured by fiber-optic strain sensors placed on the wing. The bending displacement $v(x,t)$ and torsional rotation $\theta(x,t)$ can be directly computed from these measurements.

The first unstable mode of the elastic wing occurs above an airspeed of 453 ft/s. Figure 14.2a, b show the unstable open-loop response of the wing elastic motion at 500 ft/s. Figure 14.3 shows the unstable aircraft response.

A Hamiltonian stabilizing control is designed with $\mu = 0$ and $c = 0.01\omega_f$, respectively, where ω_f is the flutter frequency. In addition, the rudder feedback control is designed by the standard LQR method to improve lateral-directional handling qualities of the aircraft. Figure 14.4a, b show the stable closed-loop response of the wing elastic motion at 500 ft/s with the Hamiltonian control. The stable closed-loop response of the aircraft is shown in Fig. 14.5a along with the control surface deflections in Fig. 14.5b.

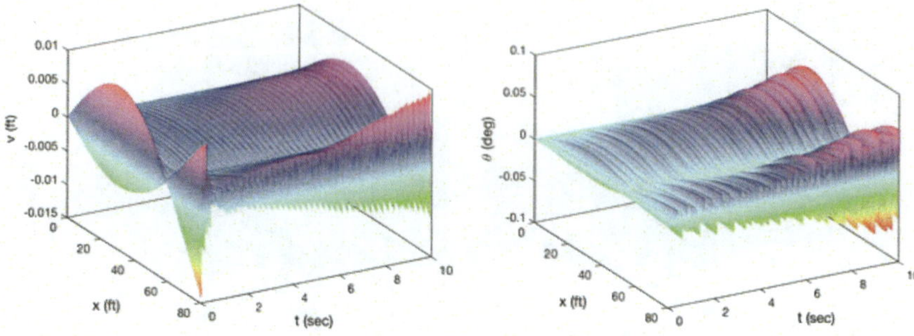

(a) (b)

Fig. 14.2 Wing bending and torsional displacement without stabilizing control

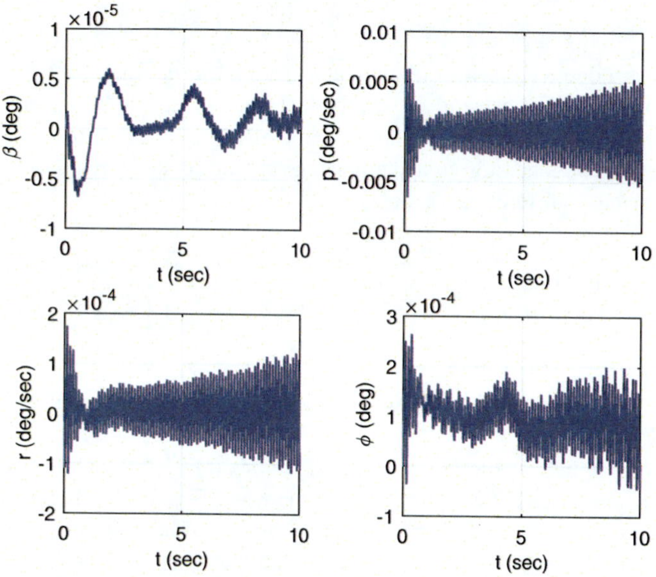

Fig. 14.3 Aircraft states without stabilizing control

A sinusoidal gust disturbance with a frequency of 30 rad/s is introduced. Figure 14.6a, b show the wing bending and torsional moments, respectively. A gust load alleviation (GLA) distributed optimal control is designed and implemented to reduce the wing bending and torsional moments. Figure 14.7a, b show the wing bending and torsional moments, respectively, with the distributed optimal control. Both the maximum bending and torsional moments are substantially reduced. The gust load reduction is substantial; 4.65 times for the maximum bending moment and 1.54 times for the maximum torsional moment.

Figure 14.8a shows the aircraft response with the distributed optimal control along with the control surface deflections in Fig. 14.8b. The effect of the distributed opti-

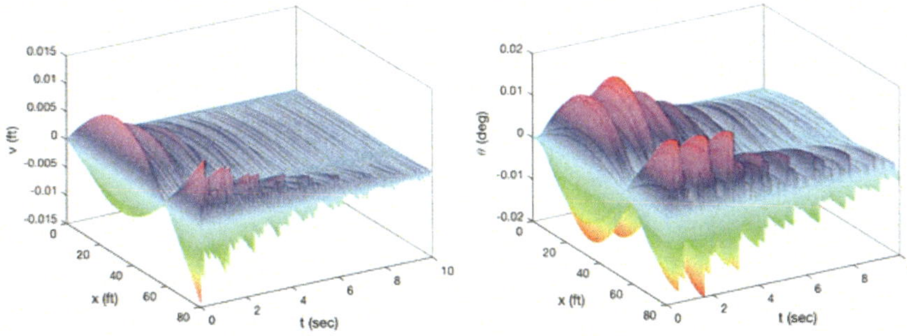

(a) (b)

Fig. 14.4 Wing bending and torsional sisplacement with Hamiltonian control

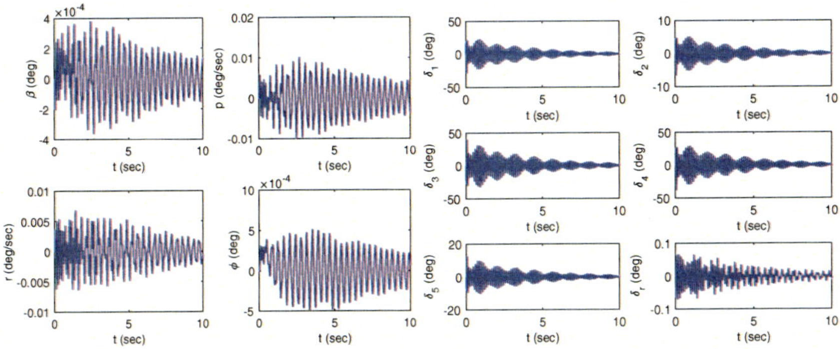

(a) (b)

Fig. 14.5 Aircraft states and control surface deflections with Hamiltonian control

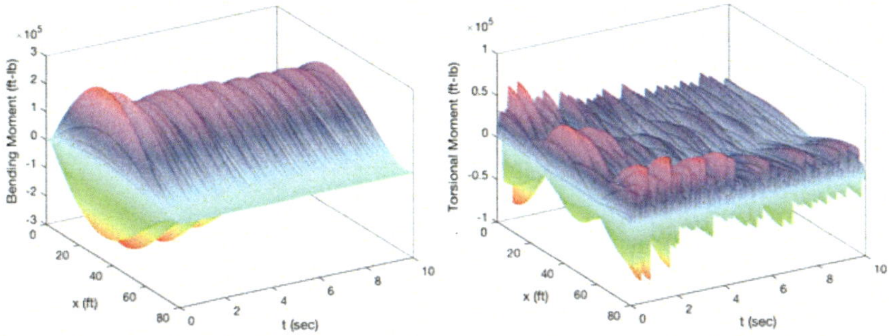

(a) (b)

Fig. 14.6 Wing bending and torsional moments with Hamiltonian control during gust encounter

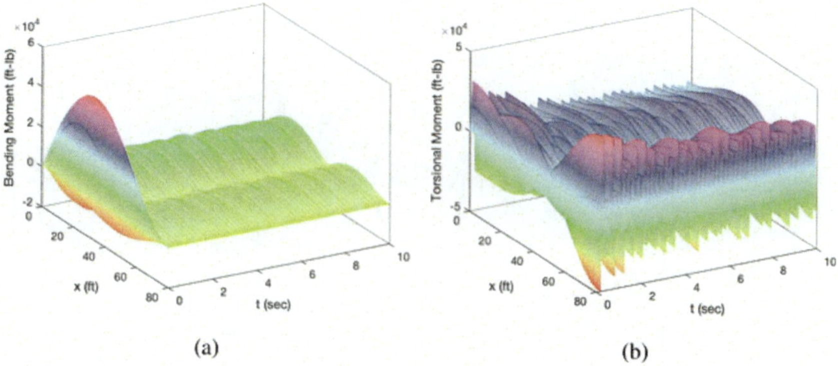

Fig. 14.7 Wing bending and torsional moments with Hamiltonian and distributed optimal control

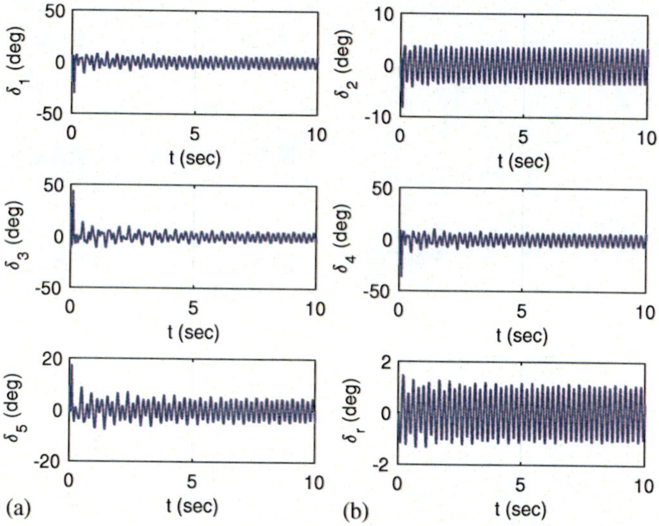

Fig. 14.8 Aircraft states and control surface deflections with Hamiltonian and distributed optimal control

mal control on the aircraft response is small due to the dominant response to the atmospheric turbulence modeled as process noise for the wing. The control surface deflections on the wing are much smaller with than without the distributed optimal control. There are small large initial transients up to 43.5° but overall the deflections are less than 10°. The maximum rudder deflection is 1.5° which is about the same as that without the GLA control. A redesign could be implemented to increase the rudder deflection. However, typically the rudder deflection is restricted to a small operating limit well less than 10° in high speed flight.

14.6 Conclusion

This paper presents a distributed Hamiltonian optimal control method for a class of distributed Lagrangian systems coupled to nonlinear finite-dimensional systems. A virtual control concept in conjunction with an optimal control allocation strategy is introduced to provide an infinite-dimensional distributed Hamiltonian control solution without any approximation as was employed in the previous work. The Hamiltonian method provides a stabilizing control for the distributed Lagrangian system. A distributed optimal control method previously developed is applied to the coupled systems using a semi-group abstraction. A flexible aircraft flight control application is presented to illustrate the theory.

References

1. Butkovskiy, A.G.: Distributed Control Systems. American Elsevier Publishing Company, New York (1969)
2. Lions, J.L.: Optimal Control of Systems Governed by Partial Differential Equations, Grundlehren der Mathematischen Wissenschaften 170. Springer, Berlin, New York (1971)
3. Raymond, J.P., Zidani, H.: Pontryagin's principle for state-constrained control problems governed by parabolic equations with unbounded controls. SIAM J. Control Optim. **36**(3), 1853–1879 (1998)
4. Fursikov, A.V., Gunzburger, M.D., Hou, L.S.: Boundary value problems and optimal boundary control for the Navier-Stokes system: the two-dimensional case. SIAM J. Control Optim. **36**(3), 852–894 (1998)
5. Kazemi, M.A.: A gradient technique for an optimal control problem governed by a system of nonlinear first order partial differential equations. J. Aust. Math. Soc. Ser. B **36**, 261–273 (1994)
6. Krstic, M., Smyshlyaev, A.: Adaptive control of PDEs. Ann. Rev. Control **32**, 149–160 (2008)
7. Ishihara, A., Nguyen, N.: Distributed parameter e-modification for an aeroelastic torsion model. In: Conference on Decision and Control, December 2014
8. Ishihara, A., Nguyen, N., Balas, M.: A stability result for distributed control of the beam. In: American Control Conference, June 2013
9. Menon, A., Chakravarthy, A., Nguyen, N.: Adaptive control for hybrid PDE models inspired from morphing aircraft. In: AIAA Guidance, Navigation, and Control Conference, AIAA-2018-0872, January 2018
10. Nguyen, N., Ting, E., Ishihara, A., Swei, S.: Distributed parameter optimal control by adjoint aeroelastic differential operators for mode suppression control. In: AIAA Guidance, Navigation, and Control Conference, AIAA-2013-4859, August 2013
11. Nguyen, N., Ardema, M.: Predictive optimal control of a hyperbolic distributed model for a wind tunnel. AIAA J. Guidance Control Dyn. **29**(3) (2006)
12. Nguyen, N., Bright, M., Culley, D.: Adaptive feedback optimal control of flow separation on stators by air injection. AIAA J. **45**(6) (2007)
13. Nguyen, N., Hashemi, K., Arabi, E., Yucelen, T.: Model-reference adaptive control of distributed Lagrangian infinite-dimensional systems using Hamilton's principle. In: AIAA Guidance, Navigation, and Control Conference, AIAA-2020-0840, January 2020
14. Nguyen, N.: Hamiltonian adaptive and optimal control of Lagrangian infinite-dimensional systems with gust load alleviation application. In: AIAA Guidance, Navigation, and Control Conference, AIAA-2021-1129, January 2021

15. Nguyen, N.: Distributed optimal control of Lagrangian infinite-dimensional systems with Kalman filter estimation. In: AIAA Guidance Navigation and Control Conference, AIAA-2022-2027, January 2022
16. Nguyen, N.: Distributed optimal control and estimation of Lagrangian infinite-dimensional systems coupled to finite-dimensional systems. In: AIAA Guidance, Navigation, and Control Conference, AIAA-2023-1633, January 2023
17. Brown, R.E., Fedde, M.R.: Airflow sensors in the avian wing. J. Exp. Biol. **179**(1) (1993)
18. Bai, H., Li, S., Barreiros, J., Tu, Y., Pollock, C.R.: Stretchable distributed fiber-optic sensors. Science **370**(6518) (2020)
19. Evans, L.: Partial Differential Equations. American Mathematical Society (2010)

Chapter 15
Stabilization of Optimal Trajectories of Dynamical Systems

Ivan E. Kaspirovich and Robert G. Mukharlyamov

Abstract The conditions of stability and stabilization of invariant sets of a dynamical system corresponding to its partial integrals are being determined. The solution of the control problem is presented as an inverse problem of dynamics. The requirement to limit deviations from the optimal trajectory is considered as a problem of stabilizing the constraints corresponding to the known integrals of the dynamical equations of a closed system. The conditions of stability of the optimal trajectory of a variable mass point in the central field of forces are established. Partial integrals of the dynamical equations of a closed system, determined by the known first integrals corresponding to the given initial conditions, are accepted as the constraint equations.

Keywords Optimal trajectory · Dynamical system · Constraint stabilization

15.1 Introduction

The methods of control theory and stability approach allow us to construct dynamic equations describing the movements of a system with predetermined properties. Depending on the formulation of the problem, differential equations of motion can admit invariant sets, correspond to the conditions of optimality, stability and provide the necessary accuracy of the numerical solution. All this variety of necessary properties of the functioning of the system allows us to present the problems of mathematical modeling of dynamics processes as inverse problems of the theory of differential equations and dynamics. A system of ordinary differential equations with a given integral curve was constructed by Erugin [1]. In [2], the structure of systems of differential equations constructed by known partial integrals is formed. Almukhamedov constructed systems of differential equations having singular points of a given type [3] and given limit cycles on the plane [4]. Some possible solution of the inverse problem of the qualitative theory of differential equations is presented in [5]. The

I. E. Kaspirovich (✉) · R. G. Mukharlyamov
Peoples' Friendship University of Russia, Moscow, Russia
e-mail: kaspirovich_ie@pfur.ru

D. Azimov (ed.), *Proceedings of the IUTAM Symposium on Optimal Guidance and Control for Autonomous Systems 2023*, IUTAM Bookseries 40,
https://doi.org/10.1007/978-3-031-39303-7_14

construction of a non-autonomous system of differential equations based on known partial integrals [6] made it possible to solve the problem of motion control with the bypass of moving obstacles.

The first formulation of the inverse problem of dynamics in classical mechanics was formulated by Newton [7] as the definition of the expression of the force that ensures the movement of a point mass along a trajectory corresponding to Kepler's laws. Expressions of the force corresponding to the movement of a point along a conic section are established in the works of Bertrand [8], Darboux [9], Imshenetsky [10]. Various formulations of inverse problems of the dynamics of a point mass and a mechanical system were investigated in the works of Suslov [11], Zhukovsky [12], Shebehely [13, 14], Bosis and Ihtiaroglu [15], Santilli [16, 17], Galiullina [18], Mukharlyamova [19], Pyatnitsky [20], Matyukhina [21], Ramírez [22], Tleubergenov [23].

Meshchersky considered the problems of determining the reactive force acting on a point of variable mass [24]. Since the inverse problem does not have a unique solution, the expression of the force corresponding to a given trajectory can be used to solve stability or optimization problems in some sense. Galiullin [18] gives a classification of inverse problems of dynamics, presenting them as problems of determining control actions, problems of constructing a part of the equations of motion and problems of constructing a set of systems that correspond to the differential equations whose solutions have given properties [1]. In paper [2], Mukharlyamov investigates the dynamics of systems with variable mass is studied. Inverse problems of classical mechanics and Helmholtz systems mechanics are described in [16, 17, 25–27]. Methods of solving inverse dynamics problems are widely used in solving applied problems in control theory, in the development of technical, economic, financial and other systems.

Research in the field of space navigation requires the development of modern methods of optimal control theory, stabilization and stability, mathematical modeling and numerical solution of differential equations. The problem of determining optimal trajectories in the problems of rocket dynamics and space navigation was posed in the work of Lawden [28]. Methods of analytical solution of these problems are proposed in [29]. In some cases, the construction of an analytical solution allows us to establish the first integrals of the equations of motion [30]. A detailed description of the development of the theory of constructing optimal trajectories and determining the first integrals is contained in the works of Asimov [31].

In the problems of modeling the dynamics of controlled systems, it is essential to ensure the stability of motion. Mukharlyamov [32] formulated stability conditions for invariant sets and developed methods for constructing dynamic equations having an asymptotically stable integral manifold. However, the asymptotic stability of invariant sets of a closed system of dynamics equations is not sufficient to ensure stability in the numerical solution [33]. Inaccuracies that can be caused by the deviations of initial conditions and errors in the numerical solution of differential equations lead to an increase in deviations from the constraint equations.

In the process of solving the optimal control problem, control forces are determined that can provide the required indicator of the quality of the system only if the initial conditions, the constraint equations and methods for solving the equations of dynamics of a closed system are strictly observed. The solution of control problems of systems admitting integrals should be accompanied by restrictions on deviations from invariant sets determined by the constraint equations, due to possible errors in the representation of the initial data and errors in the solution of differential equations. To solve the problem of constraint stabilization, Baumgarte [34] proposed to introduce additional controls-constraint reactions determined from the constraint perturbation equations determined by a linear combination of deviations from the constraint equations together with their derivatives. In the general formulation, the problem of constraint stabilization is reduced to the construction of systems of differential equations having given partial integrals [35] and refers to inverse problems of dynamics.

15.2 Problem Statement

The work is aimed at solving the problem of determining additional control forces acting on a closed optimal control system to stabilize partial integrals. Assuming the equations of the dynamics of a closed system and some first integrals to be known, it is required to develop a method for stabilizing the constraints determined by partial integrals corresponding to the given initial conditions.

The change in the state of the controlled system is described by the equation:

$$\dot{\boldsymbol{q}} = \boldsymbol{f}(\boldsymbol{q}, \boldsymbol{u}), \dot{\boldsymbol{q}} = \frac{d\boldsymbol{q}}{dt} \tag{15.1}$$

where $\boldsymbol{q} = \left(q^1, \ldots, q^n\right)$—state vector of a system, $\boldsymbol{u} = \left(u^1, \ldots, u^m\right)$—vector of control forces. Components u^μ, $\mu = 1, \ldots, m$, of vector $\boldsymbol{u}$ are selected from the set of piecewise continuous functions contained in the set U of the change of this vector: $\boldsymbol{u} \in U$. It is assumed that the set U can be open or closed, and components u^μ of vector $\boldsymbol{u}$ are piecewise continuous functions t, $t_0 \le t \le t_1$. The control aim [36, 37] is to move the system for period of time $t_1 - t_0$ from initial state $\boldsymbol{q}_0 = \boldsymbol{q}(t_0)$ to final state $\boldsymbol{q}_1 = \boldsymbol{q}(t_1)$, maintaining the extrema of the functional

$$J = \int_{t_0}^{t_1} f^0(\boldsymbol{q}, \boldsymbol{u})dt, \; f^0(\boldsymbol{q}, \boldsymbol{u}) > 0 \tag{15.2}$$

To solve this optimization problem, we introduce a Hamiltonian:

$$H(\lambda_0, \boldsymbol{\lambda}, \boldsymbol{q}, \boldsymbol{u}) = \sum_{i=0}^{n} \lambda_i \boldsymbol{f}^i(\boldsymbol{q}, \boldsymbol{u}) \tag{15.3}$$

and vector $\boldsymbol{\lambda}$, which components are presented as nontrivial solution of a system:

$$\dot{\lambda}_i = -\frac{\partial H}{\partial q^i}, \quad i = 1, \ldots, n. \tag{15.4}$$

From expression (15.3) follows the equality

$$\frac{\partial H}{\partial \lambda_0} = f^0(\boldsymbol{q}, \boldsymbol{u})$$

and Eq. (15.1) written in coordinate form

$$\dot{q}^i = \frac{\partial H}{\partial \lambda_i}, \quad i = 1, \ldots, n. \tag{15.5}$$

A necessary condition for solving the control problem is the existence of such a constant value $\lambda_0 \leq 0$ and such a nontrivial solution of the system (15.4), that for all t,that corresponds to the point of continuity of the function $\boldsymbol{u}(t)$, the following condition takes place:

$$\sup_{\boldsymbol{u} \in U} \boldsymbol{H}(\lambda_0, \boldsymbol{\lambda}, \boldsymbol{q}, \boldsymbol{u}) = \boldsymbol{H}(\lambda_0, \boldsymbol{\lambda}, \boldsymbol{q}, \boldsymbol{u}(t)) \equiv 0. \tag{15.6}$$

Equality (15.6) allows us to express the control vector $\boldsymbol{u}(t)$, corresponding to the necessary optimality conditions, expressed through vector expressions $\boldsymbol{\lambda} = \boldsymbol{\lambda}(t)$, $\boldsymbol{q} = \boldsymbol{q}(t)$, determined by the solution of a system of differential equations

$$\dot{q}^i = \frac{\partial H(\lambda_0, \boldsymbol{\lambda}, \boldsymbol{q}, \boldsymbol{u}(t))}{\partial \lambda_i}, \; \dot{\lambda}_i = -\frac{\partial H(\lambda_0, \boldsymbol{\lambda}, \boldsymbol{q}, \boldsymbol{u}(t))}{\partial q^i}, \tag{15.7}$$

under boundary conditions imposed on the vector $\boldsymbol{q}(t)$:

$$\boldsymbol{q}(t_0) = \boldsymbol{q}_0, \; \boldsymbol{q}(t_1) = \boldsymbol{q}_1.$$

15.3 Stabilization of Partial Integrals of Equations of Dynamics of a Controlled System

The equations of dynamics of a closed system can be presented in vector form:

$$\begin{aligned} \dot{\boldsymbol{r}} &= \boldsymbol{v}(t) \\ \dot{\boldsymbol{v}} &= \boldsymbol{a}(\boldsymbol{r}, \boldsymbol{v}, t) \end{aligned}, \tag{15.8}$$

Let some first integrals of (15.8) be known:

$$\boldsymbol{g}(\boldsymbol{r}, \boldsymbol{v}, t) = \boldsymbol{c}, \; \boldsymbol{g} = (g_1, \ldots g_a), a < 2n. \tag{15.9}$$

We assume that the initial conditions

$$\begin{aligned} \boldsymbol{r}(t_0) &= \overline{\boldsymbol{r}}_0 \\ \boldsymbol{v}(t_0) &= \overline{\boldsymbol{v}}_0 \end{aligned} \tag{15.10}$$

correspond to the partial integrals of Eq. (15.8)

$$\boldsymbol{g}(\boldsymbol{r}, \boldsymbol{v}, t) = 0. \tag{15.11}$$

This means that if the condition is satisfied

$$\boldsymbol{g}(\overline{\boldsymbol{r}}_0, \overline{\boldsymbol{v}}_0, t_0) = 0, \tag{15.12}$$

then equality (15.11) will be valid along the corresponding solution $\boldsymbol{r} = \boldsymbol{r}(t, \overline{\boldsymbol{r}}_0, \overline{\boldsymbol{v}}_0, t_0)$, $\boldsymbol{v} = \boldsymbol{v}(t, \overline{\boldsymbol{r}}_0, \overline{\boldsymbol{v}}_0, t_0)$. Obviously, due to inaccuracies in the initial conditions (15.10) and errors in the process of numerical integration of the system (15.8), deviations from the constraint equations occur:

$$\boldsymbol{y}(t_0) = \boldsymbol{y}^0, \ \boldsymbol{y}(t) = \boldsymbol{g}(\boldsymbol{r}, \boldsymbol{v}, t). \tag{15.13}$$

To limit the deviations (15.13), it is necessary to make changes to the right side of Eq. (15.8) by adding an extra term:

$$\begin{aligned} \dot{\boldsymbol{r}} &= \boldsymbol{v}(t) \\ \dot{\boldsymbol{v}} &= \boldsymbol{a}(\boldsymbol{r}, \boldsymbol{v}, t) + \boldsymbol{R} \end{aligned}, \quad \boldsymbol{R} = \left(R^1, \ldots, R^n\right). \tag{15.14}$$

Equation (15.14) should be composed so that, under initial conditions (15.10), its solution coincides with the solution of Eq. (15.8). This means that Eq. (15.14) must also have partial integrals defined by Eq. (15.11), so vector $\boldsymbol{R}$ should be selected depending on the magnitude of the deviations $\boldsymbol{y}:\ \boldsymbol{R} = \boldsymbol{R}(\boldsymbol{y}, \boldsymbol{r}, \boldsymbol{v}, t)$ and so that equality $\boldsymbol{R}(0, \boldsymbol{r}, \boldsymbol{v}, t) = 0$ is satisfied. Equality (15.11) represents partial integrals of Eq. (15.14) if the following equality takes place:

$$\dot{\boldsymbol{y}} = \boldsymbol{Y}(\boldsymbol{y}, \boldsymbol{r}, \boldsymbol{v}, t), \ \boldsymbol{Y}(0, \boldsymbol{r}, \boldsymbol{v}, t) = 0, \tag{15.15}$$

from which follows the equation for determining the vector $\boldsymbol{R}$:

$$\boldsymbol{G}\boldsymbol{R} = \boldsymbol{Y}, \boldsymbol{G} = (g_{\alpha i}), \ g_{\alpha i} = \frac{\partial g_\alpha}{\partial x^i}, \alpha = 1, \ldots, a. \tag{15.16}$$

Let's rewrite (15.14) considering the expression $\boldsymbol{R} = \boldsymbol{G}^T\left(\boldsymbol{G}\boldsymbol{G}^T\right)^{-1}\boldsymbol{Y}$:

$$\begin{aligned} \dot{\boldsymbol{r}} &= \boldsymbol{v}(t) \\ \dot{\boldsymbol{v}} &= \boldsymbol{a}(\boldsymbol{r}, \boldsymbol{v}, t) + \boldsymbol{G}^T\left(\boldsymbol{G}\boldsymbol{G}^T\right)^{-1}\boldsymbol{Y}. \end{aligned} \tag{15.17}$$

To stabilize the constraints given by Eq. (15.11), it is necessary that the trivial solution $y = 0$ of the equation of constraint perturbations (15.15), considered together with (15.17), was stable asymptotically. Otherwise, when $Y = Ky$ and matrix K is stable, for the asymptotic stability of a trivial solution, it is sufficient that the roots of the characteristic equation $\dot{y} = Ky$ had negative real parts. In general, the Lyapunov function method can be used to study stability [32]. So, if we take for the Lyapunov function a positive definite quadratic form defined by the equality $2V = y^T Ay$ with a constant matrix of coefficients, then trivial solution $y = 0$ will be stable asymptotically if its derivative $\dot{V} = y^T AY(y, r, v, t)$ will be negatively defined and a vector function $y = g(r, v, t)$ admits an infinitesimal higher limit. System (15.17) can be rewritten by introducing state vector $X = (r, v)$:

$$\dot{x} = X(x, t) + \tilde{R}, \ \tilde{R} = (0, R).$$

The stabilization of the constraints (15.11) in the numerical solution of Eq. (15.17) requires the fulfillment of additional conditions that must be imposed on the vector function $Y(y, x, t)$. In the case of a linear equation of perturbations of constraints of the form $\dot{y} = Ky$ with a constant matrix of coefficients when constructing a solution to Eq. (15.17) using the Euler method

$$x_{k+1} = x_k + \left(X(x, t) + \tau G^T\left(GG^T\right)^{-1}Y\right)_k, \ x_0 = x(t_0), \ k = 1, \dots, K,$$

condition $||y_k|| \le \varepsilon$ will be satisfied for all $k = 1, \dots, K$, if all the roots of the characteristic equation are different and the following inequalities take place

$$||y_0|| \le \varepsilon, \ ||(E + \tau K)|| \le (1 - \alpha)\varepsilon, \ ||g_k^{(2)}|| \le \alpha\varepsilon, \ 0 \le \alpha \le 1.$$

Here $g_k^{(2)}$ represents the residual terms of the expansion into a series of function (15.13) by degrees $\tau = t_{k+1} - t_k$. If the characteristic equation has equal roots with negative real parts [38], then the numerical solution at a certain interval of the argument change may lead to an increase in deviations from the constraint equations. In a more general case, the function can be taken as an estimate of deviations from the constraint equations $V = 0,5y^T Ay$. Suppose $V_0 = 0,5\left(y^0\right)^T Ay^0 \le \varepsilon$. Let's assume that the inequality takes place

$$V_k = 0,5\left(y^k\right)^T Ay^k \le \varepsilon, \ y^k = g(x(t_k), t_k).$$

Let's expand V_{k+1} to series by powers of τ:

$$\begin{aligned}2V_{k+1} &= 2V_k + 2\tau\left(y^k\right)^T A\left(G\dot{x}_k + g_t\right) + V_k^{(2)} = 2V_k \\ &+ 2\tau\left(y^k\right)^T A\left(G_k\left(X(x, t) + G^T\left(GG^T\right)^{-1}Y\right)_k + g_t\right) \\ &+ V_k^{(2)} = 2V_k + \tau\left(y^k\right)^T AY^k + V_k^{(2)}\end{aligned} \qquad (15.18)$$

Let's consider (15.4), (15.5) as a linear system with constant coefficients: $\dot{y} = \boldsymbol{K}y$, $\boldsymbol{K} = (k_{\mu\kappa})$, $k - const$. Then equality (15.18) takes place

$$V_{k+1} = V_k + \tau\left(y^k\right)^T \boldsymbol{A}\boldsymbol{K}y^k + V_k^{(2)} = \left(y^k\right)^T \boldsymbol{A}y^k + \tau\left(y^k\right)^T \boldsymbol{A}\boldsymbol{K}y^k$$
$$+ V_k^{(2)} = \left(y^k\right)^T \boldsymbol{A}(\boldsymbol{E} + \tau\boldsymbol{K})y^k + V_k^{(2)} = \left(y^k\right)^T \boldsymbol{B}y^k + V_k^{(2)}. \quad (15.19)$$

The following conclusion can be drawn from equality (15.19). If the following inequalities take place

$$||V_0|| \le \varepsilon,\ ||\boldsymbol{B}|| \le (1-\alpha)\varepsilon,\ ||V_k^{(2)}|| \le \alpha\varepsilon,\ 0 \le \alpha \le 1.$$

then condition $||V_0|| \le \varepsilon$, will be satisfied for all $k = 1, \ldots, K$.

15.4 Motion Control of a Point Mass in the Central Force Field

In this section the problem of controlling the motion of a rocket with variable mass $m(t)$ is considered. Let the position of which is determined by the vector $\boldsymbol{r} = (x, y, z)$, $\boldsymbol{v} = \mathrm{d}\boldsymbol{r}/\mathrm{d}t = (v_x, v_y, v_z)$—velocity vector. We will assume that the rocket is moving in the central force field with the potential $U = -\frac{\mu}{|r|}$. The dynamics of a body with a variable mass is described by Meshchersky's equations

$$\frac{\mathrm{d}\boldsymbol{r}}{\mathrm{d}t} = \boldsymbol{v},\ \frac{\mathrm{d}\boldsymbol{v}}{\mathrm{d}t} = \boldsymbol{F}_{\boldsymbol{r}} - \nabla_{\boldsymbol{r}}U,$$

where $\boldsymbol{F}_{\boldsymbol{r}}$—vector of reactive force, $\nabla_{\boldsymbol{r}} = \left(\frac{\partial}{\partial x}, \frac{\partial}{\partial y}, \frac{\partial}{\partial z}\right)$—gradient vector.

To solve the problem of the rocket's motion control, a limited function of the second fuel consumption can be selected as a control function $\dot{m}(t) = -\beta$ and a unit thrust vector $\boldsymbol{e}$, where $\boldsymbol{F}_{\boldsymbol{r}} = \frac{2P}{I_{sp}g}\boldsymbol{e}$, I_{sp}—specific impulse, $P = \frac{1}{2}\beta I_{sp}^2 g^2$—specified engine power. The differential equations of motion can then be written in the form of the Lowden equations [39]:

$$\dot{\boldsymbol{v}} = \frac{2P}{I_{sp}gm}\boldsymbol{e} - \frac{\mu}{r^3}\boldsymbol{r},\ \dot{\boldsymbol{r}} = \boldsymbol{v},\ \dot{m} = -\frac{2P}{I_{sp}^2 g^2}. \quad (15.20)$$

Unit thrust vector defines the constraint:

$$f_1 = \boldsymbol{e}^2 - 1 = 0, \quad (15.21)$$

Limited magnitudes of both impulse and engine power also establish the following restrictions:

$$f_2 = P(P_{max} - P) - \gamma^2 = 0, \tag{15.22}$$

$$f_3 = \left(I_{max} - I_{sp}\right)\left(I_{sp} - I_{min}\right) - \eta^2 = 0, \tag{15.23}$$

where γ and η—are control functions.

Let's introduce a 7-dimensional state vector of a rocket: $\boldsymbol{x} = (\boldsymbol{r}, \boldsymbol{v}, m)$ and its conjugated $\boldsymbol{\lambda} = (\boldsymbol{\lambda}_r, \boldsymbol{\lambda}_v, \lambda_m)$. The rocket dynamics equations in canonical variables will be written with Hamiltonian H:

$$H = (\boldsymbol{\lambda}_r, \boldsymbol{v}) + \left(\boldsymbol{\lambda}_r, \frac{2P}{I_{sp} g m}\boldsymbol{e} - \frac{\mu}{r^3}\boldsymbol{r}\right)$$
$$- \frac{2P\lambda_m}{I_{sp}^2 g^2} + \sigma_1 f_1 + \sigma_2 f_2 + \sigma_3 f_3, \tag{15.24}$$

where $\boldsymbol{\sigma}$—Lagrange multiplier for constraints (15.22) and (15.23).

To determine the vector of the control functions $\boldsymbol{u} = \left[\boldsymbol{e}, I_{sp}, P, \gamma, \eta\right]^T$ local extrema principle is considered

$$\nabla_{\boldsymbol{u}} H = 0. \tag{15.25}$$

The analysis of the obtained expressions (15.25) and the Weierstrass conditions makes it possible to divide the trajectory to sections according to possible power P and specific impulse I_{sp} magnitudes. The paper considers cases of maximum thrust $P = P_{max}$, $I_{sp} = I_{min}$ and minimum thrust $P = P_{max}$, $I_{sp} = I_{max}$.

The problem of determining the trajectory is solving at the interval $t \in [t_0, t_1]$ with conditions $\psi_s(\boldsymbol{x}_0) = 0$, $s = 1, \ldots, 7$, $\boldsymbol{x}_0 = \boldsymbol{x}(t_0)$ and $\Theta_r(\boldsymbol{x}_1, t) = 0$, $r \leq 7$, $\boldsymbol{x}(t_1) = \boldsymbol{x}_1$. At the same time we need to provide the minimization of a functional $J\left(x_{1,r+1}, \ldots, x_{1,7}\right)$. Then initial and boundary conditions for vector $\boldsymbol{\lambda}$ are determined with the help of transversality conditions:

$$\lambda_{0,i} = \lambda_i(t_0) = \nu_s \frac{\partial \psi_s}{\partial x_{0,i}}; \tag{15.26}$$

$$-\lambda_{1,i} = -\lambda_i(t_1) = \frac{\partial J}{\partial x_{1,i}} + \tilde{\nu}_r \frac{\partial \Theta_r}{\partial x_{1,i}}, \quad i = 1, \ldots, 7; \tag{15.27}$$

where $\boldsymbol{\nu}$, $\tilde{\boldsymbol{\nu}}$—are Lagrange multipliers, and we assume summation under repeated indexes. The system of canonical equations corresponding to the Hamiltonian (15.24) with the local extrema principle, boundary conditions and transversality conditions allow us to determine the expressions of the control vector and the state vector of the rocket.

15.5 Stabilization of First Integrals

Consider the case when k partial integrals corresponding to the given initial conditions $g_l(\boldsymbol{x}, t) = C_l$, $l = 1, \ldots, k \leq 14$ are imposed on the system. We assume that in modeling, the initial conditions are introduced with an error of the deviation vector $\boldsymbol{\varepsilon}$. Then deviated initial conditions can be written as:

$$\boldsymbol{x}'(\boldsymbol{t}_0) = \boldsymbol{x}_0(1 + \varepsilon), \tag{15.28}$$

where ε: $0 < \varepsilon < 1$. Obviously, that the deviations will also affect the vector $\boldsymbol{\lambda}$, since the transversality conditions (15.5) associate their initial values with the initial values $\boldsymbol{x}'(\boldsymbol{t}_0)$.

We define the value of deviations from partial integrals as the difference $\tilde{\boldsymbol{g}} = \boldsymbol{g}\left(\boldsymbol{x}', \boldsymbol{\lambda}', t\right) - \boldsymbol{C}$. With numerical integration of the equations, the values of deviation from the expected trajectory may increase due to the accumulation of errors. Stabilization of the partial values of the first integrals in the numerical solution of the dynamical equations in the general case will be possible if the structure of the control forces corresponds to the partial integrals of the equations of dynamics given by constraint equations. In the simplest case, control forces can be selected based on the system of equations of constraint perturbations

$$\frac{\mathrm{d}\tilde{\boldsymbol{g}}}{\mathrm{d}t} = \boldsymbol{F}\left(\tilde{\boldsymbol{g}}, \boldsymbol{x}', \boldsymbol{\lambda}', t\right), \tag{15.29}$$

where vector-function $\boldsymbol{F}$—is homogeneous with respect for $\tilde{\boldsymbol{g}}$, and it is chosen in such a way that it provides stability with respect to deviations from the first integrals. The right side of Eq. (15.29) should be chosen so that the trivial solution is stable asymptotically.

The system of motion equations will be written in the following form:

$$\begin{aligned} \dot{\boldsymbol{x}} &= \tfrac{\partial H}{\partial \boldsymbol{\lambda}} + \gamma_l \tfrac{\partial \tilde{g}_l}{\partial \boldsymbol{x}}; \\ \dot{\boldsymbol{\lambda}} &= -\tfrac{\partial H'}{\partial \boldsymbol{x}} + \gamma_l \tfrac{\partial \tilde{g}_l}{\partial \boldsymbol{\lambda}}; \\ \tfrac{d\tilde{\boldsymbol{g}}}{dt} &= \boldsymbol{F}\left(\tilde{\boldsymbol{g}}, \boldsymbol{x}', \boldsymbol{\lambda}', t\right); \end{aligned} \tag{15.30}$$

where—γ—are arbitrary multipliers. As $\tilde{\boldsymbol{g}}$ is a first integrals, so $\frac{d\tilde{\boldsymbol{g}}}{dt} = 0$, it is possible to uniquely define arbitrary multipliers as functions of $\boldsymbol{x}'$, $\boldsymbol{\lambda}'$ and t. To do this, we must solve the following system:

$$\left(\frac{\partial \tilde{g}_l}{\partial x^i}\frac{\partial \tilde{g}_k}{\partial x^i} + \frac{\partial \tilde{g}_l}{\partial \lambda^i}\frac{\partial \tilde{g}_l}{\partial \lambda^k}\right)\gamma_k = F_l\left(\tilde{\boldsymbol{g}}, \boldsymbol{x}', \boldsymbol{\lambda}', t\right). \tag{15.31}$$

15.6 Stabilization of the First Two Integrals of Optimal Motion

Let's rewrite the Hamiltonian (15.24) for plane motion in polar coordinates, taking into account the values of arbitrary multipliers for sections of maximum or minimum thrust of constant power:

$$H = -\frac{\mu}{r^2}\lambda_1 + \lambda_4 v_1 + v_2\left(\frac{\lambda_1 v_2}{r} - \frac{\lambda_2 v_1}{r} + \frac{\lambda_5}{r}\right) + \frac{P_{max}}{2\lambda_7 m^2}\left(\lambda_1^2 + \lambda_2^2\right), \quad (15.32)$$

where $\boldsymbol{x} = (v_1, v_2, r, \theta, m)$—state vector and its conjugated $\boldsymbol{\lambda} = (\lambda_1, \lambda_2, \lambda_4, \lambda_5, \lambda_7)$.

The canonical system of equations with respect to the Hamiltonian (15.32) has the at least two first integrals $\boldsymbol{g} = (g_1, g_2)$. First—is a full energy: $H = E_0$, the other—following expression:

$$\lambda_1 v_1 + \lambda_2 v_2 - 2r\lambda_4 - \frac{5}{2\lambda_7 m^2}\int\left(\lambda_1^2 + \lambda_2^2\right)\mathrm{d}t + 3E_0 t = E_1,$$

where E_0, E_1—are integrational constants.

Let's consider linear (15.29): stabilization function:

$$\frac{d}{dt}\begin{pmatrix}\tilde{g}_1\\ \tilde{g}_2\end{pmatrix} = -\begin{pmatrix}k_1 & 0\\ 0 & k_2\end{pmatrix}\begin{pmatrix}\tilde{g}_1\\ \tilde{g}_2\end{pmatrix}, \quad (15.33)$$

where k_1, k_2—given constants, determining stability with respect to the first integrals.

The solution of the system of Eqs. (15.30) for the Hamiltonian (15.32) taking into account the stabilization function (15.33) leads to the following results:

All graphs, Figs. 15.1, 15.2 and 15.3 show the functional dependence between first integral standard deviation and time coordinate. On Fig. 15.1 the parameters of stabilization equation k_1, k_2 were fixed and each line correspond to the specific value of deviation from initial data ε. On Figs. 15.2 and 15.3, on contrary the value of initial deviation was fixed and numerical integration were computed varying values of stabilization parameters k_1, k_2 correspondingly.

The deviation represented along the ordinate axis is considered as the norm of the vector of deviations from the first integrals $\sigma = \sqrt{(\tilde{g}_1 - E_0)^2 + (\tilde{g}_2 - E_1)^2}$. The figure shows that when the rocket moves, the deviation values tend to zero, that is, the constraint stabilization method ensures the stability of the numerical solution, despite possible deviations in the initial data and errors in numerical integration.

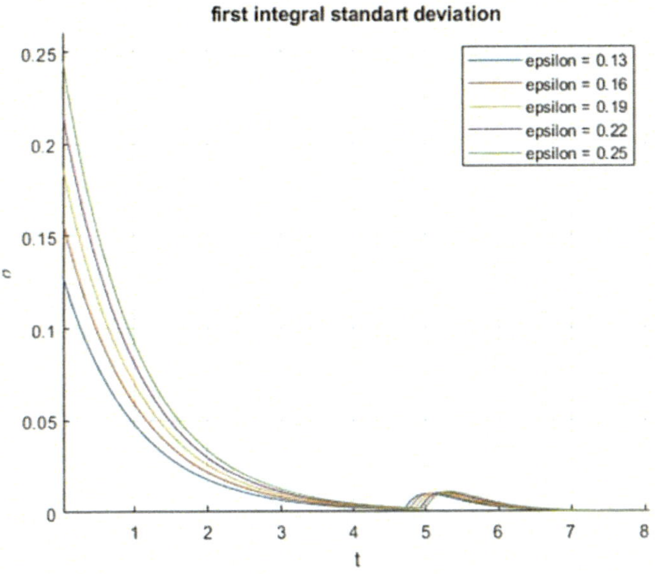

Fig. 15.1 Deviation from the first integrals from the initial deviations

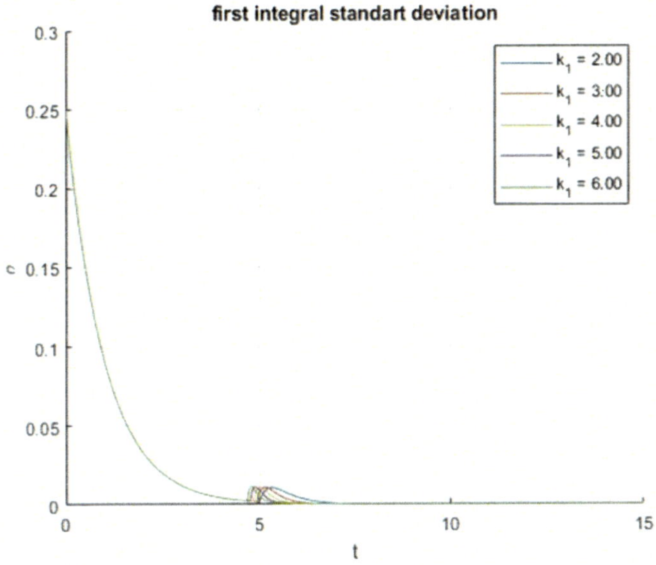

Fig. 15.2 Deviation from the first integrals of the magnitude k_1

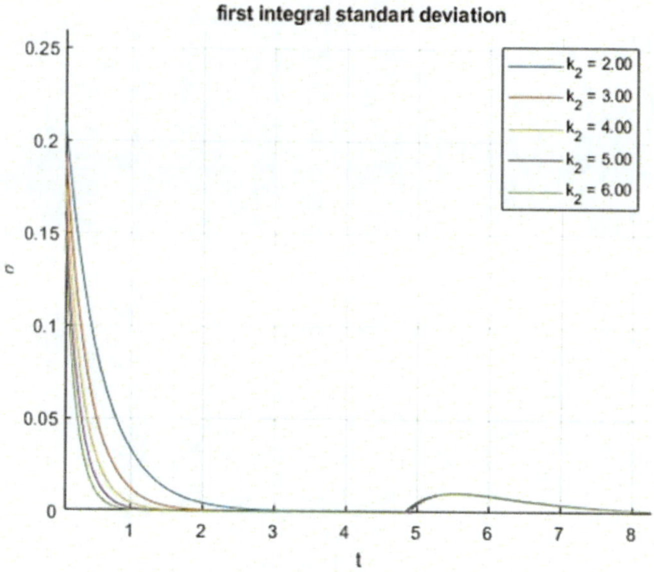

Fig. 15.3 Deviation from the first integrals of the magnitude k_2

15.7 Conclusion

The direct solution of the optimal control problem of systems whose phase coordinates are constrained can lead to instability of the numerical solution of the equations of dynamics of a closed system. The methods of constraint stabilization proposed in the paper allow us to ensure the necessary accuracy of the implementation of the constraint equations in the process of solving the control problem.

Acknowledgements The present work has been possible thanks to the funding granted by the Russian Science Foundation, project code 23-21-10065.

References

1. Erugin, N.P.: Construction of the entire set of systems of differential equations having a given integral curve. PMM **21**(6), 659–670 (1952)
2. Mukharlyamov, R.G.: On the construction of differential equations of optimal motion for a given variety. Differ. Eq. **7**(10), 1825–1834 (1971)
3. Almukhamedov, M.I.: On the construction of a differential equation having its limit cycles given curves. Izv. Vuzov. Matematika **1**(44), 12–16
4. Almukhamedov, M.I.: Inverse problem of qualitative theory of differential equations. Izv. Higher Schools Ser. Math. **8**(4), 3–6
5. Mukharlyamov, R.G.: On inverse problems of the qualitative theory of differential equations. Differ. Eq. **3**(10), C. 1673–1681 (1967)

6. Ibusheva, O.V., Mukharlyamov, R.G.: Construction of a non-autonomous system of differential equations for a given set of partial integrals in a multidimensional space. Uchen. Zap. Cauldron. Ser. Phys. Matem. Sci. **150**(3), 133–139 (2008)
7. Newton, I.: Mathematical principles of natural philosophy, 688 pp. Translated from Latin and notes by A. N. Krylov. M.: Nauka (1989)
8. Bertrand, M.G.: The theorem on the relativity of the motion of one point in clothing without fixing the center. Compte. Rendus. **77**(16), 849–853 (1873)
9. Darboux, M.G.: Investigation of the question of how to deal with the central forces to determine the trajectory of motion in order to avoid collision. Compte. Rendus. **LXXXIV**(16), 760–762 (1877)
10. Imshenetsky, V.G.: Determination of the Force Moving a Material Point Along a Conic Section in a Function of Its Coordinates, issue 1, pp. 5–15. Reports of the Kharkov Mathematical Society under the Emperor. Kharkov. Univ. (1879)
11. Suslov, G.K.: On a Force Function Admitting Given Integrals, 114 pp. Kiev Publishing House of Kiev University (1890)
12. Zhukovsky, N.E.: Determination of the force function for this family of trajectories. In: Complete Collection of SOC, vol. I, pp. 293–308 (1937) (M.-L., The Main editorial office of aviation literature)
13. Shebehey, V.: Generalized inverse problem of orbit calculation, in the process of the 2nd international. In: Space Science Symposium, vol. 3, p. 318. North Holland Publishing House, Amsterdam (1961)
14. Shebehey, V.: Open problems on the eve of the next millennium. Celes. Mech. Din. Astron. **65**, 205–211 (1997)
15. Bosis, G., Ihtiaroglu, S.: Existence and construction of dynamical systems having a given integral of motion—inverse problem. Inverse Prob. **3**, 213–227 (1987)
16. Santilli, R.M.: Fundamentals of Theoretical Mechanics. I: Inverse Problem in Newtonian Mechanics, 266 pp. Springer-Verlag, New York, Heidelberg, Berlin (1978)
17. Santilli, R.M.: Fundamentals of Theoretical Mechanics. II: Birkhoff Generalization of Hamiltonian Mechanics, 370 pp. Springer-Verlag, New York, Heidelberg, Berlin (1982)
18. Galiullin, A.S.: Methods of Solving Inverse Problems of Dynamics, 224 pp. M. Nauka (1986)
19. Mukharlyamov, R.G.: Inverse Problems of Dynamics. Stability of Motion. Analytical Mechanics. Motion Control, pp. 217–223. M. Publishing House “Science” (1981)
20. Polyak, B.: Selected Works of E.S. Pyatnitskii in 3 Volumes. Moscow: Fizmatlit, 2004, vol. 1; 2005, vol. 2. Autom. Remote Control. **66**, 1701–1702 (2005). https://doi.org/10.1007/s10513-005-0203-1
21. Matyukhin, V.I.: Control of Mechanical Systems, 320 pp. M. Fizmatlit (2009)
22. Libre, J., Ramirez, R.: Inverse problems of ordinary differential equations and their applications, p. 266. Springer International Publishing House, Switzerland (2016)
23. Marat, T., Vasilina, G., Azhymbayev, D.: Stochastic Helmholtz problem and convergence in distribution. Filomat. **36**(7), 2451–2460 (2022). https://doi.org/10.2298/FIL2207451T
24. Meshchersky, I.V.: Works on Mechanics of Bodies of Variable Mass, 280 pp. M.-L. Gostekhizdat (1952)
25. Galiullin, A.S., Gafarov, G.G., Malaishka, R.P., Khvan A.M.: Analytical dynamics of Helmholtz, Birkhoff, Nambu systems, 324 pp. Editorial Board of the Journal “Successes of Physical Sciences” (1997)
26. Mukharlyamov, R.G.: Reduction to a given structure of equations of dynamics of systems with connections. PMM. **71**(3), 401-410 (2007)
27. Kaspirovich, I.E., Mukharlyamov, R.G.: On methods for constructing dynamic equations taking into account bond stabilization. Proc. Russian Acad. Sci. Solid State Mech. **3**, 124–135 (2019)
28. Lawdon, D.F.: Optimal trajectories for space navigation, 152 pp. Publishing House “Mir”. M (1966)
29. Azizov, A.G., Korshunova, N.A.: On the analytical solution of the optimal trajectory problem in the gravitational field. Celest. Mech. **38**(4), 297–306 (1986)

30. Azizov, A.G., Korshunova, N.A.: Application of the Levi-Civita method in the analysis of optimal trajectories. Space Res. **17**(3), 378–386 (1979)
31. Asimov, D.M.: Analytical Solutions for Extreme Space Trajectories, 317 pp. Book (2018)
32. Mukharlyamov R.G. On the construction of a set of systems of differential equations of stable motion on an integral manifold. Equations. **5**(C), 688–699 (1969)
33. Mukharlyamov, R.G.: On solving systems of nonlinear equations. ZHVM MF. **11**(4), 829–836 (1971)
34. Baumgarte, J.: Stabilization of constraints and integrals of motion in dynamical systems. Comp. Math. Appl. Mech. Eng. **1**, 1–16 (1972)
35. Mukharlyamov, R.G.: On the equations of motion of mechanical systems. Differ. Eq. **19**(12), 2048–2056 (1983)
36. Pontryagin, L.S., Boltyansky, V.G., Gamkrelidze, R.V., Mishchenko., E.F.: Mathematical Theory of Optimal Processes, 400 pp. M. Ed. "Science" (1961)
37. Boltyansky, V.G.: Mathematical Methods of Optimal Control, 308 pp. M. Publishing House "Science" (1966)
38. Asher, U.M., Hongsheng, C., Reich, S.: Stabilization of DAES and invariant manifolds. Mathematics. **67**(2), 131–149 (1994)
39. Jo, M., Azimov, D.: Semi-analytical and extreme solutions for the design and synthesis of descent and landing trajectories with a drive. Int. J. Space Sci. Eng. **6**(3), 227–246 (2021)

Part V
Modeling and Data Analytics

Chapter 16
Koopman Operator Based Modeling and Control of Quadrotors

Simone Martini, Alessandro Rizzo, Margareta Stefanovic, Patrizia Livreri, Matthew J. Rutherford, and Kimon P. Valavanis

Abstract A novel set of Koopman observables is introduced to transform the nonlinear quadrotor dynamics into a linear representation. The proposed approach is a combination of feedback linearization and Koopman observables, and it offers the advantage of overcoming the quadrotor underactuation problem. Compared to previous approaches that use dynamic inversion for both position and attitude dynamics, in this work, only the attitude dynamics are feedback linearized. The formulation is shown to be considerably more compact and capable of achieving lower model mismatch error when compared to already published similar work. The advantage of this formulation is that it allows for designing control architectures exploiting well-known linear control theory. As such, a Koopman based linear quadratic (LQ) controller is tested via numerical simulations to show applicability and implementability of the approach.

Keywords Koopman modeling · Koopman control · Quadrotor

S. Martini (✉) · M. Stefanovic · M. J. Rutherford · K. P. Valavanis
Ritchie School of Engineering and Computer Science, University of Denver, Denver, CO 80208, USA
e-mail: simone.martini@du.edu

M. Stefanovic
e-mail: margareta.stefanovic@du.edu

M. J. Rutherford
e-mail: matthew.rutherford@du.edu

K. P. Valavanis
e-mail: kimon.valavanis@du.edu

A. Rizzo
Politecnico di Torino, Torino, TO 10129, Italy
e-mail: alessandro.rizzo@polito.it

P. Livreri
Department of Engineering, University of Palermo, Palermo, PA 90128, Italy
e-mail: patrizia.livreri@unipa.it

D. Azimov (ed.), *Proceedings of the IUTAM Symposium on Optimal Guidance and Control for Autonomous Systems 2023*, IUTAM Bookseries 40,
https://doi.org/10.1007/978-3-031-39303-7_15

16.1 Introduction

Koopman Theory [7, 8], has recently gained momentum as a methodology to model and control complex nonlinear systems, such as robotic systems [1, 5, 6, 17] and quadrotors [11, 12, 21]. The main advantage of Koopman based modeling relates to representing nonlinear system behaviours as linear dynamics exploiting a (possibly infinite) set of state *observables*, which form the basis of the linear lifted space [4]. Hence, the resulting system representation allows for controlling the underlying system using well-known linear control architectures [19]. However, the biggest challenge is to find a suitable set of observables that is able to capture the full nonlinear dynamics of the system.

The most well-known approach found in literature, which is adopted to overcome the challenge of finding the right set of observables is based on data driven estimation of Koopman observables [2, 5, 20]. State of the art approaches employ deep learning and neural network architectures to identify a nonlinear set of coordinates based on which the resulting dynamic model becomes linear [11, 16]. All such methods rely on 'learning' or 'guessing' a proper finite set of observables that can capture the nonlinear system behaviours, however, the accuracy of the results depends heavily on the choice of the learned or guessed set. There is only limited reported research that focuses on directly deriving a set of observables in a systematic way [9, 21, 22].

In this paper a novel set of observables is introduced to transform the quadrotor nonlinear dynamics into a linear representation. Observables are derived in a systematic way, and they may be truncated to a finite subset of observables while maintaining an accurate representation of the full nonlinear dynamics of the quadrotor. The set is derived by mixing the equivalent quadrotor Newton-Euler dynamics in $SO(3)$ with the Euler-Lagrange formulation of position dynamics. The resulting set can achieve smaller relative error and it is considerably more compact than other reported formulations [9, 21] because it does not employ matrix vectorization and it can be truncated to a smaller set of observables while maintaining high fidelity of the nonlinear behaviour. The resulting formulation is then used to design and test a linear quadratic (LQ) controller.

The rest of the paper is organized as follows: In Sect. 16.2 the adopted notation is introduced along with a brief summary of Koopman theory, followed by derivation of the quadrotor nonlinear model. Section 16.3 presents the derivation of observables, and it shows how the resulting linear system can be used to design a controller. In Sect. 16.4 the fidelity of the derived lifted dynamics is analyzed with respect to the original nonlinear model, numerical simulations of a LQ controller are presented, and results are compared against literature studies. Lastly, conclusions are presented in Sect. 16.5.

16.2 Preliminaries

16.2.1 Notation

Let 0_n, I_n denote the $n \times n$ zero matrix and identity matrix, respectively, while $0_{n\times m}$ represents the non-square zero matrix with dimensions $n \times m$. Considering a matrix X, $X^\dagger$ refers to its Moore-Penrose inverse. Moreover, $blkdiag(X_1, ..., X_n)$ indicates the block diagonal matrix composed of the matrices $X_1, ..., X_n$. $SO(3)$ denotes the special orthogonal group in $\mathbb{R}^{3\times 3}$ such that [9]

$$SO(3) = \{R \in \mathbb{R}^{3\times 3} | det(R) = 1, R^T R = RR^T = I_3\}.$$

Given the vectors $a, b \in \mathbb{R}^3$, denote $a^\times b = a \times b$, where $a^\times$ is the 3×3 skew symmetric matrix composed of the elements of a. Last, $[i, j]$ defines the interval from i to j.

16.2.2 The Koopman Operator

A dynamic system may be described by a set of ordinary differential equations

$$\frac{dx(t)}{dt} = f(x), \tag{16.1}$$

where $x \in \mathbb{R}^n$ represents the state vector. The corresponding discrete time dynamics are given by

$$x_{k+1} = F(x_k), \tag{16.2}$$

where F is the mapping from x_k to x_{k+1}, and

$$F(x(t_0)) = x(t_0) + \int_{t_0}^{t_0+t} f(x(\tau))d\tau. \tag{16.3}$$

Given a countably infinite set of observables $\mathbf{b} = (b_1(x(t)), b_2(x(t)), ...)$ of the state vector, the Koopman operator is a infinite-dimensional linear operator that advances measurements linearly to the next 'snapshot' $(k + 1)$ in the observable space [13] as follows

$$\mathcal{K}\mathbf{b}(x_k) = \mathbf{b}(F(x_k)) = \mathbf{b}(x_{k+1}). \tag{16.4}$$

Since the Koopman operator $\mathcal{K}$ is possibly infinite dimensional, the goal is to find, within the countably infinite set of observables that span the linear lifted space, a finite collection of observables, usually named *lifted space*, which is able to approximate the nonlinear dynamic evolution of the system.

As such, the concept of the Koopman operator is adopted to model and evaluate performance of a quadrotor, which is a nonlinear underactuated system.

16.2.3 Quadrotor Nonlinear Dynamics

The Newton-Euler quadrotor dynamics in $SO(3)$ are presented as [15]

$$\dot{R} = RS, \tag{16.5}$$

$$J\dot{\omega} = M + SJ\omega, \tag{16.6}$$

$$\dot{p} = Rv, \tag{16.7}$$

$$\dot{v} = \frac{1}{m}T\mathbf{e}_3 - Sv - gR^T\mathbf{e}_3, \tag{16.8}$$

where $R \in \mathbb{R}^{3\times3}$ is the rotation matrix from body frame to inertial frame, p is the position vector in the inertial frame, $S = \omega^\times$ is the 3×3 skew symmetric matrix of the angular velocities ω, v is the quadrotor linear velocity vector in the body reference frame, $J \in \mathbb{R}^{3\times3}$ is the inertia matrix, M is the external torque produced by the propellers, m is the quadrotor mass, g is the gravitational acceleration, T is the total thrust, and $\mathbf{e}_3 = [0, 0, 1]^T$. By performing feedback linearization, (16.6) can be rewritten as

$$\dot{\omega} = J^{-1}\tilde{u}, \tag{16.9}$$

where $\tilde{u} = M + SJ\omega$ is the virtual control input. Moreover, the Euler-Lagrange formulation already introduced in [18] is

$$\ddot{\eta} = J_R^{-1}(M - C\dot{\eta}), \tag{16.10}$$

$$\ddot{p} = \frac{1}{m}TR\mathbf{e}_3 - g\mathbf{e}_3, \tag{16.11}$$

with J_R being the rotated inertia matrix and C the matrix accounting for centrifugal and Coriolis effects. $\eta = [\phi, \theta, \psi]^T$ is the vector of Euler angles and $p = [x, y, z]^T$ is the position vector. The position dynamics of both formulations are equivalent and they can be obtained from one another given the transformation $v = R^T\dot{p}$. Note that the quadrotor system is underactuated since it has four inputs, $u = [\tilde{u}, T]^T$, and six generalized coordinates $(x, y, z, \phi, \theta, \psi)$ to control.

16.3 Koopman Lifted Linear Dynamics

16.3.1 Derivation of Observables

A systematic approach to derive a set of observables that span the lifted space of the nonlinear quadrotor dynamics is introduced as follows

$$\mathbf{b} = (b_1, b_2, ...) = (\omega, \{g_k\}_{k=0}^{\infty}, \{v_k\}_{k=0}^{\infty}, \{p_k\}_{k=0}^{\infty}), \tag{16.12}$$

where $g_k = (S^T)^k g_0$, $v_k = (S^T)^k v_0$, $p_k = (S^T)^k p_0$ and $g_0 = R^T[0, 0, g]^T$, $v_0 = R^T \dot{p}$, $p_0 = R^T p$, for $k \in \mathbb{N}$. This selection of observables allows for writing higher order derivatives in a systematic way. This is verified by recursively deriving the state observables as presented next.

Derivation of $\mathbf{g_k}$

Since $\dot{R}^T = S^T R^T$, then

$$\dot{g}_0 = \dot{R}^T[0, 0, g]^T = S^T R^T[0, 0, g]^T = \underbrace{S^T g_0}_{g_1},$$

$$\dot{g}_1 = \dot{S}^T g_0 + S^T \dot{g}_0 = \underbrace{S^T g_1}_{g_2} + \dot{S}^T g_0,$$

$$\dot{g}_2 = \dot{S}^T g_1 + S^T \dot{g}_1 = \underbrace{S^T g_2}_{g_3} + S^T \dot{S}^T g_0 + \dot{S}^T S^T g_0.$$

A pattern emerges in this recursive procedure, which can be summarized as follows

$$\dot{g}_k = g_{k+1} + \sum_{i=1}^{k} (S^T)^{i-1} \dot{S}^T (S^T)^{k-i} g_0, \tag{16.13}$$

for $k \in [1, N_1]$.

Derivation of $\mathbf{v_k}$

When calculating the derivative of v_0, the Euler Lagrange dynamics (16.11) is substituted as follows

$$\dot{v}_0 = R^{\dot{T}}\dot{p} = \dot{R}^T\dot{p} + R^T\ddot{p} = S^T v_0 + R^T(\frac{1}{m}R[0,0,T]^T + [0,0,g]^T)$$
$$= \underbrace{S^T v_0}_{v_1} + \underbrace{R^T[0,0,g]^T}_{g_0} + \frac{1}{m}[0,0,T]^T,$$

$$\dot{v}_1 = \dot{S}^T v_0 + S^T\dot{v}_0 = \underbrace{S^T v_1}_{v_2} + \underbrace{S^T g_0}_{g_1} + S^T\frac{1}{m}[0,0,T]^T + \dot{S}^T v_0,$$

$$\dot{v}_2 = \dot{S}^T v_1 + S^T\dot{v}_1 = \underbrace{S^T v_2}_{v_3} + \underbrace{S^T g_1}_{g_2} + (S^T)^2\frac{1}{m}[0,0,T]^T + S^T\dot{S}^T v_0 + \dot{S}^T S^T v_0.$$

Similar to g_k, the resulting k-th term is written as

$$\dot{v}_k = v_{k+1} + g_k + \frac{1}{m}(S^T)^k[0,0,T]^T + \sum_{i=1}^{k}(S^T)^{i-1}\dot{S}^T(S^T)^{k-i}v_0, \quad (16.14)$$

for $k \in [1, N_2]$. Note that during the recursive procedure, besides the dependence on v_{k+1}, the previously derived observables g_k appear linearly.

Derivation of $\mathbf{p_k}$

$$\dot{p}_0 = R^{\dot{T}}p = \dot{R}^T p + R^T\dot{p} = \underbrace{S^T p_0}_{p_1} + \underbrace{R^T\dot{p}}_{v_0},$$
$$\dot{p}_1 = \dot{S}^T p_0 + S^T\dot{p}_0 = \underbrace{S^T p_1}_{p_2} + \underbrace{S^T v_0}_{v_1} + \dot{S}^T p_0,$$
$$\dot{p}_2 = \dot{S}^T p_1 + S^T\dot{p}_1 = \underbrace{S^T p_2}_{p_3} + \underbrace{S^T v_1}_{v_2} + S^T\dot{S}^T p_0 + \dot{S}^T S^T p_0.$$

Leading to

$$\dot{p}_k = p_{k+1} + v_k + \sum_{i=1}^{k}(S^T)^{i-1}\dot{S}^T(S^T)^{k-i}p_0, \quad (16.15)$$

for $k \in [1, N_3]$. Similar to the previous case, we can observe a linear dependence on both p_{k+1} and v_k.

By considering $N_1, N_2, N_3 \to \infty$, an infinite set of observables is computed, the linear evolution of which embeds the nonlinear dynamics of the quadrotor. Moreover, the truncated subset of observables is composed as

$$\mathbf{b} = (b_1, ..., b_N) = (\omega, \{g_k\}_{k=0}^{N_1}, \{v_k\}_{k=0}^{N_2}, \{p_k\}_{k=0}^{N_3}), \quad (16.16)$$

where $[b_1, b_2, b_3]^T = \omega$, $[b_4, b_5, b_6]^T = g_0$, $[b_{3+3N_1+1}, ..., b_{3+3N_1+3}]^T = v_0$, and $[b_{3+3N_1+3N_2+1}, ..., b_{3+3N_1+3N_2+3}]^T = p_0$ with $N = 3 + 3N_1 + 3N_2 + 3N_3$.

When compared to the approach presented in [21], position and attitude are associated with p_k and g_k, respectively. In particular, ϕ and θ can be obtained from $R^T[0, 0, g]^T$ while ψ is intrinsically obtained from ω. On the other hand, the position vector is computed as $p = Rp_0$. These transformations to the original generalized coordinates are defined for $-\frac{\pi}{2} < (\phi, \theta) < \frac{\pi}{2}$, which is the same condition of (16.10).

16.3.2 Koopman Linearized System

The complete set of equations governing the evolution of the system is

$$\begin{cases} \dot{\omega} = J^{-1}\tilde{u} \\ \dot{g}_0 = g_1 \\ \dot{g}_k = g_{k+1} + \sum_{i=1}^{k}(S^T)^{i-1}\dot{S}^T(S^T)^{k-i}g_0 \\ \dot{v}_0 = v_1 + g_0 + \frac{1}{m}[0, 0, T]^T \\ \dot{v}_k = v_{k+1} + g_k + \frac{1}{m}(S^T)^k[0, 0, T]^T + \sum_{i=1}^{k}(S^T)^{i-1}\dot{S}^T(S^T)^{k-i}v_0 \\ \dot{p}_0 = p_1 + v_0 \\ \dot{p}_k = p_{k+1} + v_k + \sum_{i=1}^{k}(S^T)^{i-1}\dot{S}^T(S^T)^{k-i}p_0 \end{cases} . \quad (16.17)$$

Considering the state vector and respective state and control matrices as

$$X = (\omega, g_0, \{g_k\}_{k=1}^{N_1-1}, v_0, \{v_k\}_{k=1}^{N_2-1}, p_0, \{p_k\}_{k=1}^{N_3-1}),$$

$$A = \begin{bmatrix} [0_3] & 0 & 0 & 0 \\ 0 & [A_1] & 0 & 0 \\ 0 & [I_{l_2} \; A_2] & 0 \\ 0 & 0 & [I_{l_3} \; A_3] \end{bmatrix}, \quad A_i = \begin{bmatrix} 0_{l_i^* \times 3} & I_{l_i^*} \\ 0_3 & 0_{3 \times l_i^*} \end{bmatrix},$$

$$l_i = 3N_i, l_i^* = 3(N_i - 1)$$

$$B(X)u = \begin{bmatrix} J^{-1}u \\ 0_3 \\ \sum_{i=1}^{N_1-1}(S^T)^{i-1}\dot{S}^T(S^T)^{k-i}g_0 \\ \frac{1}{m}[0, 0, T]^T \\ \sum_{i=1}^{N_2-1}(S^T)^{i-1}\dot{S}^T(S^T)^{k-i}v_0 \\ 0_3 \\ \sum_{i=1}^{N_3-1}(S^T)^{i-1}\dot{S}^T(S^T)^{k-i}p_0 \end{bmatrix},$$

since in $B(X)u$ the control appears linearly, the state dependent control matrix $B(X)$ can be obtained by using MATLAB to write the set of linear equations in matrix form using *equationToMatrix(* $B(X)u, u$ *).*

The resulting dynamics can be written as

$$\dot{X} = AX + B(X)u, \tag{16.18}$$

where $A \in \mathbb{R}^{N\times N}$, $B(X) \in \mathbb{R}^{N\times 4}$, $X \in \mathbb{R}^N$, and $u \in \mathbb{R}^4$. It is important to note that the unforced dynamics are linear, but the control matrix B is state dependent.

16.3.3 Koopman Control

Following a more or less similar approach as in [9, 21], for control purposes, the system is rewritten as

$$\dot{X} = AX + B^{\star}U^{\star}(X), \tag{16.19}$$

where $U^{\star}(X) = B(X)u$ and $B^{\star}_{N\times N} = blkdiag(I_3, 0_3, I_{l1}, 0_2, 1, I_{l1}, 0_3, I_{l1})$, which guarantees the system $(A, B^{\star})$ to be controllable. To obtain the actual control input, we solve the following least square optimization problem:

$$\textbf{min:}(B(X)u - B^{\star}U^{\star}(X))^T(B(X)u - B^{\star}U^{\star}(X)),$$

the solution of which is given by

$$u = B^{\dagger}(X)B^{\star}U^{\star}(X). \tag{16.20}$$

16.4 Simulations

In this section numerical simulations are presented to demonstrate the validity of the proposed approach. First, the quality of the approximation of the derived formulation is analyzed with respect to the quadrotor nonlinear dynamics. Second, the feasibility of one controller implementation using the Koopman linear dynamics is shown. Last, results are compared against existing literature.

16.4.1 Model Comparison

To test the Koopman model accuracy we provide both (linear and nonlinear) models with the same input and evaluate the relative error for different combination of N_1, N_2, and N_3. In view of comparing results with existing literature work, the input is chosen as in [21]:

$$u = 0.001 * \gamma(t)\sin(0.1t),$$

with $\gamma(t) \in [-5, 5]$. The initial condition are chosen as $\omega(0) = [0, 0, 0]^T$, $v(0) = [0.1, 0.1, 0.1]^T$, $R(0) = I_3$, and $p(0) = [3, 3, 3]^T$. Figure 16.1 shows the relative error computed as $|\mathbf{a} - \mathbf{b}|/\mathbf{b}$ where $\mathbf{a}$ is the vector variable obtained by integrating the linear dynamics in Koopman observables, while $\mathbf{b}$ is the respective variable computed by integrating the quadrotor nonlinear dynamics. The relative error is evaluated for multiple combination of N_1, N_2, and N_3 to show how the accuracy of the proposed formulation increases by allowing a larger number of observables. The notation $N^*_{i,j,k} = N$ refers to the total number of observables N obtained with the combination of $N_1 = i$, $N_2 = j$, and $N_3 = k$, with $i, j, k \in \mathbb{N}$. Note that $\dot{p}_{N_2-1}$ depends v_{N_2-1} and $\dot{v}_{N_1-1}$ depends on g_{N_1-1}, which are both affected by error due to the truncation. Therefore, to avoid further integration of error, the values N_i are often chosen so that $N_1 > N_2 > N_3$.

16.4.2 Controlled Dynamics

Here we present a numerical simulation of the Koopman controlled linear and non-linear dynamics. To this end, a linear quadratic (LQ) trajectory tracking controller is designed in the Koopman framework. The control action is obtained as the solution of an optimization problem, which aims at minimizing the following cost function [14]:

$$J = \frac{1}{2}\int_0^\infty (X^T Q X + U^{\star T} R U^\star)\, dt, \tag{16.21}$$

with Q, R design weight matrices. The solution of this well known optimization problem, for the linear model in (16.19), is given as

$$U^\star = -R^{-1} B^{\star T} K^\star(t) X(t) = -F^\star X(t), \tag{16.22}$$

where $K^\star$ is the solution of the algebraic Riccati equation (ARE)

$$0 = -K^\star A - A^T K^\star - Q + K^\star B^\star R^{-1} B^{\star T} K^\star. \tag{16.23}$$

The resulting controller is known as infinite horizon LQR controller.

To generate the desired trajectory p_d, $\dot{p}_d$, $\ddot{p}_d$, we use ninth order polynomial. Next, Eq. (16.11) is manipulated as in [23] to obtain the the following relations between the Euler angles, total thrust, and linear accelerations:

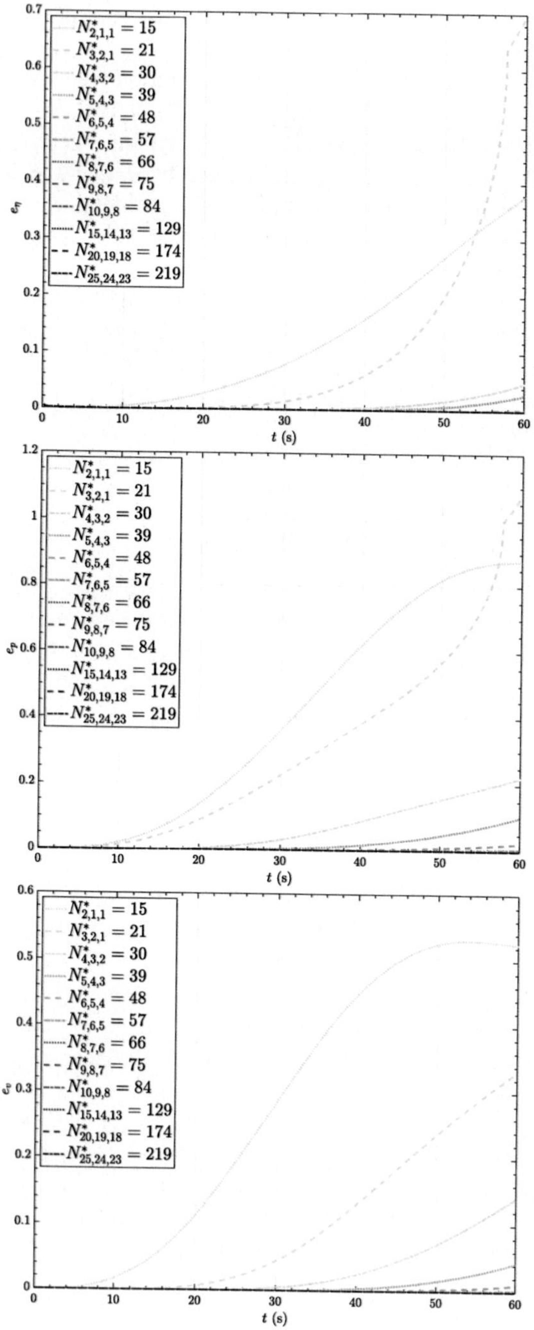

Fig. 16.1 Relative error between quadrotor nonlinear model and Koopman truncated linear model with different combination of N_1, N_2, N_3. The variables are η, p, and v, respectively

$$\theta_d = \arctan\left(\frac{U_1 \cos\psi_d + U_2 \sin\psi_d}{U_3 + g}\right), \tag{16.24}$$

$$\phi_d = \arcsin\left(\frac{U_1 \sin\psi_d - U_2 \cos\psi_d}{\sqrt{U_1^2 + U_2^2 + (U_3+g)^2}}\right), \tag{16.25}$$

$$\begin{aligned} T_d =& m[U_1(\sin\theta_d \cos\psi_d \cos\phi_d + \sin\psi_d \sin\phi_d) \\ &+ U_2(\sin\theta_d \sin\psi_d \cos\phi_d - \cos\psi_d \sin\phi_d), \\ &+ (U_3 + g)\cos\theta_d \cos\phi_d] \end{aligned} \tag{16.26}$$

where $[U_1, U_2, U_3]^T = \ddot{p}_d$ and $\psi_d = 0$. It is now possible to compute $R_d(\phi_d, \theta_d, \psi_d)$, and therefore, $g_{0,d} = R_d^T[0, 0, g]^T$, $p_{0,d} = R_d^T p$, $v_{0,d} = R_d^T \dot{p}$. The LQ controller is obtained by modifying the LQR formulation for trajectory tracking as in [3, 10], leading to the control input

$$u = -B^{\dagger} B^{\star} F^{\star}(X - X_d) + u_d, \tag{16.27}$$

where $X_d = [0_{3\times1}, g_{0,d}, 0_{l_1^*\times1}, v_{0,d}, 0_{l_2^*\times1}, p_{0,d}, 0_{l_3^*\times1}]^T$ and $u_d = [0, 0, 0, T_d]^T$.

The trajectory is generated so that the quadrotor will move from zero position and velocity to the desired position $p_d = [10, 10, 10]^T$ in 60 s with zero final velocity. However, to show that the controller is able to manage a large initial error, the actual starting position and velocity are $p_s = [3, 3, 3]^T$ and $v_s = [0.1, 0.1, 0.1]^T$, respectively. The design optimization matrix are chosen as $Q = blkdiag(I_{3+l1}, 10 \times I_2, 1, I_{l_2^*+l_3})$ and $R = I_N$. The resulting controller is applied to both the Koopman linear dynamic and the nonlinear quadrotor dynamics. As shown in Fig. 16.2, the Koopman based controller is able to reach the desired state for both systems.

It is important to state how the proposed controller is designed exploiting linear control theory, and yet, thanks to the Koopman framework of nonlinear observables of the state, the resulting algorithm can succesfully control the nonlinear quadrotor model. However, due to error integration in the proposed formulation, further analysis will be necessary test stability properties.

16.4.3 Comparison with Prior Methods

Compared to the set of observables presented in [21], our formulation is able to achieve smaller relative error in a considerably more compact form. In fact, as shown in Table 16.1, in our approach, a truncation at $N_{8,7,6}^*$, which result in $N = 66$, is able to achieve a better approximation of the nonlinear dynamics compared to the one in [21] for both $N_1 = N_2 = 25$ and $N_1 = N_2 = 15$, which result in $N = 479$ and $N = 289$, respectively. Moreover, for $N_{25,24,23}^*$, corresponding to $N = 219$, the relative error is four orders of magnitude smaller than the one obtained in [21] for a similar number of observables. The relatively great difference in error to observables number ratio is

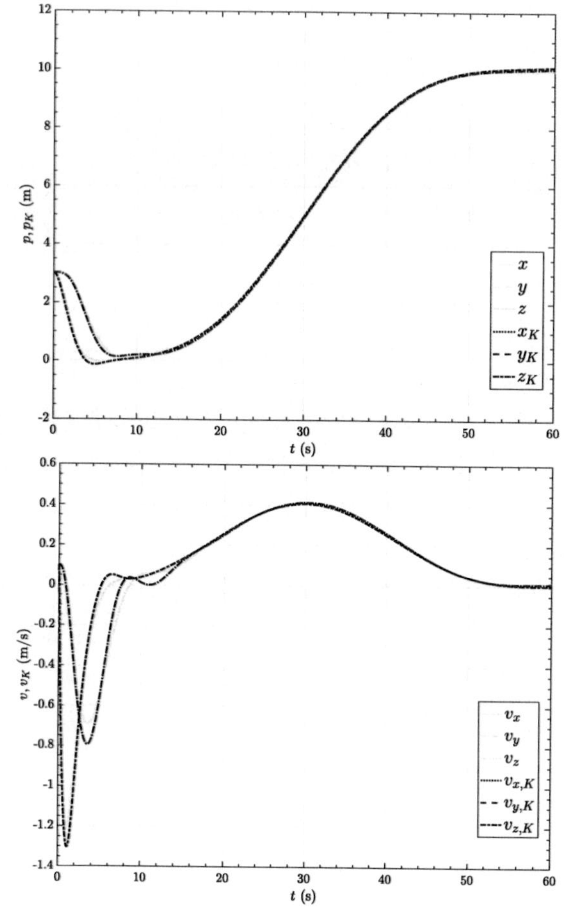

Fig. 16.2 Time evolution of position and linear velocity of the controlled quadrotor nonlinear dynamics (p, v) and the Koopman linear dynamics (p_k, v_k)

Table 16.1 Comparison with prior methods

Relative error	$N^*_{25,24,23}$	$N^*_{8,7,6}$	Method [21] $N_i = 25$	Method [21] $N_i = 15$
p	7.206×10^{-6}	9.888×10^{-4}	4.923×10^{-3}	1.697×10^{-2}
v	3.202×10^{-6}	3.341×10^{-4}	4.167×10^{-3}	1.893×10^{-2}
η	2.579×10^{-6}	5.850×10^{-5}	5.247×10^{-4}	2.457×10^{-3}

given by the fact that formulation in [21] requires 4×4 matrix vectorization for the observables associated with position and attitude, leading to $16k$ observables for each k iteration. In contrast, in this work, the observables associated to the position and attitude are p_k and g_k respectively, which generate $3k$ observables, each, per iteration.

16.5 Conclusions

A Koopman based linear dynamics model for a quadrotor UAV has been derived. A novel set of observables has been introduced, which spans the lifted space of the quadrotor nonlinear dynamics. Numerical simulations have showed that a truncated subset of observables may still embed most of the quadrotor nonlinear dynamics, and a higher number of observables is associated with an increasingly better approximation. Compared to existing studies, the resulting linear evolution of the presented observables is able to achieve a smaller mismatch to the original nonlinear dynamics in a much more compact formulation. An implementation and numerical simulation of a Koopman based controller obtained from the Koopman linearized quadrotor model is also presented, and preliminary findings show promising results.

Acknowledgements This research is conducted at the University of Denver Unmanned Systems Research Institute (DU^{2}SRI) in collaboration with the University of Palermo and Politecnico di Torino. P. Livreri and S. Martini are partially supported by Progetto "SEAVIEW", Azione 1.1.5, PO FESR 2014/2020. A. Rizzo is partially supported by the MOST—PNRR Next Generation EU. This manuscript reflects only the authors' views and opinions, neither the European Union nor the European Commission can be considered responsible for them.

References

1. Abraham, I., De La Torre, G., Murphey, T.D.: Model-based control using Koopman operators. arXiv:1709.01568 (2017)
2. Abraham, I., Murphey, T.D.: Active learning of dynamics for data-driven control using Koopman operators. IEEE Trans. Robot. **35**(5), 1071–1083 (2019)
3. Araar, O., Aouf, N.: Full linear control of a quadrotor uav, lq vs h∞. In: 2014 UKACC International Conference on Control (CONTROL), pp. 133–138. IEEE (2014)
4. Bevanda, P., Sosnowski, S., Hirche, S.: Koopman operator dynamical models: learning, analysis and control. Ann. Rev. Control **52**, 197–212 (2021)
5. Bruder, D., Gillespie, B., Remy, C.D., Vasudevan, R.: Modeling and control of soft robots using the Koopman operator and model predictive control. arXiv:1902.02827 (2019)
6. Bruder, D., Remy, C.D., Vasudevan, R.: Nonlinear system identification of soft robot dynamics using Koopman operator theory. In: 2019 International Conference on Robotics and Automation (ICRA), pp. 6244–6250. IEEE (2019)
7. Brunton, S.L., Brunton, B.W., Proctor, J.L., Kutz, J.N.: Koopman invariant subspaces and finite linear representations of nonlinear dynamical systems for control. PloS One **11**(2), e0150171 (2016)
8. Budišić, M., Mohr, R., Mezić, I.: Applied Koopmanism. Chaos Interdisc. J. Nonlin. Sci. **22**(4), 047510 (2012)
9. Chen, T., Shan, J., Wen, H.: Koopman-operator-based attitude dynamics and control on so (3). In: Distributed Attitude Consensus of Multiple Flexible Spacecraft, pp. 177–210. Springer (2022)
10. Cowling, I.D., Yakimenko, O.A., Whidborne, J.F., Cooke, A.K.: A prototype of an autonomous controller for a quadrotor UAV. In: 2007 European Control Conference (ECC), pp. 4001–4008. IEEE (2007)
11. Folkestad, C., Burdick, J.W.: Koopman NMPC: Koopman-based learning and nonlinear model predictive control of control-affine systems. In: 2021 IEEE International Conference on Robotics and Automation (ICRA), pp. 7350–7356. IEEE (2021)

12. Folkestad, C., Wei, S.X., Burdick, J.W.: Koopnet: Joint learning of Koopman bilinear models and function dictionaries with application to quadrotor trajectory tracking. In: 2022 International Conference on Robotics and Automation (ICRA), pp. 1344–1350. IEEE (2022)
13. Kaiser, E., Kutz, J.N., Brunton, S.L.: Data-driven discovery of Koopman eigenfunctions for control. Mach. Learn. Sci. Technol. **2**(3), 035023 (2021)
14. Kirk, D.E.: Optimal Control Theory: An Introduction. Courier Corporation (2004)
15. Lee, T., Leok, M., McClamroch, N.H.: Geometric tracking control of a quadrotor UAV on se (3). In: 49th IEEE Conference on Decision and Control (CDC), pp. 5420–5425. IEEE (2010)
16. Lusch, B., Kutz, J.N., Brunton, S.L.: Deep learning for universal linear embeddings of nonlinear dynamics. Nat. Commun. **9**(1), 4950 (2018)
17. Mamakoukas, G., Castano, M., Tan, X., Murphey, T.: Local Koopman operators for data-driven control of robotic systems. In: Robotics: Science and Systems (2019)
18. Martini, S., Sönmez, S., Rizzo, A., Stefanovic, M., Rutherford, M.J., Valavanis, K.P.: Euler-Lagrange modeling and control of quadrotor UAV with aerodynamic compensation. In: 2022 International Conference on Unmanned Aircraft Systems (ICUAS), pp. 369–377. IEEE (2022)
19. Mauroy, A., Susuki, Y., Mezić, I.: Koopman Operator in Systems and Control. Springer (2020)
20. Shi, L., Karydis, K.: ACD-EDMD: analytical construction for dictionaries of lifting functions in Koopman operator-based nonlinear robotic systems. IEEE Robot. Autom. Lett. **7**(2), 906–913 (2021)
21. Zinage, V., Bakolas, E.: Koopman operator based modeling for quadrotor control on se (3). IEEE Control Syst. Lett. **6**, 752–757 (2021)
22. Zinage, V., Bakolas, E.: Koopman operator based modeling and control of rigid body motion represented by dual quaternions. In: 2022 American Control Conference (ACC), pp. 3997–4002. IEEE (2022)
23. Zuo, Z.: Trajectory tracking control design with command-filtered compensation for a quadrotor. IET Control Theor. Appl. **4**(11), 2343–2355 (2010)

Chapter 17
On the Vertical and Horizontal Integration of Robotics Within Engineering and Computing Education

Vanessa Barth, Colleen Berg, Paulo Costa, Missy L. Cummings, William Denham, Robert Handler, Monson Hayes, Dinesh Karri, Nathan Kathir, Leigh McCue, Elise Miller Hooks, Cameron Nowzari, Colin Reagle, David Rosenblum, Violetta Rostobaya, Quentin Sanders, Daigo Shishika, John Shortle, and Emina Sinanovic

Abstract In this paper, we discuss the state of the art of robotics education in the US, the role of partnerships with area schools, cross-disciplinary research and education in robotics at the undergraduate and graduate levels, the importance of a holistic, multi-disciplinary perspective in the recruitment of faculty who facilitate this cross-disciplinary integration, and the future of engineering and computing education in robotics. Interwoven through the paper are case studies showcasing longitudinal and vertical integration of robotics education at Mason.

Keywords Robotics · Engineering education · Workforce development

17.1 Introduction

A recent white paper by Christensen et al., states: "Robotics has the potential to magnify the skilled workforce of the nation by complementing our workforce with automation: teams of people and robots will be able to do more than either could alone. The economic engine of the U.S. runs on the productivity of our people. The rise of automation offers new opportunities to enhance the work of our citizens and drive the innovation and prosperity of our industries. Robotics research creates new options and possibilities for robotics technologies to be explored, evaluated, and transitioned. Most critically, we need research to understand how future robot technologies can best complement our workforce to get the best of both human and

V. Barth · C. Berg · P. Costa · M. L. Cummings · W. Denham · R. Handler · M. Hayes · D. Karri · N. Kathir · L. McCue (✉) · E. M. Hooks · C. Nowzari · C. Reagle · D. Rosenblum · V. Rostobaya · Q. Sanders · D. Shishika · J. Shortle · E. Sinanovic
George Mason University, Fairfax, VA 22030, USA
e-mail: lmccuewe@gmu.edu

D. Azimov (ed.), *Proceedings of the IUTAM Symposium on Optimal Guidance and Control for Autonomous Systems 2023*, IUTAM Bookseries 40,
https://doi.org/10.1007/978-3-031-39303-7_16

automated labor in a collaborative team. Investments made in robotics research and workforce development will lead to increased GDP, an increased export–import ratio, a growing middle class of skilled workers, and a U.S.-based supply chain that can withstand global pandemics and other disruptions. In order to make the United States a leader in robotics, we need to invest in basic research, technology development, K-16 education, and lifelong learning" [1].

It is with this backdrop that we discuss the state of the art in robotics education in the United States, for example, universities such as Georgia Tech [2], Worcester Polytechnic Institute [3], Carnegie Mellon [4], and University Michigan [5] creating PhD programs, and in the case of Michigan, an entire department for robotics. At the same time, students arrive at a University as increasingly comfortable users of robotics, whether it is an Alexa in their home ("[i]n 2022, 62% of Americans 18 + use a voice assistant on any device" [6]), a robot cleaning their floor (14% of Americans own a robot vacuum [7]), or a drone they have flown (8% of Americans own a drone, and 15% of Americans report having flown one [8]). It is also common for American college students to have participate in robotics summer camps or school clubs during their K-12 years. Furthermore, whether knowingly or not, many students are exposed to ethical concerns associated with artificial intelligence, for example through tools such as ChatGPT opening new doors to opportunistic cheating [9, 10].

Cummings et al. found that "there is a significant gap between the education and research aims of academia and what is needed in industry and government" [11]. To fill this gap and maintain international superiority in aerospace autonomy, as well as autonomy in other domains, the US needs to promote the convergence in the fields of computer science and domain-specific engineering programs, as well as safety, cybersecurity, and testing. In this paper, we address ways to fill this gap, including the vertical and horizontal integration of robotics education in the engineering and computing curriculum at George Mason University. This integration spans grade levels and disciplinary boundaries. This includes active partnerships at the K-12 level with area institutions directly leading to a pipeline of students joining robotics clubs at Mason, as well as no-cost robotics camps to reach students who may not have access to these types of programs. At the undergraduate curricular level, we interweave collaborative course offerings, capstone projects, and student competitions that develop cross-disciplinary teams of students spanning Computer Science, Electrical and Computer Engineering, Mechanical Engineering, and Systems Engineering and Operations Research. The importance of cross-disciplinary emphasis has been specifically called out by industry [11] and is attractive to students. Our focus on interdisciplinarity results in the retention of many of our best and brightest undergraduates to stay at Mason for graduate research related to robotics, again often crossing traditional disciplinary boundaries from bioengineering to cybersecurity.

Lastly, robotics education provides unique opportunities for translation of university initiatives with respect to diversity, equity, and inclusion (DEI) into pedagogy. For example, it has long been established that facial recognition algorithms can exhibit demographic biases [12, 13]. A study that looked at facial recognition algorithms developed by U.S.-based researchers versus East Asian researchers concluded that "the performance of state-of-the-art face recognition algorithms varies as a joint function of the demographic origin of the algorithm and the demographic structure of the test population" [14]. Algorithms reflect the demographics of where they originate. Self-driving cars are another oft-discussed example of robotics engineering with tremendous potential for positive impact on society. At the same time, the models these vehicles use to identify pedestrians show inequities related to their ability to detect pedestrians with darker skin tones [15–17], or individuals in wheelchairs, mobility scooters, or wearing long skirts [17]. Likewise, algorithms used to automate decision processes across diverse applications, ranging from determining how to dynamically dispatch new public transit options or who should be approved for a home loan, suggest decisions with unintended biases, possibly disproportionately affecting access and economic wellbeing. Mason is one of the most diverse schools in the nation [18], and our strong focus on DEI is included at all levels of robotics education, which leads to building the diverse teams for the robotics solutions of the future.

17.2 K-12 Outreach

17.2.1 The Recruitment Pipeline

A unique benefit to Mason's geographic location with three campuses in Northern Virginia is a dense population of students and educators in our immediate region. Mason's Prince William campus is home to a so-called "Governor's School," a partial-day program for gifted and talented STEM students in the Manassas region. Students in this program have dual enrollment opportunities at Mason for introductory math, science, and engineering courses [19]. Mason also has a unique partnership with Northern Virginia Community College to provide a seamless, dual-enrollment pathway from community college to bachelor's degree [20].

17.2.2 Collaborations with Area Schools

To effectively leverage these relationships into vertically integrated pathways, we prioritize regular dialogue with area STEM educators to ensure we are providing offerings that are both desirable to the current generation of K-12 students and build upon those students' middle and high school experiences. For example, a recent STEM outreach effort at Mason, supported by the Office of Naval Research, sought to develop a kit and standards-aligned curriculum for high-school aged learners in lighter-than-air, biologically inspired, robotics. A key component to that effort was the recruitment of focus group participants, many of whom were area educators, whose feedback was solicited to inform curricular revisions [21]. These purposeful connections to area educators contextualize the multi-disciplinary efforts we are making in robotics education at Mason, as described by the following first-person student experience.

17.2.3 Student Experience: William Denham

As a student in local area schools myself, I was able to discover a career field that I am truly passionate about through my participation in competitive robotics programs. The programs allowed me to find my passion for STEM and led me to pursue a degree in electrical and computer engineering at George Mason University. At George Mason, I was provided with the STEM professionals and company connections to form a competitive robotics team known as Scitech Robotics. This team was created for students such as myself to interact with our coursework in a way that is both interactive and fun. The Vex Robotics University Competition [22] releases a new challenge every year, where student engineers from colleges and universities across the world are challenged to compete against one another, with the best team earning the prestige and honor of being the Vex Robotics World Champion. This group has been very successful in the past few years both on the competition field and each of our members successful in the classroom as well. The program has become known as a platform for engineering excellence in everything that we do, being recognized globally for our efforts and successes. Something that would not have been possible without the STEM Mentors and Educators that we were able to learn and grow with. Now that I have neared the end of my degree completion, I have accepted a position with the local school division, Prince William County Schools as a Robotics Coach and Educator, directly inspiring and helping students find their niche in STEM. As someone who always had strong mentors and teachers to allow me to find, light up, and guide me along my path, it only felt natural to be that mentor and educator giving back to today's students.

Through K-12 STEM outreach in local area schools, I have been able to form meaningful relationships with K-12 students and provide them with exposure to STEM, through competitive robotics. I have witnessed firsthand as my students have become more confident and comfortable in mechanical design, technical documentation, and software development, as a direct result of their participation in robotics programs such as the Vex Robotics Competition. These programs are designed to be an interactive, engaging way to learn seemingly difficult technical skills in a nurturing, and safe environment. These students have been able to find a passion in the robotics field and many of them have decided to pursue a robotics-related undergraduate degree, whether in mechanical engineering, electrical engineering, or computer science. If we're able to harness the students' innate curiosity for solving problems, and critical thinking skills; we can create a talent pipeline of individuals entering the STEM workforce. In the end, attracting more students to STEM comes down to forming relationships, inspiration, and skill.

17.2.4 Camps

To further reinforce the connections between area students, educators and Mason's robotics programming, we have begun to offer robotics camps leveraging RoboNation's SeaPerch and SeaGlide kits [23] for age-appropriate curricula targeting middle school and high school aged learners respectively. These camps provide hands-on maritime robotics experiences coupled with exposure to undergraduate and graduate level research on Mason's Prince William campus, and are offered at no cost to campers, in an effort to provide this opportunity to students who may not otherwise be able to afford such experiences. For further detail, the interested reader is referred to [24].

17.3 Undergraduate Education

From the first AUVSI aerial robotics competition held at Georgia Tech in 1991 to a team of Rutgers students successfully sending an underwater glider across the Atlantic in 2009, the last three decades have seen rapid advancement in the implementation of autonomy into undergraduate education. Figures 17.1 and 17.2, reproduced from [25] and discussed further in [26] provide brief timelines highlighting key moments for academics in aerial and maritime robotics. Notably, as a young academic institution, Mason's College of Engineering and Computing was born just prior to this revolution in autonomous systems. The College of Engineering and Computing was founded in 1985 as the School of Information Technology and Engineering [27]. Some departments within the College date back to those origins, such as the Department of Electrical and Computer Engineering, whereas others are quite young, such as Mechanical Engineering which was formally established in 2015. All departments within the College have grown out of this era of autonomous systems, and in that regard, have developed curricula in which robotics and mechatronics are as foundational as fluid dynamics and circuits.

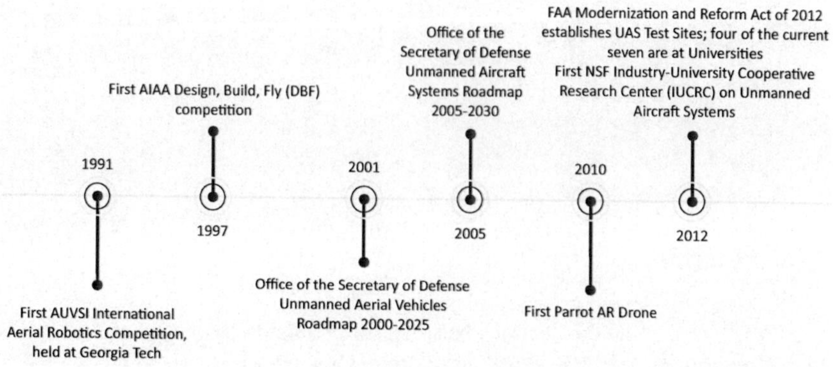

Fig. 17.1 Selected key moments in academic unmanned aerial vehicle history

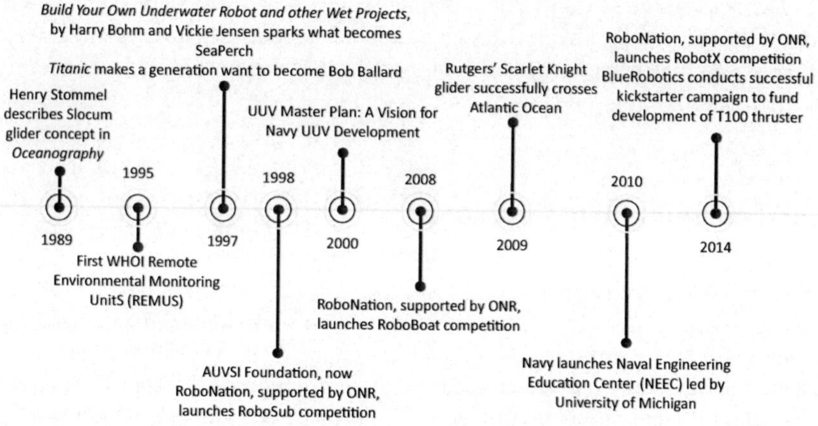

Fig. 17.2 Selected key moments in academic unmanned marine vehicle history

17.3.1 Example Curricular Revisions to Support Undergraduate Education in Robotics

The Mechanical Engineering department identified and implemented three changes that have better prepared students to practice robotics. In 2018, Mechanical Engineering worked together with Electrical and Computer Engineering to revise a two-course sequence in DC and AC circuits into a combined course, ECE 330—Circuit Theory. This gave students the background for more advanced courses in Systems Dynamics and Controls and freed up 3 credit hours to implement the second change, a required course ME 331—Mechatronics. This hands-on course focuses on sensors, microcontrollers, and actuators and is intended to be taken right before senior design commences. The final change was in the 2017 development and subsequently tweaking of ME 151—Practicum in Engineering. This replaced a more general Intro to Engineering course with one focused on algorithmic thinking and solid modelling. This required every student in the Mechanical Engineering program to use code to solve engineering problems and CAD software to create models before beginning their upper-level coursework.

17.3.2 Mechanical Design

Senior design project, also known as capstone, is a curricular requirement for the senior students in engineering programs at Mason in support of ABET's requirement for "a culminating major engineering design experience that (1) incorporates appropriate engineering standards and multiple constraints, and (2) is based on the knowledge and skills acquired in earlier course work" [28].

While Mason's Department of Mechanical Engineering was established in 2015, as early as the 2016–2017 academic year, robotics related projects were provided as capstone design opportunities for students. Robotics and mechatronics projects have included sponsor driven activities, such as developing control systems for motorcycle assistance or building robots to inspect confined spaces for corrosion, storm sewer blockage, or pipeline coating status. And similarly, multidisciplinary projects have facilitated participation in competition, be it maritime robotics focused or toward the development of novel unmanned aerial vehicles. Details on specific capstone experiences can be found in annual brochures posted online which highlight the multidisciplinary student collaborations and faculty mentorship provided for these teams [29].

As robotics at Mason continues to grow, we have found these types of corporate or competition driven culminating design experiences provide students the unique and valuable opportunity to work on multidisciplinary teams drawing from students and faculty across the College of Engineering and Computing. For those students interested in pursuing careers in robotics, this provides a unique opportunity to broaden their exposure beyond the coursework core to their primary discipline. For all participating students this emulates the work environment in which multidisciplinary perspectives are key to successful engineering design and innovation.

17.3.3 Student Experience: Emina Sinanovic

I was first introduced to robotics when I joined For Inspiration and Recognition of Science and Technology (FIRST) Robotics Team during high school in 2017. At FIRST, I joined the software team and extended my involvement to the mechanical and electrical subteams. I realized I enjoyed my experience being part of the software team. Because of this experience and my enjoyment of robotics, I decided to pursue my Bachelor's Degree in Computer Science at George Mason University (GMU). Due to my demanding coursework, I was unable to grow my interest in robotics further because of time constraints.

When I came to GMU in 2020, I found a robotics club called BOT Robotics that I was interested to be a part of. Unfortunately at the time, they did not have a software component because it was geared towards the engineering students. I was curious of other opportunities that would include a software aspect to it, and a student suggested that I reach out to Dr. Cameron Nowzari to be part of a Capstone project. My experience with the Capstone project was amazing. I enjoyed my time with my teammates quite a lot. We collaborated and brainstormed ideas to establish a functional drone and an accompanying camera. This was a lot of work, but extremely rewarding in the end. The software aspect of the team was exciting: it was my first time working with a camera, and there was a limited amount of memory afforded for both the camera and software to fly the drone. Figuring out how to work with them both and ensuring that they would not interfere with each other was challenging. I am forever grateful that I was given this opportunity to be part of this project and to be part of the team.

17.3.4 Extracurricular Activities and Undergraduate Research

The following subsections provide brief examples of characteristic extracurricular and undergraduate research opportunities for Mason students across multiple disciplines interested in Robotics.

17.3.4.1 SciTech Robotics

As referred to previously in the student experience commentary by William Denham in Sect. 16.2.3, SciTech Robotics is a student-organized robotics club formed to participate in VEX-U robotics competition. While the specific rules are updated every year, the competition usually requires students to build and program robots that can search objects in the game arena and interact with them (e.g., pick them up and throw into a goal). Such game set ups naturally lead the students to face various challenges in robotics and autonomy including mechanical design, controls, computer vision, programming etc. The competition also includes a technical report that prepares the students on scientific writing. The current team consists of about ten students across Mechanical Engineering, Electrical and Computer Engineering, and Computer Science departments complementing each other to tackle various challenges listed above. Since VEX-U provides its own set of modular components (e.g., microcontroller, sensors, and actuators) the students are able to focus their attention on the overall system design without extensive domain knowledge. This trait makes the robotics club to have low entry point and allows interested freshman and sophomore to dive into robotics research.

17.3.4.2 Patriot Pilots

Another program designed to integrate all types and levels of skills are the Patriot Pilots: a team of undergraduate and graduate students led by 3 faculty members in Electrical and Computer Engineering and Mechanical Engineering participating in a twice-a-year national Lighter-Than-Air (LTA) aerial robotics competition. The 6th event is planned to be hosted for the first time at Mason April 24–28, 2023. There will be at least 6 teams including the University of Florida, Virginia Tech, Baylor University, West Virginia University, Lehigh University, and George Mason University. Each team will have a "How-It-Works" table showing our undergraduate students how all the robots work at our student celebration day Mason Day.

17.3.5 Student Experience: Violetta Rostobaya

Multidisciplinary approach to robotics is vital for the future of engineering and computer science education. As an undergraduate mechanical engineering student, I expressed interest in control theory and system dynamics, which lead me to join a research group. Being an undergraduate member of the lab seeded the idea of choosing academia as a career pathway. Participating in reading groups along with graduate students from different fields let me learn more about their experiences with robotics and graduate education. Consistent participation and socialization only solidified my further decision to apply for a PhD program.

During my time in the lab as an undergraduate student I have also expanded my knowledge of controls, robotics and, surprisingly, game theory. Early outreach by faculty members allowed me to have an early head start in research and planning for graduate program. Being part of robotics research community also showed me robotics is not just an engineering science in its most traditional way, but as a constellation of sciences coming together to explore technology through different approaches and methods.

17.4 Graduate Education

At the graduate level, we have deliberately aimed to develop multi-disciplinary pathways for students seeking to advance their knowledge in robotics and autonomous systems, be it through bachelor-accelerated masters offerings, cross-disciplinary PhDs, or introducing students to cutting edge topics like translational AI and current research.

17.4.1 Bachelors-Accelerated Masters Programs

The College of Engineering and Computing offers 63 Accelerated Master's options. The main benefit for students is the ability to double count up to 12 credits towards both their BS and MS degrees allowing completion of both degrees in as little as 5 years. While many of these are options are for completing a Master's degree in the same field as a Bachelor's degree, some allow students to earn a Master's degree in a different discipline. Mechanical Engineering established 8 such programs as an option for students. It leverages the broad background of an ME and gives them a path to MS degrees in other robotics related fields such as Bioengineering, Electrical Engineering, and Systems Engineering.

17.4.2 Cross-Disciplinary PhDs

In addition to 'traditional' pathways to PhDs that can be autonomy or robotics intensive via the Departments of Computer Science, Electrical and Computer Engineering, or Bioengineering, Mason has also created multiple interdisciplinary pathways that graduate students may pursue. Two of note include options for Mechanical Engineering concentrations to either an Information Technology PhD or a Systems Engineering and Operations Research PhD. As described in Mason's catalog, "The Information Technology (INFT) PhD is a signature degree of the College of Engineering and Computing. The program focuses on the science, engineering, and technology of information processing, an area of study ripe for innovation in a world driven more and more by data. The PhD in INFT accommodates rigorous and cross-disciplinary PhD study that does not fit with PhD program requirements of a single VSE department" [30]. Similarly, the Mechanical Engineering concentration to the Systems Engineering and Operations Research PhD allows graduate students to pursue a degree emphasizing mechanical engineering systems, such as robotic systems.

That said, we have also begun formulating a truly multidisciplinary PhD serving the needs of learners from industry, government, or academic careers in mechatronics, robotics, or automation engineering. Foundational to the degree would be a core that ensures fundamental disciplinary knowledge in computer science, cybersecurity engineering, electrical and computer engineering, and mechanical engineering. While students would then specialize in their specific subdomain, such a program would ensure that a mechanical engineering working in prosthetics, for example, has a fundamental understanding of how to cyberharden a system to prevent malicious attacks, a computer scientist developing cybersecure software understands the physical implementation requirements of mechanical systems, or that an engineer working in construction understands the role of robotics in automation. It is this multidisciplinary perspective that we see as vital long-term to creating the graduates who will develop new autonomous systems for defense applications, design human–computer teaming solutions for the supply chains of the present and future, manage the health of large civil infrastructure, deploy autonomous systems post-disaster to support community recovery, and create robotic interventions for the medical industry.

17.4.3 Student Experience: Vanessa Barth

My experience at George Mason University started as an undergraduate student in the department of Mechanical Engineering where I was exposed to the foundations of robotics, namely computer programming, circuit theory, control theory, dynamics, and mechatronics. One of my favorite undergraduate courses was aeronautics, although not a robotics course, it was what led to my interest in unmanned and autonomous systems and desire for a future career in academia. A timely opportunity came about to stay at Mason for graduate school and conduct research in maritime robotics, with a focus on unmanned and autonomous systems, under Dr. Leigh McCue. I jumped at the chance to stay at a school I knew I would thrive at and be a part of exciting research. Recently, my work involving gaze control of an underwater remotely operated vehicle was successful during in-water demonstrations, and I am now extending my gaze control method to surface boats.

Beyond school and research, I had the opportunity to take part in the 2022 RobotX competition in Sydney, Australia with a team comprised of Mason students, alumni, and faculty advisor. This was a great experience that allowed me to use what I've learned in my education and research to achieve tasks using a drone. In my research, I have used serial communication to send messages from my computer to a vehicle and vice versa. I used this prior knowledge during the competition to achieve the task of using sockets to send messages to a server. In the end, although our drone crashed due to high winds, we walked away proud that we had made it into the finals for the search and rescue task. Being a part of this team proved to me once more that robotics is a blend of different disciplines and would not exist if one were missing.

17.4.4 Translational AI

While the rise of artificial intelligence (AI) across many domains has been wide and swift with many positive benefits, increasingly companies and government agencies are learning that there is often significant drawback to this technology. A major limitation to the deployment of systems with embedded AI is a lack of understanding of foundational principles in the design and development of such systems, especially in a safety–critical context. Often, agencies and companies focus on integrating AI into hardware or traditional rule-based software systems and struggle to adequately consider the strengths and weaknesses of AI and user interactions, as well as testing and policy implications. They also have difficulty in adapting their work practices to adequately consider the profound changes that embedding nondeterministic AI software have on both products and work processes, as well as workforce development.

A recent survey of companies and government agencies investing in autonomous systems ranked their top research and development areas as (1) safety, (2) cybersecurity, (3) system integration, (4) certification, and (5) human-autonomous system interaction. Unfortunately, this same survey showed that most of these topics were not important in the eyes of senior academics [11]. Thus, there is a gap in the skill sets industry needs and those courses available to students that provide them with the necessary multidisciplinary background to translate AI into actual systems that have the capacity to significantly impact human welfare.

To help fill this gap, Mason is considering development of graduate programs in translational AI, which will integrate foundational elements of AI (such as machine learning, optimization, and computer vision) in a risk-based systems framework. The goal is to develop practitioners that can leverage a first-principles approach to developing and implementing AI in real-world applications, and also understand holistically the strengths and weaknesses of such systems. Students will focus on constructing and deploying AI-based systems that are safe, trustworthy and equitable. The vision includes two tracks in engineering and public policy (including law and ethics), but these students will form cohorts so the two groups can learn from one another. The goal is to provide returning professionals with an updated skill set so that they can understand how AI-enabled systems meet requirements, the unique design and deployment constraints of such systems, and also are ready to tackle the future of work issues that AI-enabled systems raise.

17.4.5 Seminar Series

Departmental seminar series provide opportunities for cross-disciplinary idea pollination with discipline-specific application. For example, the Department of Civil, Environmental and Instructure Engineering has hosted speakers with expertise in connected and autonomous vehicles, structural health monitoring via drones, construction automation, and trusted AI systems for transit, which elucidate the role of automation, robotics and AI in supporting our infrastructure and infrastructure systems. In supplement to department-level activities, in the fall semester of 2022, Mason launched a new seminar series that invites internal and external speakers conducting research in the field of robotics and controls. The main motivation behind the initiation of this series was to provide a central place where robotics-related faculty and their students from all departments could regularly meet. The weekly seminars are contributing to the horizontal integration of robotics research in the College of Engineering and Computing. The technical talks, followed by a Q&A session and casual discussion over lunch, are providing opportunities for researchers from Computer Science, Electrical and Computer Engineering, Mechanical Engineering and other departments to start new collaborations beginning from a simple exchange of ideas to development of joint grant proposals.

17.4.6 Student Experience: Dinesh Karri

Robotics has become an integral part in the society in the past few decades. I got interested in robotics as a kid playing with several toys and decided to pursue my degrees in this field. Unfortunately, most universities, especially in India, where I did my undergrad from, overlooked the importance on building interdisciplinary skills on development of robotic systems and how it incorporates different fields of engineering like computer science, mechanical, electrical engineering, electronics, and communication engineering.

Defend the Republic is a research-based competition between multiple research groups to play a game of soccer with multiple balls in adversarial 3D environments autonomously. But to accomplish this the challenge took place in stages. The First stage was to create a system which can be tele-operated to accomplish the given tasks and gain intuition on how the autonomy should look like. Joining Defend the Republic (DTR) project 3 years ago where we had to build a teleoperated system to participate in this competition, I realized how important it is to develop skills in the above stated fields to build one single robotic system. I started the project as a master's student under the guidance of the professors where I built a robot from the ground up. Me being a controls engineer, I did not have much knowledge in the mechanical, computer science and communication aspects of it. But in this project, I got to work with several PhD students from different disciplines to learn and brainstorm ideas for the system which I was successfully able to build, test, and deploy.

Then as a PhD student I lead a group of students when the rules of the competition moved toward autonomy. In the team, we had PhD, masters, undergraduate students and highschoolers with backgrounds in electrical, computer, embedded systems, mechanical, and computer science who became integral parts of the development of the system.

In projects like this, everyone on the team, irrespective of type of engineering background, have a lot learn and a lot of gain from by collaborating and making decisions on the future career choices. Most of the students who worked with us, have either made career choices with respect to the newfound interest in robotics and others have opted to perceive academic paths in robotics to be a part of this interdisciplinary field. This was helpful even more so for high school students to get a taste of robotics hands-on experience which sparked interest in pursuing the field.

17.5 Faculty Recruitment and Development

17.5.1 Faculty Recruitment

Robotics is a thoroughly cross-disciplinary area of technology, requiring expertise in a large variety of both STEM and non-STEM disciplines. For a university to achieve a suitable critical mass of activity in both research and education in robotics, it is necessary for the university and its relevant departments to make significant investments in recruitment of faculty across the different disciplines of robotics. But recruiting outstanding faculty is not enough. In order to fully realize the significant benefits and impacts such a critical mass of faculty can achieve, it is necessary for the leadership within the university at all levels to be fully invested in the spirit of cross-disciplinary collaboration and to make a commitment to sustain it over a long period of time.

While many faculty recruits in robotics will specialize primarily in some area of a "home discipline," such as mechanical engineering, computer science, or cognitive psychology, many newly-minted PhDs nowadays come equipped with substantial cross-disciplinary expertise of their own. Universities that are experienced with large cluster recruitments or are effective at leveraging joint appointments of faculty in multiple departments can build up a critical mass of faculty in robotics in a short period of time. Other universities may prefer to make gradual additions to faculty as budgets permit, while also recruiting existing faculty into nascent efforts around robotics. In the following subsections two recent robotics hires, one junior and one senior, provide their personal insight.

17.5.2 Faculty Experience: Quentin Sanders

Prior to joining Mason, I had the opportunity to work at a Silicon Valley based company that prided itself on innovation through interdisciplinary collaboration. Our project in particular was focused on developing a robotic exoskeleton to assist people with mobility challenges. The project by nature is inherently interdisciplinary, but it is quite common to find teams working on similar projects composed solely of engineers. This often leads to solutions that achieve a goal (i.e. technological innovation) but do not meet the need of the user. The composition of our team consisted of biomechanics specialists, roboticists, apparel designers, and user experience researchers. It is the interdisciplinary nature of our team that allowed us to develop a novel robotic device that's performance well exceeded current state of the art devices. I have seen how interdisciplinary collaboration can lead to highly impactful devices, and this is especially true in the area of robotics. I was attracted to Mason because they sought to create a similar culture of interdisciplinary collaboration. I think this is reflected by the University's willingness to hire me into a joint appointment position in the bioengineering and mechanical engineering departments.

17.5.3 Faculty Experience: Missy Cummings

Professor Mary "Missy" Cummings, a recent new hire to George Mason, is an example of Mason's growing focus on interdisciplinarity. She was appointed jointly across three departments, Mechanical, Electrical and Computer, and Computer Science. While two-department joint hires routinely occur at other universities, three is unusual, but critical according to Missy. Robotics is inherently interdisciplinary, with no challenge belonging to any single domain. Students and junior faculty want the ability to work across departments and even colleges, so breaking down barriers between departments, schools and colleges will be critical for successful robotics workforce development. Mason's youth as a university allows it to be more agile and less mired in stale traditions that current-day students find archaic. Indeed, industry shares this same sentiment, so interdisciplinarity is the key to recruiting and producing the kind of talent industry (and academia) need.

17.6 Conclusions

Taken on a point-by-point level, these efforts are not unique. Taken in whole, however, this unified effort leads to comprehensive, multi-disciplinary education in robotics facilitating pathways for students from K-12 to PhD covering northern Virginia, including some of the largest school districts in the Commonwealth of Virginia (181,000 students in Fairfax County Public Schools [31], which is the 11th largest school district in the United States and largest in Virginia [32], 89,468 students in Prince William County Schools [33], which is the 34th largest school district in the United States and second largest in Virginia [32, 33], and 27,586 students in Arlington Public Schools [34]) coupled with being a pathway to bachelor's degrees for students beginning their college education at Northern Virginia Community College (the second largest community college in the United States [35]). For reference, the locations in which Mason's Bioengineering (BENG), Civil, Environmental, and Infrastructure Engineering (CEIE), Computer Science (CS), Cyber Security Engineering (CYSE), Electrical and Computer Engineering (ECE), Mechanical Engineering (ME), Systems Engineering and Operations Research (SEOR) departments offer activities at each of these levels is summarized in Fig. 17.3—emphasizing the strong geographic reach of Mason throughout Northern Virginia and specifically in Fairfax, Prince William County (SciTech Campus), and Arlington. This figure represents a snapshot at the time of this writing. Departmental presences on all three campuses is dynamic and continues to grow with the growth of the University at large.

This geographic reach, coupled with horizontal integration across disciplinary domains and vertical educational integration from K-12 to faculty levels, is an enabler for meeting the industry need for a growing workforce in this field. Looking to a few industry examples in particular, we note that automation and robotics in warehouses and the supply chain are quickly transforming the industry and becoming a major economic driver. "Robotic warehouse automation is currently a small portion

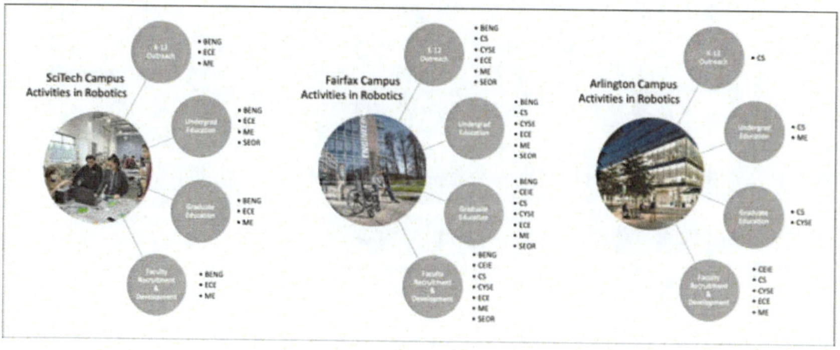

Fig. 17.3 Mapping of departmental involvement in K-12, undergraduate, graduate, and faculty development activities in robotics across Mason's three Virginia campuses

of overall warehouse automation spending, but it is growing rapidly. Advisory firm Interact Analysis predicts the mobile robotics market could grow from approximately $3.6 billion in 2021 to $18 billion in 2025. The number of warehouse facilities with mobile robots is also projected to expand from 9000 in 2020 to 53,000 in 2025. In addition, Interact Analysis forecasts a total of 2.1 million robots in service by the end of 2025 with 860,000 of these shipped during that year alone. Interact Analysis includes mobile robots as a component of the broader warehouse automation market, which it expects to grow from $49.6 billion in 2020 to $69 billion in 2025. Fixed automation systems such as automated storage and retrieval systems (AS/RS), conveyor belts and sorters will eclipse other forms of automation for the foreseeable future" [36]. Looking specifically to Amazon, "[t]oday, Amazon has more than 200,000 mobile robots working inside its warehouse network, alongside hundreds of thousands of human workers.... This robotics race—led by Amazon—will have a seismic impact on the warehouse industry, which employs more than 1.1 million Americans today" [37].

Furthermore, continued development of autonomous and remotely operated vehicles remains a growth sector, as noted in a Businesswire report which states: "The global market for Offshore AUV [autonomous underwater vehicle] and ROV [remotely operated vehicle] estimated at US$3.3 Billion in the year 2020, is projected to reach a revised size of US$7.2 Billion by 2026, growing at a CAGR of 14.3% over the analysis period" [38]. The report goes on to describe the contributions of these types of robotic and autonomous systems toward human safety, science, defense, energy, infrastructure, and food by stating: "The rising use of ROVs in a variety of industries, including search and rescue, marine biology, military, oil and gas, submerged infrastructure, and aquaculture, as well as advancements in ROV technology, is anticipated to drive growth in the segment. The need for big AUVs for military and defense applications, as well as oil and gas exploration, is driving the expansion of the AUV market segment" [38].

As another industry sector experiencing rapid growth, surgical robots have in relatively short time gone from science fiction to practice. "Within 20 years since the regulatory approval of the first surgical robot, more than 3000 units were installed worldwide, and more than half a million robotic surgical procedures were carried out in the past year alone. The exceptionally high speeds of market penetration and expansion to new surgical areas had raised technical, clinical, and ethical concerns. However, from a technological perspective, surgical robots today are far from perfect, with a list of improvements expected for the next-generation systems. On the other hand, robotic technologies are flourishing at ever-faster paces. Without the inherent conservation and safety requirements in medicine, general robotic research could be substantially more agile and explorative. As a result, various technical innovations in robotics developed in recent years could potentially be grafted into surgical applications and ignite the next major advancement in robotic surgery" [39].

Particular opportunity for robots and autonomous systems for economic impact has been highlighted by the COVID-19 pandemic; "The COVID-19 pandemic has revealed many needs that could be solved in part through the use of robotics, with sufficient investment in research, development, and job training. For example, in

2017 hospitality was a $1.6 trillion sector, one that now has additional needs for disinfection, cleaning, maintenance, and socially distanced services. Similar needs are present in healthcare to provide safety to healthcare workers and patients alike. In agriculture, robotics could provide assistance where there is now a lack of seasonal workers to harvest crops. Robotics can address the supply chain issues that were revealed by COVID-19, including the need to manufacture critical PPE and other necessary items in the U.S. Autonomous trucks could increase the efficiency of our supply chain while also increasing road safety" [1]. Research and Markets reports that in 2022 the global logistics robotics market was valued at $15.2 billion with a prediction for this to grow to $58.6 billion in 2028 [40]. They go on to report, "An increase in the number of logistics and warehousing companies that are incorporating robots to improve speed and efficiency and remain competitive in the market are propelling the demand for logistics robots worldwide. Moreover, the adoption of advanced technologies, such as robotic warehousing and logistics technologies, is growing on account of the sudden outbreak of the coronavirus disease (COVID-19) and the consequent lockdowns imposed by governments of various countries to prevent the transmission of the pandemic" [40]. In a market analysis by Astute Analytica, they note that "Covid-19 triggered the deployment of food delivery robots when social distancing and human interface became difficult and avoidable" while forecasting that "The Global Indoor Delivery Robots Market is anticipated to witness a major jump in its revenue from US$ 16.63 Mn in 2021 to US$ 253.76 Mn by 2030" [41].

It is with this common understanding of industry workforce need that we highlight the critical importance of the horizontally and vertically integrated educational structure described here, weaving together K-12, undergraduate, graduate, and faculty activities in robotics and autonomous systems, across multiple departments. This may provide an outline for institutions working to develop a similarly multi-disciplinary institutional structure to support robotics research and education.

Acknowledgements The authors of this paper gratefully acknowledge support for programs referred to in this paper from multiple sources, including NSF grant CMMI 2135619 and ONR grant N00014-21-1-2968. Any opinions, findings, and conclusions or recommendations expressed in this material are those of the author(s) and do not necessarily reflect the views of the National Science Foundation or the Office of Naval Research.

References

1. Christensen, H., Gini, M., Jenkins, C., Yanco, H.: Robotics enabling the workforce. https://cra.org/ccc/resources/ccc-led-whitepapers/#2020-quadrennial-papers
2. Georgia Tech: Ph.D. program in robotics. https://research.gatech.edu/robotics/phd-program-robotics. Accessed Feb 2022
3. WPI: PhD in robotics engineering. https://www.wpi.edu/academics/study/robotics-engineering-phd. Accessed Feb 2022.
4. Carnegie Mellon: Doctoral program in robotics (PhD). https://www.ri.cmu.edu/education/academic-programs/doctoral-robotics-program/. Accessed Feb 2022

5. University of Michigan: U-M founds first robotics department among top 10 engineering schools. https://www.engin.umich.edu/2021/12/u-m-founds-first-robotics-department-among-top-ten-engineering-schools/. Accessed Feb 2022
6. NPR: NPR & Edison research: smart speaker ownership reaches 35% of Americans. https://www.npr.org/about-npr/1105579648/npr-edison-research-smart-speaker-ownership-reaches-35-of-americans. Accessed Jan 2023
7. Interpret Weekly Insights, Mobile & CE: Robot vacuum cleaners are still mostly for the early adopter tech crowd. https://interpret.la/robot-vacuum-cleaners-are-still-mostly-for-the-early-adopter-tech-crowd/. Accessed 10 Feb 2021
8. Vuleta, B.: Drone statistics: commercial and governmental use. https://seedscientific.com/drone-statistics/. Accessed 11 Nov 2021
9. Thompson, B.: AI homework. https://stratechery.com/2022/ai-homework/. Accessed 5 Dec 2022
10. Huang, K.: https://www.nytimes.com/2023/01/16/technology/chatgpt-artificial-intelligence-universities.html. Accessed 16 Jan 2023
11. Cummings, M., Morgansen, K., Argrow, B., Singh, S.: Transforming aerospace autonomy education and research. In: 2021 IEEE Aerospace Conference (50100), pp. 1–10. Big Sky, MT, USA (2021). https://doi.org/10.1109/AERO50100.2021.9438150
12. National Institute of Standards and Technology: NIST study evaluates effects of race, age, sex on face recognition software. https://www.nist.gov/news-events/news/2019/12/nist-study-evaluates-effects-race-age-sex-face-recognition-software. Accessed 19 Dec 2019
13. Snow, J.: Amazon's face recognition falsely matched 28 members of congress with mugshots. https://www.aclu.org/blog/privacy-technology/surveillance-technologies/amazons-face-recognition-falsely-matched-28?page=7. Accessed 26 July 2018
14. Phillips, P. Jonathan, F.J., Abhijit, N., Julianne, A., O'Toole, A.J.: An Other-Race Effect for Face Recognition Algorithms. ACM Transactions on Applied Perception (2009). https://tsapps.nist.gov/publication/get_pdf.cfm?pub_id=906254
15. Wilson, B., Judy, H., Jaime, M.: Predictive inequity in object detection (2019). https://arxiv.org/pdf/1902.11097.pdf
16. Miley, J.: Autonomous cars can't recognize pedestrians with darker skin tones. In: Interesting Engineering. https://interestingengineering.com/autonomous-cars-cant-recognise-pedestrians-with-darker-skin-tones. Accessed 21 April 2021
17. Law Commission: Automated vehicles: consultation paper 3—a regulatory framework for automated vehicles. In: Law Commission Consultation Paper No 252, https://s3-eu-west-2.amazonaws.com/lawcom-prod-storage-11jsxou24uy7q/uploads/2021/01/AV-CP3.pdf. Accessed 18 Dec 2020
18. Williams, P.: Mason now a top 10 public university for diversity, innovation, and cybersecurity education, U.S. News rankings show. https://www.gmu.edu/news/2022-09/mason-now-top-10-public-university-diversity-innovation-and-cybersecurity-education-us. Accessed 12 Sep 2022
19. Prince William County Public Schools: The governor's school @ innovation park. https://governors.pwcs.edu/about_us/index. Accessed Jan 2023
20. Ball, K.S., Barton, Jr. O., Caraballo, S.A., Durant, L.W., Marks, M., Jarrouj, A., Parker, R.R.: ADVANCE: a community college and university partnership model for expanding and diversifying the talent pipeline to the jobs of tomorrow. In: Paper Presented at 2019 CoNECD—The Collaborative Network for Engineering and Computing Diversity, Crystal City, Virginia (2019). https://peer.asee.org/31740
21. McCue, L., Hagarty, A., Nelson, J., Nowzari, C., Raz, A., Riggi, M., Rosenberg, J., Shishika, D., Smith, C.: Lessons learned in the development of a STEM outreach program for biologically inspired underwater robotics. In: Abstract Submitted to the Ocean and Marine Engineering Division, American Society for Engineering Education Annual Conference, Baltimore, MD (2023)
22. Innovation First, Inc: Vex robotics. https://www.vexrobotics.com. Accessed Feb 2023
23. Robonation: Programs. https://robonation.org/programs/. Accessed Jan 2023

24. McCue, L., Barth, V., Hall, J.: SeaPerch and seaglide camp implementation. In: ASEE Southeastern Section Conference, Fairfax, VA (2023)
25. McCue, L.: Remarks delivered as part of an invited panel. In: MTS Tech Forum: Intelligence and Observations: The Nexus of Undersea Technology (2021)
26. McCue, L., Nowzari, C.: Building the UxS Workforce of the future. In: The Disruptor, issue 4 (2022)
27. College of Engineering and Computing: About. https://cec.gmu.edu/about-0. Accessed Jan 2023
28. ABET: Criteria for accrediting engineering programs, 2022–2023. https://www.abet.org/accreditation/accreditation-criteria/criteria-for-accrediting-engineering-programs-2022-2023/. Accessed Feb 2023
29. George Mason Department of Mechanical Engineering: Senior design capstone. https://mechanical.gmu.edu/hands/senior-design-capstone. Accessed Feb 2023
30. George Mason University: Information technology, PhD. https://catalog.gmu.edu/colleges-schools/engineering-computing/information-technology-phd/. Accessed Feb 2023
31. Fairfax County Public Schools: About Us. https://www.fcps.edu/about-fcps. Accessed Jan 2023
32. National Center for Education Statistics: Table 215.30. Enrollment, poverty, and federal funds for the 120 largest school districts, by enrollment size in 2018: 2017–18 and fiscal year 2020. https://nces.ed.gov/programs/digest/d20/tables/dt20_215.30.asp?current=yes. Accessed Jan 2023
33. Prince William County Public Schools: Fast facts. https://www.pwcs.edu/about_us/strategic_plan/fast_facts. Accessed Jan 2023
34. Arlington Public Schools: APS quick facts. https://www.apsva.us/wp-content/uploads/2022/12/APSQuickFacts-update1222.pdf. Accessed Jan 2023
35. Northern Virginia Community College (NOVA): Campuses & centers. https://www.nvcc.edu/campuses/#:~:text=NOVA%20is%20the%20largest%20public,from%20anywhere%20through%20NOVA%20Online. Accessed Feb 2023
36. Lawton, G.: How warehouse automation robotics transformed the supply chain. https://www.techtarget.com/searchenterpriseai/feature/How-warehouse-automation-robotics-transformed-the-supply-chain. Accessed 14 Dec 2021
37. Del Rey, J.: How robots are transforming Amazon warehouse jobs—for better and worse. https://www.vox.com/recode/2019/12/11/20982652/robots-amazon-warehouse-jobs-automation. Accessed 11 Dec 2019
38. Businesswire: Global offshore AUV and ROV market report 2021–2026: rising use of ROVs in search & rescue, marine biology, military, oil and gas, submerged infrastructure, and aquaculture. https://www.businesswire.com/news/home/20211104005762/en/Global-Offshore-AUV-and-ROV-Market-Report-2021-2026-Rising-Use-of-ROVs-in-Search-Rescue-Marine-Biology-Military-Oil-and-Gas-Submerged-Infrastructure-and-Aquaculture---ResearchAndMarkets.com. Accessed July 2022
39. Wang, Z., Liu, S., Peng, J., Chen, M.Z.: The next-generation surgical robots. In: Surgical Robotics. IntechOpen. https://doi.org/10.5772/67515
40. Research and Markets: Logistics robots market: global industry trends, share, size, growth, opportunity, and forecast 2023–2028. https://www.researchandmarkets.com/reports/5547170/logistics-robots-market-global-industry-trends. Accessed March 2023
41. Astute Analytica: Indoor delivery robots market—industry dynamics, market size, and opportunity forecast to 2030. https://www.astuteanalytica.com/industry-report/indoor-delivery-robots-market. Accessed July 2022

Chapter 18
Generalization of the Boundary Uniqueness Theorem for $A(z)$—Analytic Functions

Nasridin Zhabbarov and Behzod Husenov

Abstract We consider $A(z)$—analytic functions in case when $A(z)$ is antianalytic function. In this paper, the Nevanlinna class for $A(z)$—analytic functions is are introduced and for these classes, the boundary values of the function are investigated. For the Nevanlinna class of functions, an analogue of Fatou's theorem was proved as a proposition to show that the function has a value on the boundary of the domain. Also, the Privalov's ice-cream cone consruction is introduced for $A(z)$—analytic functions and Egoroff's theorems are applied for them. As the main result, the analog generalized boundary uniqueness theorem for $A(z)$—analytic functions is proven and the boundary uniqueness theorem for Nevanlinna classes of functions are given as a corollary.

Keywords $A(z)$—analytic functions · $A(z)$—lemniscate · The angular limit for $A(z)$—analytic functions · The nevanlinna class for $A(z)$—analytic functions · The analog generalized boundary uniqueness theorem for $A(z)$—analytic functions

18.1 Introduction

Let $A(z)$ be antianalytic function, i.e. $\frac{\partial A}{\partial z} = 0$ in the domain $D \subset \mathbb{C}$; moreover, let $|A(z)| \leq c < 1$ for all $\forall z \in D$, where c =const. The function $f(z)$ is said to be $A(z)$—analytic in the domain D if for any $z \in D$, the following equality holds:

$$\frac{\partial f}{\partial \bar{z}} = \overline{A}(z)\frac{\partial f}{\partial z}. \qquad (18.1)$$

N. Zhabbarov (✉)
Belorussian-Uzbek Joint Intersectoral Institute of Applied Technical Qualifications in Tashkent, 4 Karamurt-1 Street, Kibraisky 100071, Tashkent, Uzbekistan
e-mail: jabborov61@mail.ru

B. Husenov
Bukhara State University, 11 Muhammad Ikbal str., Bukhara 705018, Uzbekistan

D. Azimov (ed.), *Proceedings of the IUTAM Symposium on Optimal Guidance and Control for Autonomous Systems 2023*, IUTAM Bookseries 40,
https://doi.org/10.1007/978-3-031-39303-7_17

We denote by $O_A(D)$ the class of all $A(z)$—analytic functions defined in the domain D. Since an antianalytic function is infitely smooth, $O_A(D) \subset C^\infty(D)$ (see [1]). In this case, the following takes place:

Theorem 1 *(analogue of Cauchy's theorem [2]). If $f \in O_A(D) \cap C(\bar{D})$, where $D \subset \mathbb{C}$ is a domain with rectifiable boundary ∂D, then*

$$\int\limits_{\partial D} f(z)(dz + A(z)d\bar{z}) = 0.$$

Now we assume that the domain $D \subset \mathbb{C}$ is convex, and $\xi \in D$ is a fixed point in it. Consider the function

$$K(z,\xi) = \frac{1}{2\pi i} \frac{1}{z - \xi + \overline{\int_{\gamma(\xi,z)} \overline{A(\tau)}d\tau}},$$

where $\gamma(\xi, z)$ is a smooth curve which points of $\xi, z \in D$. Since the function $\overline{A}(z)$ is analytic and the domain D is simply-connected, the integral

$$I(z) = \overline{\int_{\gamma(\xi,z)} \overline{A(\tau)}d\tau}$$

is independent of the path of integration; it coincides with the antiderivative $I'(z) = \overline{A}(z)$ (see [1]).

Theorem 2 *(see [1]). $K(z,\xi)$ is an $A(z)$—analytic function outside of the point $z = \xi$, i.e. $K(z,\xi) \in O_A(D/\{\xi\})$. Moreover, at $z = \xi$ the function $K(z,\xi)$ has a simple pole.*

Remark 1 (see [1]). If a simply connected domain $D \subset \mathbb{C}$ is not convex, then the function

$$\psi(z,\xi) = z - \xi + \overline{\int\limits_{\gamma(\xi,z)} \overline{A(\tau)}d\tau}$$

although well defined in D, may have other isolated zeros except for $\zeta: \psi(z,\xi) = 0$ for $z \in P/\{\xi, \xi_1, \xi_2, ...\}$. Consequently, $\psi \in O_A(D)$, $\psi(z,\xi) \neq 0$ when $z \notin P$ and $K(z,\xi)$ is an $A(z)$–analytic function only in D/P, it has poles at the points of P. Due to this fact we consider the class of $A(z)$—analytic functions only in convex domains.

According to [[1], Theorem 2], the function

$$\psi(z,a) = z - a + \overline{\int_{\gamma(a,z)} \overline{A(\tau)}d\tau}$$

is an $A(z)$—analytic function.

The following set is an open subset of D:

$$L(a,r)=\left\{z\in D\colon |\psi(z,a)|=|z-a+\overline{\int_{\gamma(a,z)}\overline{A(\tau)}d\tau}|<r\right\}$$

For suffiently small $r>0$, this set compactly lies in D (we denote this fact by $L(a,r)\subset\subset D$) and contains the point a. The set $L(a,r)$ is called an $A(z)$—lemniscate with the centered at the point a. The lemniscate $L(a,r)$ is a simply-connected set (see [1]).

Theorem 3 *(see [3] Cauchy formula). Let $D\subset\mathbb{C}$ be a convex domain and $G\subset\subset D$ be an arbitrary subdomain with a smooth or piecewise smooth boundary ∂G that compactly lies in D. Then for any function $f(z)\in O_A(G)\bigcap C(\bar{G})$, the following formula holds:*

$$f(z)=\int_{\partial G} f(\xi)K(z,\xi)\left(d\xi+A(\xi)d\bar{\xi}\right),\quad z\in G. \tag{18.2}$$

18.2 Some Classes and Concepts for $A(z)$—Analytic Functions

18.2.1 Angular Limit for $A(z)$—Analytic Functions

Initially, we introduce an angular limit for $A(z)$—analytic functions. Let $L(a,r)\subset\subset D$ and $f(z)\in O_A(L(a,r))$.

Definition 1 Let's put $\psi(z,a)=\rho e^{i\theta}$ and $\psi(\zeta_0,a)=re^{i\phi_0}$, where $0<\rho<r,\ 0<\theta<2\pi$ and $\phi_0\in(0,2\pi)$—fixed angle. Then there is an "angular" limit of

$$\lim_{z\to\zeta_{0\angle}} f(z)$$

with $\rho\to r$ on $\zeta_0\in\partial L(a,r)$ and we denote them $f^*(\zeta_0)$, where $|\theta-\phi_0|<d(r-\rho),\ d=$const.

Remark 2 Thus, it is prescribed that point z tends to ζ_0, remaining inside the sector of solution$<\pi$ with a vertex at point ζ_0, symmetrical with respect to the radius leading from a to ζ_0. In this case, they say that $f(z)\to f^*(\zeta_0)$ with z, tending to ζ_0 in non-tangential directions. We will write it down like this:

$$f(z)\to f^*(\zeta_0)\ by\ z\to\zeta_{0\angle}.$$

18.2.2 The Nevanlinna Classes for $A(z)$—Analytic Functions

Now, we introduce the Nevanlinna class for $A(z)$—analytic functions. $A(z)$—analytic function $f(z)$ is not identically equal to 0 in lemniscate $L(a,r)$, belongs to class N if integral

$$\int\limits_{|\psi(z,a)|=\rho} \ln|f(z)|^p|dz + A(z)d\bar{z}|$$

was bounded at $z \in L(a,\rho)$.

This class in the domain of D is $A(z)$—analytic functions is denoted as $N_A(D)$. We will also look at the following properties of the function class N_A:

The function $f(z)$ is represented in the lemniscate $L(a,r)$ as the ratio of two bounded functions:

$$f(z) = \frac{g(z)}{h(z)} \tag{18.3}$$

functions $g(z)$ and $h(z)$ can always be considered bounded in lemniscate $L(a,r)$ modulo one. Class N_A can be characterized differently based on the following statement by Nevanlinna.

Statement 1. *In order for the function $f(z) \not\equiv 0$ to belong to class N_A, it is necessary and sufficient that the integral*

$$\int\limits_{|\psi(z,a)|=\rho} \ln^+|f(z)||dz + A(z)d\bar{z}| \tag{18.4}$$

is bounded at $0 < \rho < r$ by some finite number E, independent of ρ, where if $\tilde{d} \geq 1$, then $\ln^+\tilde{d} = \ln\tilde{d}$.

Proof If function $f(z) \not\equiv 0$ belongs to class N_A, i.e. it is representable as a relation (18.4) with $|g(z)| \leq 1$, $\;|h(z)| \leq 1$ in $|\psi(z,a)| < r$, then since $|f(z)| \leq \frac{1}{|h(z)|}$ in $|\psi(z,a)| < r$:

$$\int\limits_{|\psi(z,a)|=\rho} \ln^+|f(z)||dz + A(z)d\bar{z}| \leq -\int\limits_{|\psi(z,a)|=\rho} \ln|h(z)||dz + A(z)d\bar{z}|. \tag{18.5}$$

Now, if $h(z) = \sum\limits_{k=m}^{\infty} c_k\psi^k(z,a)$, $\;m \geq 0$, then according to Jensen's formula applied to function $\frac{h(z)}{\psi^m(z,a)}$, we have:

$$\begin{aligned}\frac{1}{2\pi\rho} \int\limits_{|\psi(z,a)|=\rho} \ln|h(z)||dz \qquad\qquad\qquad\qquad +A(z)d\bar{z}| \\ = \ln|c_m| + \sum_{0<|\psi_k(z,a)|\le\rho} \ln\frac{\rho}{|\psi_k(z,a)|} + m\ln\rho;\end{aligned} \tag{18.6}$$

here the zeros of function $\psi_k(z,a)$ in $h(z)$ are denoted by $0 < |\psi(z,a)| < r$, and the sum is taken for all zeros of function $h(z)$ lying in $0 < |\psi(z,a)| < r$. Since the right part in (18.6) is a non-decreasing function from ρ to $0 < \rho < r$, the right part in (18.5) will be a non-increasing function from ρ and, therefore, will be bounded from above at ρ. This proves that the integral (18.4) will also be bounded in $0 < \rho < r$.

Let now, inversely, the function $f(z) \not\equiv 0$ be such that the integral (18.4) is bounded at ρ. Then by the Jensen-Schwartz formula in $|\psi(z,a)| < r$ and $\xi \in \partial L(a,\rho)$ we have:

$$\begin{aligned}\ln f(z) = &\sum_{|\psi_k(z,a)|<\rho} \ln\frac{\rho(\psi(z,a)-\psi_k(z,a))}{\rho^2-\bar{\psi}_k(z,a)\psi(z,a)} \\ &+ \frac{1}{2\pi\rho}\int\limits_{|\psi(\xi,a)|=\rho} \ln|f(z)|\frac{\psi(\xi,a)+\psi(z,a)}{\psi(\xi,a)-\psi(z,a)}|d\xi + A(\xi)d\bar{\xi}| + ic_1,\end{aligned}$$

where c_1 is a real constant. This can be rewritten as:

$$f(z) = \frac{g_\rho(z)}{h_\rho(z)} \tag{18.7}$$

where

$$g_\rho(z) = \prod_{|\psi_k(z,a)|<\rho} \frac{\rho(\psi(z,a)-\psi_k(z,a))}{\rho^2-\bar{\psi}_k(z,a)\psi(z,a)} e^{-\frac{1}{2\pi\rho}\int_{|\psi(\xi,a)|=\rho}\ln^+\frac{1}{|f(z)|}\frac{\psi(\xi,a)+\psi(z,a)}{\psi(\xi,a)-\psi(z,a)}|d\xi+A(\xi)d\bar{\xi}|+ic_2}$$

and

$$h_\rho(z) = e^{-\frac{1}{2\pi\rho}\int_{|\psi(\xi,a)|=\rho}\ln^+|f(z)|\frac{\psi(\xi,a)+\psi(z,a)}{\psi(\xi,a)-\psi(z,a)}|d\xi+A(\xi)d\bar{\xi}|}$$

is put, where c_2 is also real constant.

Functions $g_\rho(z), h_\rho(z) \in O_A(L(a,r))$ and $|g_\rho(z)| \le 1$, $|h_\rho(z)| \le 1$ in $L(a,r)$. Taking the sequence of numbers $\rho_k \to r$, according to the principle of condensation, a subsequence $h_{\rho'_k}(z)$ can be distinguished from the sequence of function $h_{\rho_k}(z)$, which converges uniformly inside the lemniscate $L(a,r)$ to the $A(z)$–analytic function $h(z)$, and $|h(z)| < 1$ to $|\psi(z,a)| < r$, where $k \in \mathbb{N}$. Since the values of $|h_{\rho_k}(a)|$ are bounded from below by a positive value independent of k, then $h(z) \not\equiv 0$. From (18.7) it follows that function $g_{\rho'_k}$ converges in $\{|\psi(z,a)| < r\}$ to some function $g(z)$, $A(z)$—analytical in $\{|\psi(z,a)| < r\}$, and $|g(z)| \le 1$ in $\{|\psi(z,a)| \le r\}$.

The statement is proved. □

Since from the representation (18.4) for the function $f(z) \not\equiv 0$ of class N_A we have:

$$\int\limits_{|\psi(z,a)|=\rho} |\ln|f(z)|||dz + A(z)d\bar{z}| \le \int\limits_{|\psi(z,a)|=\rho} |\ln|g(z)|||dz + A(z)d\bar{z}| + \int\limits_{|\psi(z,a)|=\rho} |\ln|h(z)|||dz + A(z)d\bar{z}|$$
$$= -\int\limits_{|\psi(z,a)|=\rho} \ln|g(z)||dz + A(z)d\bar{z}| - \int\limits_{|\psi(z,a)|=\rho} \ln|h(z)||dz + A(z)d\bar{z}|,$$

and according to the Statement 1 proved in the first part of the proof, the last two integrals do not decrease by $0 < \rho < r$, then not only the integral (18.4), but also the integral

$$\int\limits_{|\psi(z,a)|=\rho} |\ln|f(z)|||dz + A(z)d\bar{z}| \tag{18.8}$$

at $0 < \rho < r$ will be bounded to a horse value independent of ρ. This property of class N_A functions will be used now. That is, in order to show the existence of limit values for the class of functions N_A, we will prove Fatou's theorems as a proposition.

Proposition 1 *If the function $f(z) \not\equiv 0$ in $L(a, r)$ and belongs to class $N_A(L(a, r))$, then it has almost everywhere on the boundary of the lemniscate $\partial L(a, r)$ certain limit values of $f^*(\zeta)$ along all non-tangential paths (angular limit), and $|\ln|f^*(\zeta)||$ is summed by $|\psi(\zeta, a)| = r$.*

Proof If the function $f(z) \in N_A$, $f(z) \not\equiv 0$ and bounded in $L(a, r)$ it has already been noted that it has almost everywhere on $|\psi(\zeta, a)| = r$ certain limit values of $f^*(\zeta)$ along all non-tangential paths, and in particular along radial paths. We denote by $\psi(z, a) = \rho e^{i\theta}$ we have $f(\rho) := |f(z)|$, where $0 \le \rho \le r$. According to the maximum modulus principle for $A(z)$—analytic functions, it follows that the monotonicity of functions $f^*(r) = \max\limits_{0 \le \rho \le r} f(\rho)$. Hence, by Fatou's lemma we have:

$$\int\limits_{|\psi(z,a)|=\rho} |\ln|f^*(\zeta)|||d\zeta + A(\zeta)d\bar{\zeta}| \le \lim_{\rho\to r} \int\limits_{|\psi(z,a)|=\rho} |\ln|f(z)|||dz + A(z)d\bar{z}|, \tag{18.9}$$

moreover, according to what has been said about the integral (18.9), the right part is bounded here. In this case, in the "radial" convergence, taking $\psi(z, a) = \rho e^{i\theta}$, $\psi(\zeta, a) = re^{i\theta}$ will be $\rho \to r$.

Therefore, $|\ln|f^*(\zeta)||$ is summable by $\partial L(a,r)$. But then the values of $|\ln|f^*(\zeta)||$ are almost everywhere on $\partial L(a,r)$ are finite, i.e. the values of $f^*(\zeta)$ are almost everywhere on $\partial L(a,r)$ are different from zero. But then the values of $|\ln|f^*(\zeta)||$ are almost everywhere finite by $|\psi(\zeta,a)| = r$, i.e. the values of $f^*(\zeta)$ are almost everywhere $|\psi(\zeta,a)| = r$ different from zero.

If now $f(z) \not\equiv 0$ is any function of class N_A, then in its representation (18.3) functions $g(z)$ and $h(z)$ have almost everywhere on $|\psi(\zeta,a)| = r$ certain limit values along non-rotational paths and these limit values are almost everywhere non-zero. But then $|\ln|f^*(\zeta)||$ almost everywhere on $\partial L(a,r)$ has certain limiting values of $f^*(\zeta)$; by applying again Fatou's lemma to the integral (18.8), we conclude that $|\ln|f^*(\zeta)||$ is summable by $\partial L(a,r)$.

The proposition are proven. □

The finite angular limit values of the function $f \in N_A(L(a,r))$ that exist almost everywhere on $|\psi(\zeta,a)| = r$ along non-tangential paths are now called its boundary values.

18.3 Boundary Uniqueness Theorem for $A(z)$—Analytic Functions

18.3.1 Privalov's Ice-Cream Cone Construction for $A(z)$—Analytic Functions

Definition 2 For $|\psi(\zeta,a)| = r$, let's put such an domain

$$S_\zeta := \left\{ z \colon |\psi(z,a)| > \frac{r}{\sqrt{2}}, \quad |\arg \psi(\zeta, z)| < \frac{\pi}{4} \right\}$$

Let's make a number of obvious remarks:

(a) $\bigcup_{|\psi(\zeta,a)|=r} S_\zeta$—is all of $\{r/\sqrt{2} < |\psi(z,a)| < r\}$.

(b) If $r/\sqrt{2} < |\psi(z,a)| < r$ for some z, $\{\zeta \colon |\psi(\zeta,a)| = r, \ \zeta \in S_\zeta\}$ is the (open) arc $\widetilde{\zeta_1, \zeta_2}$ of the boundary of lemniscate $\partial L(a,r)$.

(c) If $J = \widetilde{\zeta_1, \zeta_2}$ is the arc of the lemniscate boundary $\partial L(a,r)$, contracting the angle no more than $\pi/2$, then the set of points z, $r/\sqrt{2} < |\psi(z,a)| < r$, such that z is contained only in those sets S_ζ, which have $\zeta \in J$, forms a closed curved triangle T.

Now we can describe the Luzin-Privalov construction for $A(z)$—analytic functions. For a given closed set M on the lemniscate boundary $\partial L(a,r)$, let $\{J_k\}$ be the set (no more than countable) of arcs on the boundary $\partial L(a,r)$ adjacent (additional) to M. Using each arc J_k as a base, we will build a triangle or trapezoid T_k on it in accordance with the procedure described in (c). Let's take the closed domain

$$\bar{D} = \{|\psi(z,a)| \le r\} / \bigcup_{k=1}^{\infty} \breve{T}_k / \bigcup_{k=1}^{\infty} J_k$$

(the icon ˇ denotes the interior, and the dash on top denotes the closure).

Our domain $\bar{D}$ has the following important property:

Proposition 2 *Each point $z \in \bar{D}$, modulo a large $r/\sqrt{2}$, belongs to $\bar{S}_\zeta$ for some $\zeta \in M$.*

This follows directly from the observations (a) and (c) made above.

Note that $\partial \bar{D}$ is the Jordan curve. Indeed, if ζ_θ denotes the only point at which the ray going from a to $\psi(\zeta, a) = re^{i\theta}$, intersects $\partial \bar{D}$, then ζ_θ will be a continuous one-to-one mapping of the lemniscate $L(a, r)$ to $\partial \bar{D}$. The boundary $\bar{D}$ will even be a rectified Jordan curve, since for each k the perimeter T_k does not exceed $\tilde{c}|J_k|$, where $\tilde{c}$ is a constant that can be calculated for geometric reasons.

18.3.2 Use of Egoroff's Theorem for $A(z)$—Analytic Functions

Proposition 3 *Let be $A(z)$—analytic functions in $L(a, r)$, and put, for $|\psi(\zeta, a)|=r$,*

$$G_f(\zeta) = \sup_{z \in S_\zeta} |f(z)|$$

Then $G_f(\zeta)$ is measurable.

Proof For $n \ge 3$, take $r_n = r(1 - 1/n)$ and put, for $|\psi(\zeta, a)| = r$,

$$G_n(\zeta) = \sup\{|f(z)|: r/\sqrt{2} \le |\psi(\zeta, a)| \le r_n, \quad |\arg \psi((r_n/r)\zeta, z)| \le \pi/4\}.$$

Because $f(z)$ is continuous for $|\psi(z, a)| \le r_n$, $G_n(z)$ is continuous. Since clearly $G_f(\zeta) = \lim_{n\to\infty} G_n(\zeta)$ pointwise in $\zeta \in \partial L(a, r)$, $G_f(\zeta)$ is measurable.

The proposition is proved. □

Proposition 4 *Let $f(z)$ be $A(z)$—analytic in $L(a, r)$. Suppose there is a set M of positive measure on the boundary of lemniscate $\partial L(a, r)$ such that*

$$\lim_{z \to \zeta_\angle} f(z) = 0$$

for each $\zeta \in M$. Then there is a closed set M, $\mu(M) > 0$, such that $|f(z)| \to 0$ uniformly for $|\psi(z, a)| \to r$ and z in the union of the S_ζ with $\zeta \in S_\zeta$.

Proof For $n \geq 3$ and $|\psi(\zeta, a)| = r$ put

$$P_n(\zeta) = \sup\left\{|f(z)|:\ z \in S_\zeta,\ \ |\psi(z, a)| \geq \left(r - \frac{r}{n}\right)\right\}.$$

The argument used in the proof of the previous proposition shows that each $P_n(\zeta)$ is measurable; so, then, is the set M^* of ζ for which $\lim\limits_{n\to\infty} P_n(\zeta) = 0$.

By hypothesis, $\lim\limits_{n\to\infty} P_n(\zeta) = 0$ for $\zeta \in M$, and $\mu(M) > 0$. Therefore $\mu(M^*) > 0$, and Egoroff's theorem gives us a measurable $M_0 \subset M^*$, $\mu(M_0) > 0$, with $\lim\limits_{n\to\infty} P_n(\zeta) = 0$ uniformly for $\zeta \in M_0$. Also, on an open set M^*/M_0 has $\mu\left(\zeta \in M^*/M_0\text{: } \lim\limits_{n\to\infty} P_n(\zeta)\right) = 0$. Taking a closed $M \subseteq M_0$ with $\mu(M) > 0$, we have the proposition.

The proposition is proved. □

18.3.3 Generalization of the Boundary Uniqueness Theorem for $A(z)$—Analytic Functions

Now we will prove the analog generalized boundary uniqueness theorem for $A(z)$—analytic functions.

Theorem 4 *Let $f \in O_A(L(a, r))$. Suppose that M is the set of positive measure on the boundary $\partial L(a, r)$, such that $\lim\limits_{z\to\zeta_\angle} f(z) = 0$ is for $\zeta \in M$. Then the function $f(z)$ is identically equal to zero.*

Proof This theorem for analytic functions was proved by Luzin N.N. and Privalov I.I. in 1924. By Proposition 4, we can find a closed set M, $\mu(M) > 0$, on the lemniscate boundary $L(a, r)$, such that S_ζ, $\zeta \in M$ is uniform at $|\psi(z, a)| \to r$, if z belongs to the union of sets S_ζ, $\zeta \in M$. This means that if we carry out the Luzin-Privalov construction described in Subsect. 18.3.1, starting from the set M, we get a domain $G \subset L(a, r)$ on which $f(z) \to 0$ is uniformly at $|\psi(z, a)| \to r$, $z \in G$. In the construction of the Luzin-Privalov construction for $A(z)$—analytic functions, we see that ∂D consists of segments in $\{|\psi(z, a)| < r\}$, going to the points of the set M on the boundary $\partial L(a, r)$, and, in addition, from the set M itself. Therefore, if we define function $f(z)$ as zero by M, we get a continuous function by $\bar{D}$ and an $A(z)$—analytic function in D.

In accordance with what is stated in Subsect. 18.3.1, ∂D is a Jordan straighten curve. We take the homeomorphic map φ of lemniscate $\{|\psi(w, b)| < r\}$ to D and for $\{|\psi(w, b)| < r\}$ we put $f(w) = F(\varphi(w))$, where point b is the center of this lemniscate (see [4, 5]). According to the Carateodori theorem, φ actually continues up to $\{\psi(v, b) = r\}$ and displays this boundary one-to-one and continuously at ∂D, where $\zeta \in \partial L(b, r)$. This means that $f(w)$ continues continuously up to $\{|\psi(v, b)| = r\}$, since $f(z)$ continues continuously up to ∂D. Let $S = \psi^{-1}(M)$.

Then $f(\nu) = 0$ for $\nu \in S$. The subset M of the rectified curve ∂D has a positive measure: $\mu(S) = \int\limits_S |d\nu + A(\nu)d\bar{\nu}|$. We take $d\mu = d\nu + A(\nu)d\bar{\nu}$, then μ will be a measure of $\{|\psi(\nu, b)| = r\}$. Measure μ is absolutely continuous. That is, for every $\varepsilon > 0$ there is such a $\delta > 0$ that

$$\left| \int\limits_S |d\mu| \right| < \varepsilon$$

for every measurable $S \subset \partial L(b, r)$ such that $\mu(S) < \delta$. Denote the (pairwise disjoint) open intervals additional to S—the so-called adjacent intervals—by (α_k, β_k), $1 \leq k \leq n$; there are no more than a countable set of them. In other words, the arcs $\{\psi(\zeta, b) = re^{it} : \ \alpha_k < t < \beta_k\}$ do not intersect in pairs and together with $\{t : \nu \in S\}$, fill exactly the entire boundary of the lemniscate $\{|\psi(\nu, b)| = r\}$. From $\mu(S) < \delta$ it follows that

$$\lim_{n\to\infty} \left| \int\limits_{\partial L(b,r)} |d\mu| - \sum_{k=1}^{n} (\beta_k - \alpha_k) \right| = \left| \int\limits_S |d\mu| \right| = \mu(S) < \varepsilon = \delta.$$

Therefore, it follows from this that $\mu(S) > 0$. Since the function $f(w)$ $A(w)$—analytic in $L(b, r)$, is continuous on a closed lemniscate $\bar{L}(b, r)$ and is zero on S, then $f \equiv 0$. That is, the Cauchy formula for $A(z)$—analytic functions (18.2) on the piece boundary $S \subset \partial L(b, r)$ follows

$$f(w) = \int\limits_S f(\nu)K(w, \nu)(dw + A(w)d\bar{w}) = 0, \quad w \in L(b, r)$$

From where $f(w) \equiv 0$. If we perform

$$z = r^2 \frac{\psi(w, b) - \psi(a, b)}{r^2 - \bar{\psi}(a, b)\psi(w, b)}$$

isomorphism, here $z = z(w) : L(b, r) \to L(a, r)$ (see [6]), we get $f(z) \equiv 0$. □

From this theorem, we obtain the following corollary:

Corollary 1 *Let be given $f \in N_A$ function and a set M of a positive measure of a piece of boundary $\partial L(a, r)$. If the $\lim\limits_{z\to\zeta_\angle} f(z) = 0$ is for $\zeta \in M$, then $f(z)$ is identically zero.*

Proof Obviously, $f \in N_A$ and if $f \not\equiv 0$ is in $|\psi(z, a)| < r$, then, according to Proposition 1, $\ln |f^*(\zeta)|$ is almost everywhere on $|\psi(\zeta, a)| = r$ is finite, i.e. $f^*(\zeta)$ is almost everywhere on $|\psi(\zeta, a)| = r$ is finite and different from zero; this also contradicts the condition that $f^*(\zeta) = 0$ is almost everywhere on M (the points at which $f(z) \equiv 0$ are excluded). □

If we consider these statement as $f = f_1 - f_2$ we get $f_1 \equiv f_2$. This relation, on the other hand, means that the $A(z)$—analytic function belonging to all classes is uniquely defined in our $L(a, r)$ lemniscates, which we will consider.

References

1. Sadullaev, A., Zhabborov, N.M.: On a class of A—analytic functions. J. Siberian Fed. Univ. **9**(3), 374–383 (2016). https://doi.org/10.17516/1997-1397-2016-9-3-374-383
2. Arbuzov, E.V.: The Cauchy problem for second-order elliptic systems on the plane. J. Siberian Math. **44**(1), 3–20 (2003). https://doi.org/10.1023/A:1022034001292
3. Zhabborov, N.M., Otaboyev, T.U.: An analog of the integral Cauchy theorem for A—analytic functions. J. Uzbek Mat. **4**(4), 50–59 (2016). (in Russian)
4. Bojarski, B.: Homeomorphic solutions of the Beltrami systems. Proc. USSR Acad. Sci. **102**(4), 661–664 (1955). (in Russian)
5. Bojarski, B.: Generalized solutions of a system of differential equations of the first order of the first order of the elliptic type with discontinuous coefficients. Sbornik Math. **4**(43), 451–503 (1957). (in Russian)
6. Zhabborov, N.M., Otaboyev, T.U.: Khursanov: Schwarz inequality and Schwarz formula for A—analytic functions. J. Modern math. Fundamental Directions **64**(4), 637–649 (2018). https://doi.org/10.22363/2413-3639-2018-64-4-637-649

Chapter 19
Control of an Upper-Stage Rocket with Partially Filled Fuel Tanks via Takagi-Sugeno Fuzzy Model

Chokri Sendi and Aaron Inks

Abstract This paper investigates the attitude stability of an upper-stage rocket equipped with a thrust control system and a reaction wheel via a fuzzy control technique. The upper-stage rocket is subject to disturbances due to fuel sloshing in its tanks. Two types of fuzzy controllers are presented. The proposed techniques solve the problems of actuators' constraints. The Takagi-Sugeno (T-S) fuzzy model, in conjunction with the Parallel Distributed Compensation technique (PDC), is used to obtain the fuzzy feedback control law. The controller design is presented as an optimization problem in the form of Linear Matrix Inequalities (LMI). The proposed controller's performance and stability are analyzed via numerical simulation.

Keywords T-S fuzzy model · Parallel distributed compensation (PDC) · Fuel slosh · Uncertainties

19.1 Introduction

Thrust control is used by spacecraft and rockets to control the direction in airless environments where control surfaces are ineffective. The thrust is achieved by pushing gas out of a nozzle allowing an object-free rotation and translation along one or more axes. While aircraft can turn using aerodynamic control surfaces such as ailerons, the absence of air precludes this method for spacecraft, making thrust control the primary method for controlling direction.

A large portion of a spacecraft's mass comprises the propellant it carries in its tanks, which it needs to launch and maneuver at its destination site. As this fuel is consumed, it leaves a vacancy in the tanks, allowing the fuel to move around. This produces what is known as slosh dynamics. Propellant slosh dynamics present a significant challenge for rocket thrust control. Methods for slosh suppression include

C. Sendi (✉) · A. Inks
University of Alaska Anchorage, Anchorage, AK 99508, USA
e-mail: csendi@alaska.edu

A. Inks
e-mail: aainks@alaska.edu

D. Azimov (ed.), *Proceedings of the IUTAM Symposium on Optimal Guidance and Control for Autonomous Systems 2023*, IUTAM Bookseries 40,
https://doi.org/10.1007/978-3-031-39303-7_18

baffles or partitions within the fuel tanks. Although these methods have a mitigating effect, they do not eliminate the effects of propellant slosh. Control methods have previously addressed fuel slosh dynamics and their interaction with spacecraft dynamics. However, these methods employed linear control, which could have been more effective. Often such methods result in unintended transverse vehicle motion. For example, where the thrust vector of a gimballed thruster should intersect the vehicle's center of mass, it can experience large disturbance torques and velocity errors, causing its true vector to misalign with the intended direction. Different methods have been used to derive the dynamic model of a spacecraft subject to fuel sloshing. For instance, it is possible to represent the fuel sloshing dynamics using one or multiple pendulums [1, 2], or multiple spring-mass systems [3]. Either way, it is generally difficult to predict slosh motion [4], and the problem becomes even more challenging when the liquid exhibits boiling phenomena as with cryogenic fuel or when the tank has internal structures, such as baffles or capillary guides. Considering the challenges in modeling fuel slosh, many criteria need to be taken into consideration when designing a control system, including but not limited to the robustness of the designed control towards internal and external disturbances and uncertainties, computational time, control power consumption, the accuracy of the output, and capability of the spacecraft to process the designed controller. The problem of attitude control of spacecraft using nonlinear controllers has been extensively studied over the last decades. It has been demonstrated [5–7] that nonlinear controllers can achieve a better performance of such systems. For instance, Sliding mode control [8, 9], Feedback linearization [10], Adaptive control [11, 12], and Optimal control [13, 14] to mention a few. We refer the reader to [15] for a comprehensive list of publications.

In the past few decades, control design technique based on fuzzy logic has been utilized as an alternative approach to traditional control for complex nonlinear systems. As it may be, the Takagi-Sugeno (T-S) fuzzy model is considered one of the most known forms of fuzzy systems. Its recognition derives from the fact that the stability and performance characteristics of the system represented by the T-S model can be analyzed using a Lyapunov function approach [16–18]. Subsequently, many problems have been solved, and some promising results for T-S fuzzy systems have been reported in the literature. The TS fuzzy model is a linear model that allows the designer to apply fuzzy control techniques to a nonlinear system. Traditional linearization techniques create a model that loses accuracy as the states diverge from the linearization point. The T-S fuzzy model is theoretically capable of approximating the entire nonlinear system over the entire state-space to any degree of accuracy desired. Though in practice, compromises are made to reduce the computational cost. The ability to approximate nonlinear functions makes fuzzy systems well-suited to nonlinear state estimation. The versatility of fuzzy models allows designers to use the same model to estimate nonlinear states and obtain the feedback control law via linear matrix inequalities. A comprehensive list of fuzzy control techniques can be found in [19, 20].

In this paper, two different T-S fuzzy controllers are presented with application to an upper-stage rocket equipped with thrusters and reaction wheels for attitude stability. The spacecraft is subject to fuel slosh in its tank. The objective of the controllers is to reduce the effect of the disturbances due to the fuel slosh and to stabilize the attitude of the spacecraft. It should be noted that the fuzzy controller is designed with the assumption that the initial conditions are unknown, but the upper bound is known. Finally, the actuator's amplitude constraints are considered in the design.

This paper is organized as follows: Sect. 19.2 presents a brief derivation of the dynamic model based on the spherical pendulum to model the fuel slosh. The equation of motion of the 2D model is made possible using the Lagrange equations. Section 19.3 introduces the concept of the Takagi-Sugeno (T-S) fuzzy model and the parallel distributed Compensator (PDC). In Sect. 19.4, two types of fuzzy controllers are introduced, the first controller is a fuzzy controller with disturbance rejection properties, and the second is a fuzzy optimal controller. Simulation results are provided throughout the paper to demonstrate the performance of the proposed controllers. The conclusion in Sect. 19.5 provides some comments regarding the performance of the controllers.

19.2 Mathematical Model

A simplified dynamic model of the spacecraft equipped with a spherical pendulum to model the fuel slush and side thrusters moving in a fixed plane is illustrated in Fig. 19.1. The mathematical model based on this configuration has been used to design linear and nonlinear controllers [21]. The derivation of the equations of motion can be summarized as follows.

Consider v_x, v_y to be the components of the velocity of the center of the fuel tank, θ represents the attitude of the spacecraft with respect to the inertial frame, and ψ is the angle of the spherical pendulum with respect to the body reference frame. The

Fig. 19.1 Spacecraft model

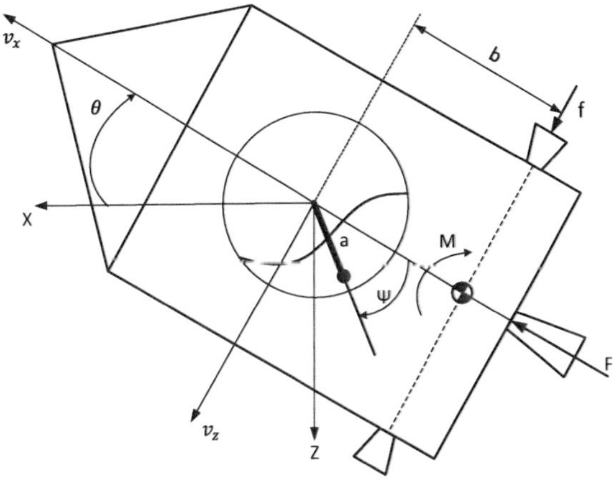

actuators consist of the main thrust F acting to propel the spacecraft forward, a side thrust f, and a moment M to control the pitch and the position of the spacecraft with respect to z-axis.

Assuming the mass m and the moment of inertia I of the spacecraft as well as the fuel mass m_f and the fuel moment of inertia I_f to be constant, and let a be the length of the pendulum. The center of mass position with respect to the body frame is expressed as:

$$r = \begin{bmatrix} x - b \\ z \end{bmatrix} \tag{19.1}$$

The velocity of the spacecraft is:

$$\dot{r} = \begin{bmatrix} \dot{x} + z\dot{\theta} \\ \dot{z} - x\dot{\theta} + b\dot{\theta} \end{bmatrix} = \begin{bmatrix} v_x \\ v_z + b\dot{\theta} \end{bmatrix} \tag{19.2}$$

The position and the velocity of the center of the fuel tank, respectively, are given by:

$$r_f = \begin{bmatrix} x - a\cos\psi \\ z + a\sin\psi \end{bmatrix} \tag{19.3}$$

and

$$\dot{r}_f = \begin{bmatrix} \dot{x} + a\dot{\psi}\sin\psi + \dot{\theta}(z + a\sin\psi) \\ \dot{z} + a\dot{\psi}\cos\psi - \dot{\theta}(x - a\cos\psi) \end{bmatrix} = \begin{bmatrix} v_x + a(\dot{\theta} + \dot{\psi})\sin\psi \\ v_z + a(\dot{\theta} + \dot{\psi})\cos\psi \end{bmatrix} \tag{19.4}$$

Assuming no gravitational potential energy is present, the total kinetic energy is written as

$$T = \frac{1}{2}m\dot{r}^2 + \frac{1}{2}m_f\dot{r}_f^2 + \frac{1}{2}I\dot{\theta}^2 + \frac{1}{2}I_f(\dot{\theta} + \dot{\psi})^2 \tag{19.5}$$

Substituting Eqs.(19.3, 19.4) into Eq.(19.5), the kinetic energy is expressed as

$$\begin{aligned} T = {} & \frac{1}{2}m\left[v_x^2 + (v_z + b\dot{\theta})^2\right] \\ & + \frac{1}{2}m_f\left[(v_x + a(\dot{\theta} + \dot{\psi})\sin\psi)^2 + (v_z + a(\dot{\theta} + \dot{\psi})\cos\psi)^2\right] \\ & + \frac{1}{2}I\dot{\theta}^2 + \frac{1}{2}I_f(\dot{\theta} + \dot{\psi})^2 \end{aligned}$$

Using Lagrange formulation [22], the equations of motion in the absence of gravitational potential energy are given as

$$\frac{d}{dt}\frac{\partial L}{\partial v} + \hat{\omega}\frac{\partial L}{\partial v} = \tau_t \tag{19.6}$$

$$\frac{d}{dt}\frac{\partial L}{\partial \omega} + \hat{\omega}\frac{\partial L}{\partial \omega} + \hat{v}\frac{\partial L}{\partial v} = \tau_r \tag{19.7}$$

$$\frac{d}{dt}\frac{\partial L}{\partial \dot{\eta}} - \frac{\partial L}{\partial \eta} + \frac{\partial R}{\partial \dot{\eta}} = 0 \tag{19.8}$$

where τ_t, and τ_r represents the actuators input on the spacecraft, and $R = \frac{1}{2}\epsilon\dot{\psi}^2$ is the Rayleigh dissipation function. Let

$$v = \begin{bmatrix} v_x \\ 0 \\ v_z \end{bmatrix}, \omega = \begin{bmatrix} 0 \\ \dot{\theta} \\ 0 \end{bmatrix}, \tau_t = \begin{bmatrix} F \\ 0 \\ f \end{bmatrix}, \tau_r = \begin{bmatrix} 0 \\ M + fb \\ 0 \end{bmatrix},$$

The equations that describe the motion of the spacecraft with a partially filled fuel tank moving in the plane are:

$$(m + m_f)(\dot{v}_x + \dot{\theta}v_z) + m_f a(\ddot{\theta} + \ddot{\psi})\sin\psi + mb\dot{\theta}^2 + m_f a(\dot{\theta} + \dot{\psi})^2\cos\psi = F \tag{19.9}$$

$$(m + m_f)(\dot{v}_z + \dot{\theta}v_x) + m_f a(\ddot{\theta} + \ddot{\psi})\cos\psi + mb\ddot{\theta} - m_f a(\dot{\theta} + \dot{\psi})^2\sin\psi = f \tag{19.10}$$

$$(I + mb^2)\ddot{\theta} + mb(\dot{v}_z - \dot{\theta}v_x) - \epsilon = M + bf \tag{19.11}$$

$$(I_f + m_f a^2)(\ddot{\theta} + \ddot{\psi}) + m_f a\Big[(\dot{v}_x + \dot{\theta}v_z)\sin\psi + (\dot{v}_z + \dot{\theta}v_x)\cos\psi\Big] + \epsilon\dot{\psi} = 0 \tag{19.12}$$

To see the detailed derivation of the dynamic model, we refer the reader to [21].

19.3 Takagi-Sugeno (T-S) Fuzzy Model

19.3.1 T-S Fuzzy Model for Systems with Disturbance

A nonlinear dynamic system with disturbances is represented by the following (T-S) fuzzy model:

Model Rule i: IF $z_1(t)$ is about $\mu_{i1}[z_1(t)]$, ..., $z_p(t)$ is about $\mu_{ip}[z_p(t)]$, THEN

$$\begin{cases} \dot{x}(t) = A_i x(t) + B_i u(t) + D_i \varphi(t) \\ y(t) = C_i x(t) \end{cases} \tag{19.13}$$

where $\mu_{ij}[z_p(t)]$ is the grade of the membership of $z_p(t)$, x represents the state vector, u is the control input vector, y is the output vector, $(i = 1, 2, \ldots, r)$ specifies the number of fuzzy rules, and $\varphi(t)$ represents the disturbance. A_i, B_i, D_i, and C_i are known constant matrices with appropriate dimensions. The firing strength of each rule can be determined using *T-norm* product as follows:

$$w_i[z(t)] = \prod_{j=1}^{p} \mu_{ij}[z_j(t)] \tag{19.14}$$

Furthermore, the normalized membership functions are computed as follows:

$$h_i[z(t)] = \frac{w_i[z(t)]}{\sum_{i=1}^{r} w_i[z(t)]} \tag{19.15}$$

Connecting all the rules, the (T-S) fuzzy model takes the following form:

$$\Sigma_{T-S} : \begin{cases} \dot{x}(t) = \sum_{i=1}^{r} h_i[A_i x(t) + B_i u(t) + D_i \varphi(t)] \\ y(t) = \sum_{i=1}^{r} h_i C_i x(t) \end{cases} \tag{19.16}$$

19.3.2 Parallel Distributed Compensation Control

The Parallel Distributed Compensation (PDC) introduced by Wang et al. [23, 24] provides a platform to design a fuzzy controller from a given T-S fuzzy model. The feedback control law for each model rule is designed from the corresponding rule of a T-S fuzzy model. The general structure of each control rule is as follows:

Control Rule i:
IF $z_1(t)$ is about $\mu_{i1}[z_1(t)]$, ..., $z_p(t)$ is about $\mu_{ip}[z_p(t)]$, THEN

$$u_i(t) = -K_i x(t) \qquad i = 1, 2, \ldots, r \tag{19.17}$$

where K_i represents the control gain. The overall control input is formulated by fuzzy blending as follows:

$$u(t) = -\sum_{i=1}^{r} h_i(z(t)) K_i x(t) \tag{19.18}$$

It should be noted that we are considering that the initial states are unknown but bounded such that $||x(0)|| < \phi$ and the amplitude of the actuators is constrained such that $||u(t)|| < \mu \; \forall \, t \geq 0$

19.4 Fuzzy Controller Design

19.4.1 *Fuzzy Controller Design for System with Disturbances*

The disturbance rejection can be realized by minimizing γ Subject to

$$\sup_{||\varphi(t)||_2 \neq 0} \frac{||y(t)||_2}{||\varphi(t)||_2} \leq \gamma \tag{19.19}$$

The fuzzy controller design is formulated as an optimization problem using linear matrix inequalities (LMIs). The control feedback gains K_i stabilizing the system given by Eq. (19.16), and satisfying the control input constraint $||u(t)|| < \mu$, and unknown bounded initial condition $||x(0)|| < \phi$ can be achieved by solving the following linear matrix inequalities (LMIs).

$\underset{x, M_1, \ldots, M_r}{\text{minimize}} \, \gamma^2$

Subject to:

$$X > 0; \begin{bmatrix} M(1,1) & -\frac{1}{2}(D_i + D_j) & \frac{1}{2}X(C_i + C_j)^T \\ -\frac{1}{2}(D_i + D_j)^T & \gamma^2 I & 0 \\ \frac{1}{2}(C_i + C_j)X & 0 & I \end{bmatrix} \geq 0 \tag{19.20}$$

$$\begin{bmatrix} 1 & x_0^T \\ x_0 & X \end{bmatrix} \geq 0; \begin{bmatrix} X & M_i^T \\ M_i & \mu^2 I \end{bmatrix} \geq 0 \tag{19.21}$$

$$\phi^2 I \leq X \tag{19.22}$$

where

$$\begin{aligned} M(1,1) = -\frac{1}{2}(XA_i^T - M_j B_i^T + A_i X - B_i M_j \\ + XA_j^T - M_i^T B_j^T + A_j X - B_j M_i) \end{aligned}$$

such that $i, j = 1, 2, \ldots, r$, $i \leq j \; s.t. \; h_i \cap h_j \neq \emptyset$, and X is a positive definite matrix. If a feasible solution exists, the feedback gain can be computed as

$$K_i = M_i X^{-1} \tag{19.23}$$

To see the detailed proof, we refer the reader to [24].

Table 19.1 Spacecraft parameters

Parameters	Value
spacecraft mass m [kg]	600
spacecraft moment of inertia I [km/m^2]	720
mass of the fuel m_f [kg]	100
moment of inertia of the fuel I_f [kg/m^2]	90
length of the pendulum a [m]	0.2
distance to the center of mass b [m]	0.3
engine thrust F [N]	2300
dissipation constant ϵ [kg m$^2/s$]	0.19

19.4.2 Numerical Simulation

The fuzzy controller Eq. (19.18) developed and implemented for a spacecraft to stabilize the pitch angle θ, the altitude z, and to suppress the sloshing motion ψ. The physical parameters of the spacecraft are listed in Table 19.1.

The efficacy of the fuzzy controller in Eq. (19.18) is tested by considering a set of two rules for each premise variable θ, $\dot{\theta}$, ψ and $\dot{\psi}$ as shown in the Fig. 19.2. A local sector nonlinearity approach was used to obtain a linear fuzzy model from the nonlinear system. Therefore, with a sufficient number of fuzzy rules, a nonlinear model can be represented precisely by a linear fuzzy model within a local interval. Since we used two rules for each premise variable, sixteen local fuzzy models are obtained and combined to describe the nonlinear model accurately.

The unknown initial state are bounded by $||x(0)|| \leq 10^4$ where for simulation purposes, $x(0) = [0 \ -230 \ 2° \ 30° \ 10000 \ 350 \ 0.57° 0]^T$. The LMI given by Eq. (19.22) allow for a simulation to start from any initial condition as long as $||x(0)|| \leq \phi = 10^4$. We use Matlab toolbox YALMIP [25] to solve the set of the LMIs in Eqs. (19.20–19.22). YALMIP is a modeling language for solving optimization problems.

Overall, the performance of the T-S fuzzy controller is very satisfactory. The simulation results are shown in Figs. 19.3, 19.4, 19.5, 19.6 and 19.7. Figure 19.3 shows the position of the spacecraft's center of mass and the velocity in the $x-$direction. A longer simulation need to be performed to observe the behavior of the position of the spacecraftin the x-direction.

Figure 19.4 shows the position and the velocity in the z-direction.

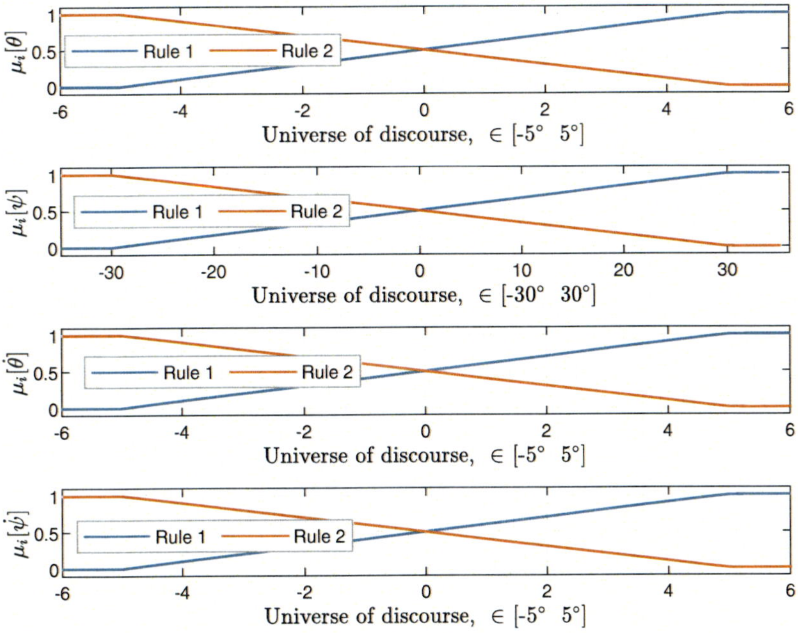

Fig. 19.2 Fuzzy membership function

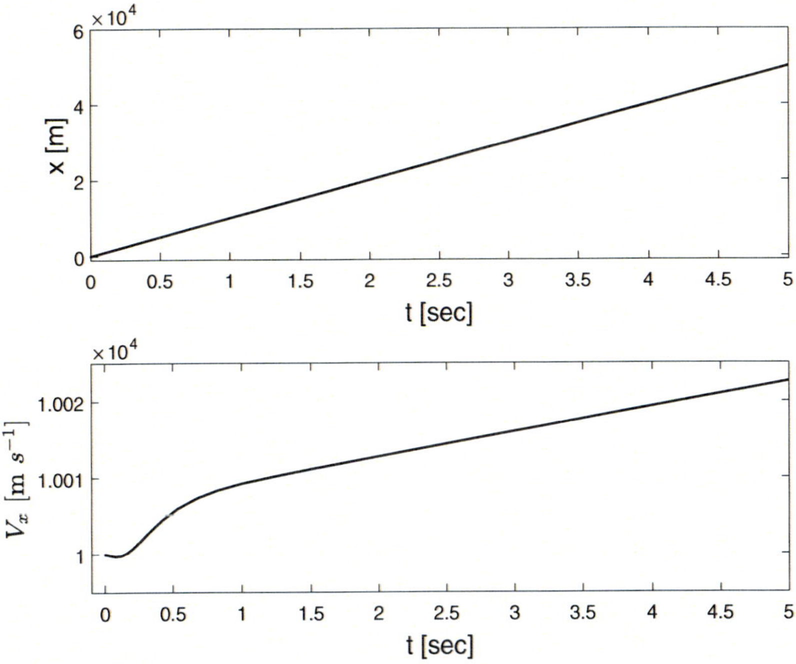

Fig. 19.3 Position and velocity in the x-direction of the spacecraft

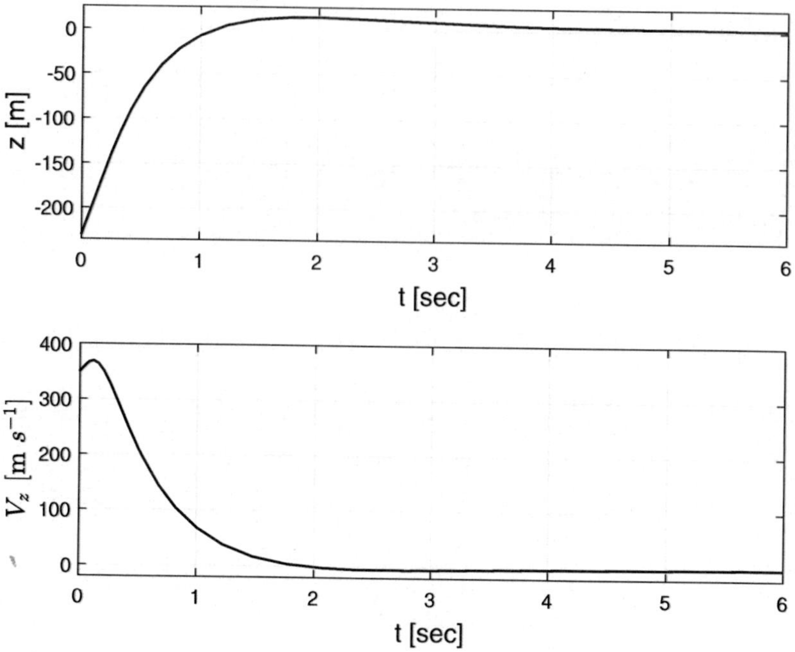

Fig. 19.4 Position and velocity in the z-direction of the spacecraft

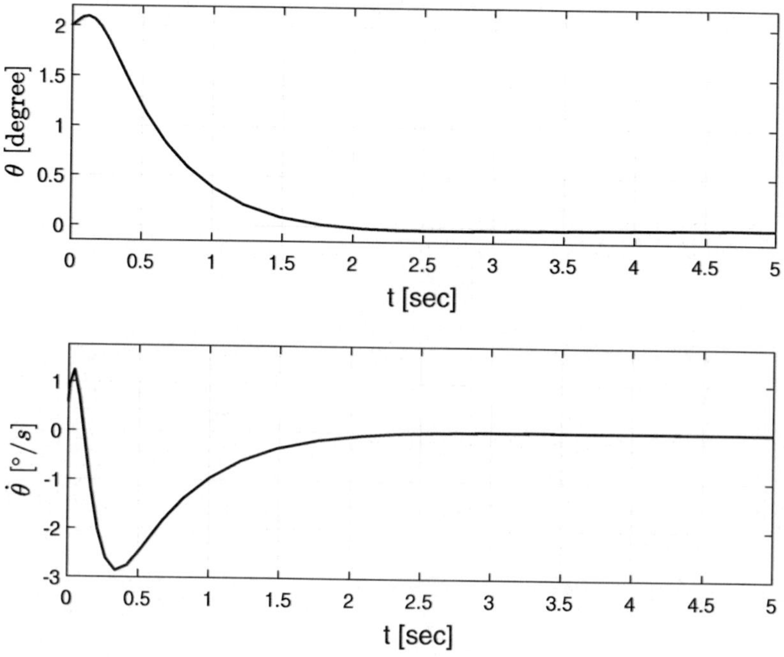

Fig. 19.5 Angular position and angular velocity of the spacecraft

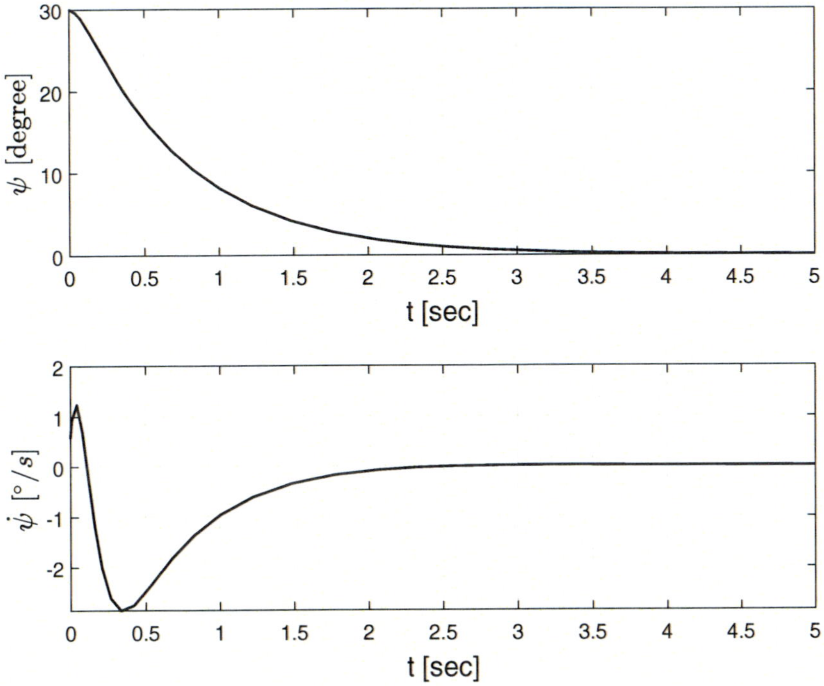

Fig. 19.6 Angular position and angular velocity of the pendulum

The fuzzy controller can stabilize the spacecraft's altitude and suppress the vertical velocity. Figures 19.5 and 19.6 shows the angular position and the angular velocity of the spacecraft and the pendulum, respectively. The fuel sloshing is completely eliminated. Finally, the amplitude of the control input is shown in Fig. 19.7.

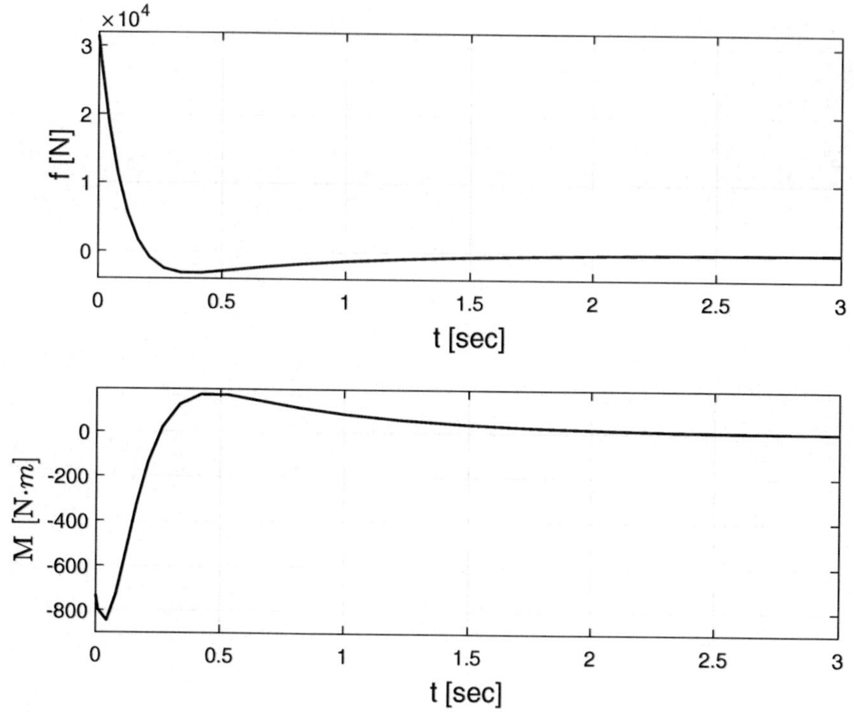

Fig. 19.7 Control input

19.4.3 Optimal Fuzzy Control

The main objective of the proposed fuzzy controller is to minimize the upper bound of the following quadratic performance cost function, assuming the initial condition x_0 is known.

$$J = \int_0^{\infty} \left[y^T \, W \, y(t) + u^T(t) \, R \, u(t) \right] dt \; < \; x_0^T \, X^{-1} \, x_0 \; < \; \lambda \tag{19.24}$$

where $y(t)$ is the measured output, $u(t)$ is the control input, W and R are weighting coefficient matrices. it can be shown [24] that the optimal fuzzy controller can be formulated as a multi-objective optimization problem as

$\underset{x, M_1, \ldots, M_r}{\text{minimize}} \, \lambda$

Subject to:

$$X > 0, \quad \begin{bmatrix} \lambda & x_0^T \\ x_0 & X \end{bmatrix} > 0 \tag{19.25}$$

$$U_{ii} = \begin{bmatrix} \begin{pmatrix} XA_i^T + A_iX \\ -B_iM_i - M_i^TB_i^T \end{pmatrix} & * & * \\ C_iX & -W^{-1} & 0 \\ -M_i & 0 & -R^{-1} \end{bmatrix} < 0 \tag{19.26}$$

$$V_{ij} = \begin{bmatrix} \begin{pmatrix} XA_i^T + A_iX - B_iM_j - M_j^TB_i^T \\ +XA_j^T + A_jX - B_jM_i - M_i^TB_j^T \end{pmatrix} & * & * & * & * \\ C_iX & -W^{-1} & 0 & 0 & 0 \\ -M_j & 0 & -R^{-1} & 0 & 0 \\ C_jX & 0 & 0 & -W^{-1} & 0 \\ -M_i & 0 & 0 & 0 & -R^{-1} \end{bmatrix} < 0 \tag{19.27}$$

$$i, j = 1, 2, \ldots, r, i < j \ s.t. \ h_i \cap h_j \neq \emptyset$$

If a feasible solution of the LMI's (19.25–19.27) exists, then the closed-loop is asymptotically stable, and the controller feedback gains can be computed as:

$$K_i = M_iX^{-1} \qquad i = 1, 2, \ldots, r. \tag{19.28}$$

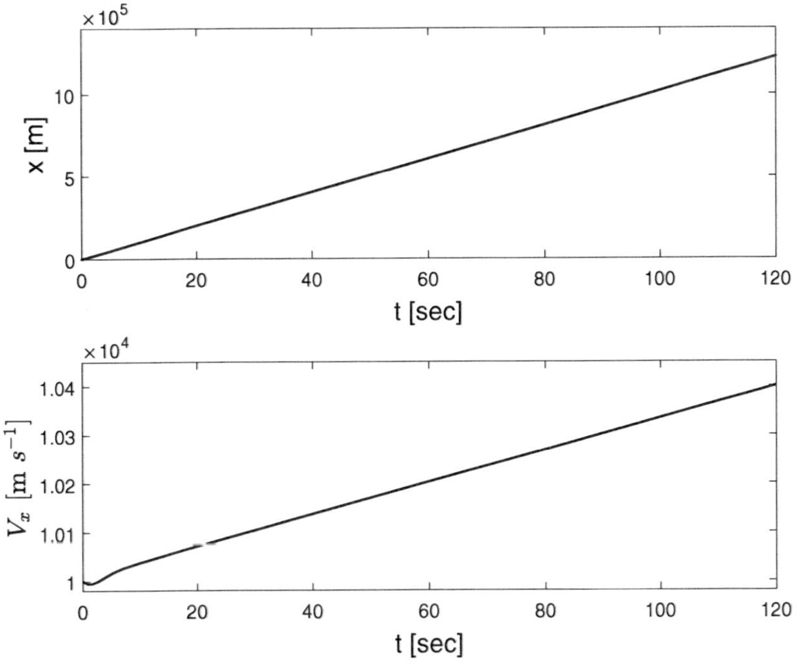

Fig. 19.8 Position and velocity in the x-direction of the spacecraft

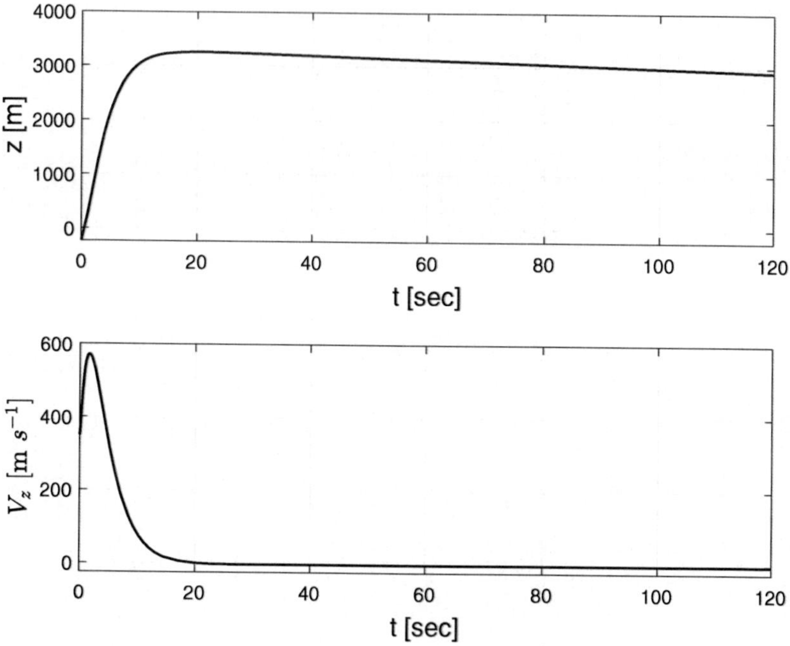

Fig. 19.9 Position and velocity in the z-direction of the spacecraft

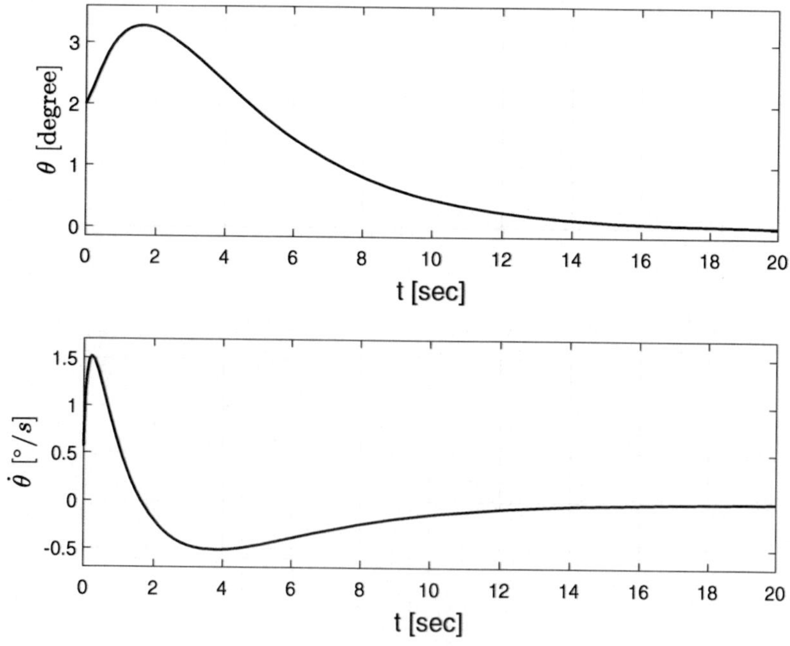

Fig. 19.10 Angular position and angular velocity of the spacecraft

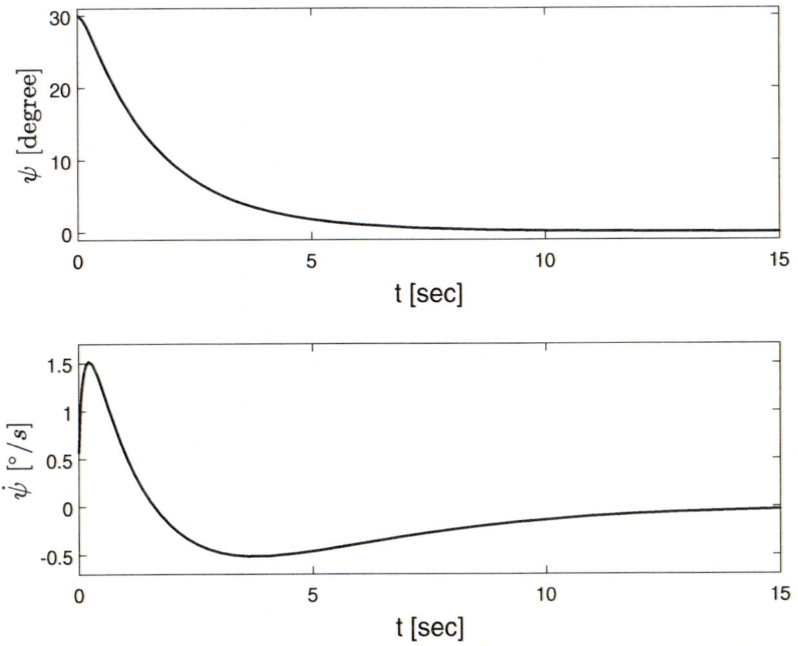

Fig. 19.11 Angular position and angular velocity of the pendulum

To see the detailed proof, we refer the reader to [24].

19.4.4 Numerical Simulation

The performance of the optimal fuzzy controller with feedback gains calculated by solving the LMI's (19.25–19.27) is tested for the same spacecraft shown in Fig. 19.1 with parameters presented in Table 19.1 and using the same set of fuzzy rules shown in Fig. 19.2.

The numerical simulation of the closed-loop is shown in Figs. 19.8, 19.9, 19.10, 19.11 and 19.12. Overall, the performance of the optimal fuzzy controller is good, and we notice that the controller stabilizes the states over a more extended period, and the amplitude of the control input f and M are less than the one obtained from the controller for the system with disturbances (Fig. 19.8).

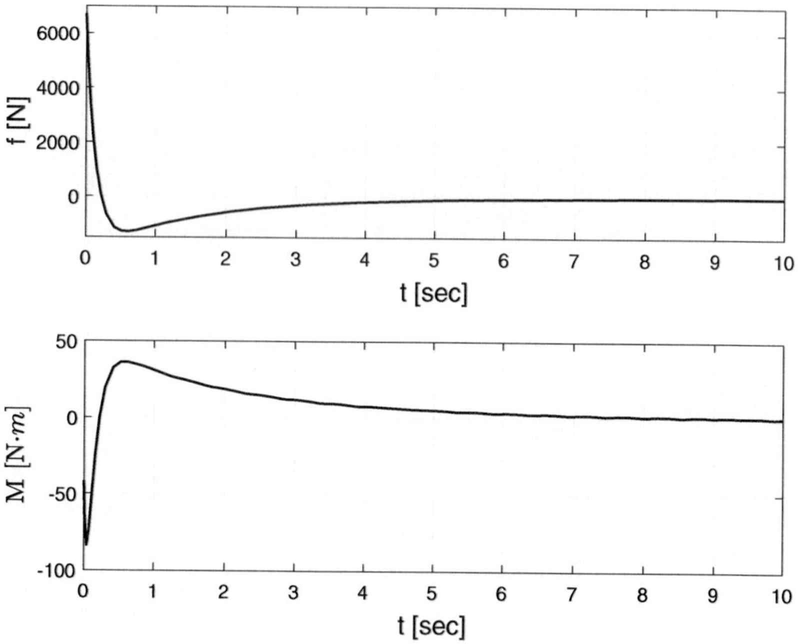

Fig. 19.12 Control input

19.5 Conclusion

The fuzzy logic controller presented is a nonlinear controller with a simple structure based on a linear model. They cover a more comprehensive range of operating conditions and are more readily customizable in natural language. They are quick to comprehend conceptually. They are tolerant of unmodeled dynamics, uncertainty, and incorrect data. In addition, the actuator amplitude saturation and unknown initial condition can be easily incorporated as linear matrix inequalities and solved simultaneously. This advantage allows the design engineer to choose the best actuator to perform the task appropriately.

Nevertheless, the feasibility of the solution is not guaranteed since it is often challenging to find a common positive definite matrix X such that all the constraints are satisfied. For instance, in this particular problem and using only two rules for each premise variable, we must solve 138 constraints with a common positive definite matrix X. Another problem is tuning the controller to obtain the best performance.

References

1. Cho, S., McClamroch, N.H., Reyhanoglu, M.: Dynamics of multibody vehicles and their formulation as nonlinear control systems. In: Proceedings of American Control Conference, pp. 3908–3912 (2000a)
2. Dodge, F.: The New Dynamic Behavior of Liquids in Moving Containers, NASA Rept. SP-106, Southwest Research Institute, San Antonio, TX, USA (2000)
3. Ibrahim, R.A.: Liquid Sloshing Dynamics: Theory and Applications. Cambridge University Press, 1–12, 294–329. Cambridge, England, U.K. (2005)
4. Vreeburg, J.P.B., Veldman, A.E.P.: Transient and sloshing motions in an unsupported container. In: Monti, R. (ed.) Physics of Fluids in Microgravity, pp. 293–321. Taylor and Francis, London (2001)
5. van de Wal, M., de Jager, B.: Control structure design: a survey. In: Proceedings of 1995 American Control Conference—ACC'95, vol. 1, pp. 225–229. Seattle, WA, USA (1995). https://doi.org/10.1109/ACC.1995.529242
6. Yin, S., Xiao, B., Ding, S.X., Zhou, D.: A review on recent development of spacecraft attitude fault tolerant control system. IEEE Trans. Industr. Electron. **63**(5), 3311–3320 (2016). https://doi.org/10.1109/TIE.2016.2530789. May
7. Khan, S.A., Shiyou, Y., Ali, A., Rao, S., Fahad, S., Jing, W., Tong, J., Tahir, M.: Active attitude control for micro spacecraft; a survey and new embedded designs. Adv. Space Res. **69**(10), 3741–3769. ISSN 0273-1177. https://doi.org/10.1016/j.asr.2022.02.020
8. Liu, H., Li, J., Hexi, B.: Sliding mode control for low-thrust Earth-orbiting spacecraft formation maneuvering. Aerosp. Sci. Technol. **10**(7), 636–643 (2006). ISSN 1270-9638. https://doi.org/10.1016/j.ast.2006.04.008
9. Chen, Y.-P., Lo, S.-C.: Design, Sliding-Mode Controller, for Spacecraft Attitude Tracking Maneuvers: American Control Conference, 195–196. San Francisco, CA, USA (1993). https://doi.org/10.23919/ACC.1993.4792835
10. Bang, H., Lee, J.S., Eun, Y.J.: Nonlinear attitude control for a rigid spacecraft by feedback linearization. KSME Int. J. **18**, 203–210 (2004). https://doi.org/10.1007/BF03184729
11. Junkins, John L., Akella, Maruthi R., Robinett, Rush D.: Nonlinear adaptive control of spacecraft maneuvers. J. Guidance Control Dyn. **20**(6), 1104–1110 (1997)
12. Wu, B., Xu, C., Zhang, Y.: Decentralized adaptive control for attitude synchronization of multiple spacecraft via quantized information exchange. Acta Astronautica **175**, 57–65. ISSN 0094-5765. https://doi.org/10.1016/j.actaastro.2020.05.013
13. Xin, M., Pan, H.: Indirect robust control of spacecraft via optimal control solution. IEEE Trans. Aerosp. Electron. Syst. **48**(2), 1798–1809 (2012). https://doi.org/10.1109/TAES.2012.6178102
14. Xin, M., Pan, H.: Nonlinear optimal control of spacecraft approaching a tumbling target. Aerosp. Sci. Technol. **15**(2), 79–89 (2011). ISSN 1270-9638. https://doi.org/10.1016/j.ast.2010.05.009
15. Hassrizal, H.B., Rossiter, J.A.: A survey of control strategies for spacecraft attitude and orientation. In: 2016 UKACC 11th International Conference on Control, IEEE, 31 Aug–2 Sept 2016, Belfast, UK. https://doi.org/10.1109-CONTROL.2016.7737543
16. Tanaka, K., Sugeno, M.: Stability analysis of fuzzy systems using Lyapunov's direct method. In: Proceedings of NAFIPS'90, Toronto, ON, Canada, June 1990, pp. 133–136
17. Langari, R., Tomizuka, M.: Analysis and synthesis of fuzzy linguistic control systems. In: Proceedings of ASME Winter Annual Meeting. Dallas, TX, Nov 1990, 35–42
18. Singh, S.: Stability analysis of discrete fuzzy control systems. In: Proceedings of 1st IEEE International Conference on Fuzzy Systems, San Diego, CA, Mar 1992, pp. 527–534
19. Sugeno, M.: An introductory survey of fuzzy control. Inf. Sci. **36**(1–2), 59–83. ISSN 0020-0255. https://doi.org/10.1016/0020-0255(85)90026-X
20. Nguyen, A.-T., Taniguchi, T., Eciolaza, L., Campos, V., Palhares, R., Sugeno, M.: Fuzzy control systems: past, present and future. IEEE Comput. Intell. Mag. **14**(1), 56–68 (2019). https://doi.org/10.1109/MCI.2018.2881644. Feb

21. Reyhanoglu, M., Rubio Hervas, J.: Nonlinear control of a spacecraft with multiple fuel slosh modes. In: 2011 50th IEEE Conference on Decision and Control and European Control Conference, Orlando, FL, USA, 2011, pp. 6192–6197. https://doi.org/10.1109/CDC.2011.6160660
22. Meirovitch, L., Kwak, K.M.: Dynamics and control of spacecraft with retargeting flexible antennas. J. Guidance Control Dyn. **13**(2) (1990)
23. Wang, H.O., Tanaka, K.: Parallel distributed compensation of nonlinear systems by Takagi-Sugeno fuzzy model. In: Proceedings of 1995 IEEE International Conference on Fuzzy Systems, vol. 2, pp. 531–538. https://doi.org/10.1109/FUZZY.1995.409737
24. Tanaka, K., Ikeda, T., Wang, H.O.: Robust stabilization of a class of uncertain nonlinear system via fuzzy control: quadratic stabilizability, H_∞ control theory, and linear matrix inequalities. IEEE Trans. Fuzzy Syst. **4**(1) (1996)
25. Löfberg, J.: YALMIP: a toolbox for modeling and optimization in MATLAB. In: Proceedings of the CACSD Conference, Taipei, Taiwan, 2004

Chapter 20
Microwave Thermal Rocket Engine-Based Orbital Launch System Implementation

Akhror Agzamov, Abdulazal Toshkhujaev, and Makhsud Yusupov

Abstract Chemical rocket engines' specific impulse (Isp) limitations hinder the introduction of fully reusable single-stage launch vehicles with high payload mass-fraction. Modern launchers rely on multi-staging, face reusability difficulties, and resulting in a low payload mass-fraction compared to other vehicles. Microwave Thermal Rocket (MTR) method was explored as a more efficient solution but was abandoned due development cost risks. There are three types of rocket engines for orbital launch: (1) Chemical engines, (2) nuclear thermal rocket (NTR) engine, and (3) beam-powered engines (microwave or laser). Chemical engines provide low Isp, but high thrust-to-weight ratio (T/W). NTR engines offer high Isp but have low T/W and cause radiation pollution. MTR engines provide similar to NTR Isp and T/W similar to chemical engines without radiation pollution. MTR differ from conventional rockets as the power source is ground-based. The engine and ground-based beam director communicate to adjust power intensity and guidance parameters for autonomous ascent. This communication is to be done without human interaction, which is the ability of autonomy. However, as strong interference may put the viability of this essential feature in doubt. We propose a solution to address this problem. Research also proposes a commercially viable microwave-powered launch method utilizing reduced frequency Gyrotrons, air-launch method, and direct methane heating design to reduce research, development, and infrastructure costs. MTR engines can provide high Isp, enabling fully reusable single-stage vehicles with higher payload mass-fraction. Our approach trades off maximum performance for cost reduction, aligning with private investing market needs and supporting the trend of commercial space expansion.

A. Agzamov (✉) · A. Toshkhujaev · M. Yusupov
Spaceborne Corporation, Turkkurgan 31, Tashkent, Uzbekistan 100138
e-mail: agzamov@spacebornecorp.com

A. Toshkhujaev
e-mail: toshkhujaev@spacebornecorp.com

M. Yusupov
e-mail: yusupov@spacebornecorp.com

D. Azimov (ed.), *Proceedings of the IUTAM Symposium on Optimal Guidance and Control for Autonomous Systems 2023*, IUTAM Bookseries 40,
https://doi.org/10.1007/978-3-031-39303-7_19

Keywords Microwave propulsion · Space launch · Beamed energy propulsion

20.1 Introduction

20.1.1 The Problem of Space Launch and the Need for a Step-Up from Chemical Propulsion

Traditional chemical propulsion-based launch vehicles lack payload lift capability due to the natural limits of their energy source—chemical fuel mixtures. Reaching an orbital velocity by burning chemical fuel mixtures requires multi-stage and semi-reusable or completely expendable vehicles to be applied. The payload ratio is meager, around 4%, especially compared to other vehicles; for example, cargo aircraft might have ten times for payload ratio [1].

Besides the lack of payload ratio, the expendable nature of rocket vehicles means that the launch cost heavily depends on manufacturing vehicles for each flight mission. Designing fully reusable rocket vehicles faces significant technological challenges due to an even more reduction of payload mass fraction; for small launch vehicles, it is barely possible at the current state-of-the-art material science [2].

Both factors mentioned above naturally lead to prohibitively high launch costs. The commercial side of the space industry focuses only on the products that can bring significant value [3] being launched on a costly launch service. Significant launch cost reduction to the levels comparable with other types of transportation will presumably open opportunities for massive use of space for commercial purposes.

The natural energy limitations of chemical propulsion are not a solvable problem at the current understanding of physics, and the typical way of improvement relies on (1) improving the dry-to-wet mass ratio of rocket stages, (2) reducing the cost of production, or (3) introducing reusability technologies. The above methods do not promise significant improvement and remain minor advancements [2]. The technology reaches its natural ceiling; the major leaps in space launch technology cannot be promised by chemical propulsion-based launch vehicles. An introduction of completely novel launch technology is needed to resolve the above-stated problems.

Nuclear Thermal Rocket (NTR) engines are among the options considered for space launch applications due to their extremely high specific impulse, reaching up to 900 s [4]. Despite this advantage, NTRs face drawbacks such as a low thrust-to-weight ratio, falling below 35 [1], and the considerable issue of radioactive pollution, which negatively impacts their overall appeal.

20.1.2 Beamed Energy Space Launch Technology

The method of powering rocket engines via external direct energy sources was first introduced by Konstantin Tsiolkovsky in 1924 [5], named "parallel beam of shortwave electromagnetic rays". However, the author admitted that the (1) microwave generator, (2) beam focusing, (3) directing, and (4) receiver material science technologies were not mature enough.

An artificially generated energy beam, either a microwave or laser, might carry various energy densities. Which indeed far exceed what can be extracted from chemical combustion energy sources. Thus, it leads to an ability to design rocket engines with significantly higher specific impulse. With the current state-of-the-art material science, thermal-type rocket engines can reach specific impulse, similar to nuclear thermal engines, which are under 900 s of specific impulse [1]. However, it depends on the reaction mass, molar weight, and temperature (Fig. 20.1).

The method is based on heating the reaction mass and then converting the energy of the heated gas into jet thrust. Various researchers suggest different types of realization of this method. It might vary from (1) pulsed bursts or (2) continuous flow of energy or be based on transferring energy on a (1) laser or (2) microwave beam. The pulsed beam concepts are typically (1) pulse detonation engines [6] or (2) ablative thrusters. On the other side, the continuous energy beam concepts are typically (1) thermal rockets, (2) light sails, or (3) photonic thrusters [7].

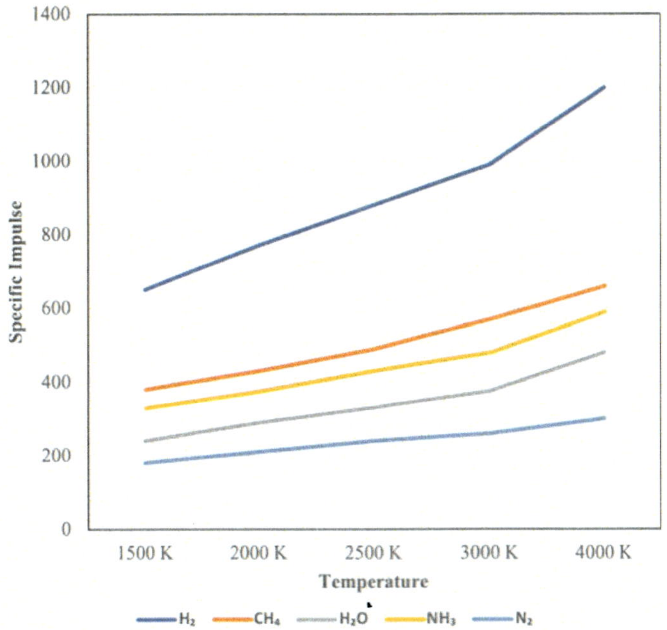

Fig. 20.1 Specific impulse parameters of different types of reaction mass

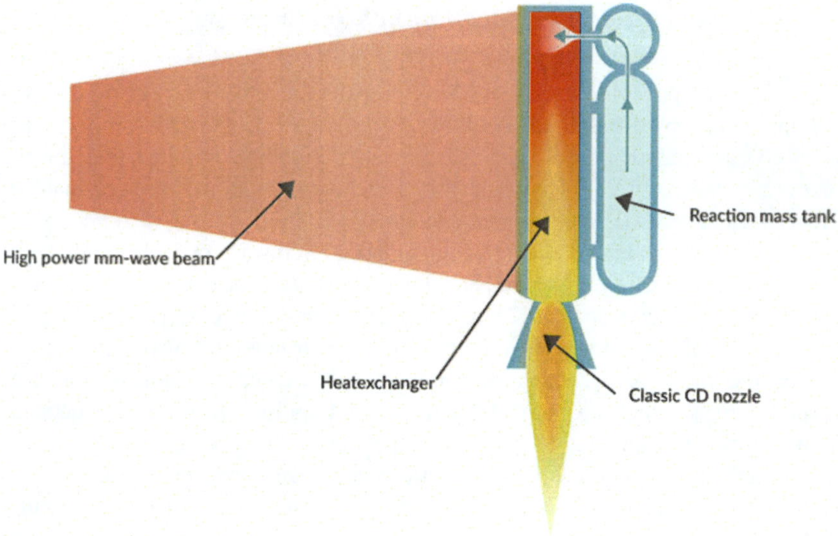

Fig. 20.2 Microwave thermal rocket engine schematic image

The (1) energy storage, (2) microwave generator, and (3) beam director are located on Earth and serve as a stationary energy source that transmits power wirelessly to the rocket engine system (Fig. 20.2) [7]. The core concept of beamed energy propulsion is based on leaving the energy source, which typically constitutes a significant mass, out of the rocket on the ground. As there is no theoretical limit to the energy source and the power of the beam, unlike in chemical fuels, the technology promises significant improvements with the improvement of material science, primarily via improvements in handling high-temperature gas flows.

20.2 Benefits of Beamed Energy Propulsion Based on Microwave Thermal Rockets

Unlike other concepts, the microwave continuous thermal engine concept has an outstanding readiness level for realization. The (1) multi-megawatt continuous wave microwave generators, (2) beam combining, and (3) receiver material science have a Technology Readiness Level of 8–9 according to NASA classification [7]. This concept is typically proposed for the space launch application, as it is most suitable due to its mm-wave diffraction limits and the ability to generate and direct gigawatt scale beams, which are required for space launch [1].

Another differential point is that the microwave thermal engine concept allows multiple methods of energy absorption. Those can be (1) direct heating of heat exchanger, or (2) placing the microwave susceptor material inside of microwave transparent cavity, or (3) direct reaction mass heating [1]. Unlike the continuous laser beam concept, where energy is absorbed directly by the outer side of the heat exchanger, above mentioned variety of energy absorption methods promises better performance and efficiency [8].

The higher specific impulse naturally drives a higher payload ratio. The thrust-to-weight ratio of such a thermal engine system is driven primarily by reaction mass density [1]. Thus, it does not disrupt the high specific impulse advantage as it happens with nuclear rocket engines [1, 7].

Considering the extra performance driven by higher specific impulse, a single-stage rocket becomes a preferred solution at some moment. From an economic point of view, adding reusability features is another option that optimizes launch costs [9].

As there is no combustion, the reaction mass is homogenious, thus requiring one type of onboard storage tank, valves, and pumps. Essentially, the factors mentioned above reduce vehicle complexity too [9].

20.3 Implementation Problems of Microwave Thermal Launch Systems and Proposed Solutions

20.3.1 Suggested Components of the Concept and System Description

The microwave thermal launch propulsion system consists of (1) ground and (2) onboard components. The ground component comprises (1) energy storage or source, (2) energy distribution, (3) microwave generators, and (4) beam director. The onboard component comprises a (1) reaction mass storage and (2) distribution system, a (3) planar power receiver, and a (4) jet nozzle [1, 7].

Most researchers suggest using hydrogen as a reaction mass due to its lowest molecular mass and significantly higher specific impulse [1, 7, 9]. The system design supposes that a thermal engine depends on a continuous external energy flow. At the stated conditions, the ground take-off scheme requires placing a (1) beam director close to the launch pad and another (2) beam director along the launch trajectory line to satisfy the propulsion system's tracking and power density requirements (Fig. 20.2) [9, 10].

The following methods are suggested for energy receiving and conversion (1) direct heating of the heat exchanger on board the launch vehicle, (2) cooling via transferring the heat to the reaction mass, and then (3) expelling it through the jet nozzle [10, 11].

20.3.2 Launch Control and Monitoring

Establishing feedback and control loop sequence for a microwave beam launch system is required for defining the critical parameters to monitor and control during the launch process. Control adjustments are critical according to these parameters as it is the key feature of adaptive launch control algorithms, for example Powered Explicit Guidance [12].

During the launch, the feedback and control system must continuously monitor the following critical parameters in order to adjust main microwave beam power intensity:

- Thrust to weight ratio
- Delta-v
- Spacecraft altitude and velocity
- Spacecraft attitude (pitch, roll, yaw) [12].

However, that's challenging to establish direct radio communication due to strong interference and highly probable damaging factor to on-board communication system from high power microwave beam that points onto the launch vehicle [7]. The satellite relay communication is viable due to the satellite communication antenna on the launch vehicle can be placed on top side of the vehicle in the shadow of high-power microwave beam, however this method possesses high latency (estimated latency might be up to 1000 ms) which considerably limits the guidance systems ability to respond accordingly to quickly changing conditions. Studied literature on microwave beamed launch method do not provide any viable solutions to the problem, mostly focusing on other problems.

Proposed solution

Using a combination of ground-based radar [13] and infrared telescopes [14, 15] to indirectly monitor the launch vehicle's performance and additional relay satellite-based data link [16] can be a viable approach. The proposed method would involve the following steps:

1. Ground-based radar: Tracks the launch vehicle's velocity, altitude, acceleration, and attitude using a radar system. This data would provide insights into the vehicle's trajectory and overall performance. Through knowing the acceleration, the system can analytically estimate the thrust to weight ratio:

$$\frac{T}{W} = \frac{a}{g} \tag{20.1}$$

where $\frac{T}{W}$ is the thrust-to-weight ratio, a is the acceleration, and g is the gravitational acceleration (approximately 9.81 m/s^2).
2. Infrared telescope: Monitors the engine heat signature using an infrared telescope to estimate specific impulse and energy consumption. This information would help assess the efficiency of the microwave beam energy absorption and

conversion into thrust. Infrared telescope is capable to retrieve the temperature data from the engine plume [14], which allows to estimate the Specific Impulse:

$$Isp = V_e g, \tag{20.2}$$

where

$$v_e = \sqrt{\left(\frac{2k}{k-1}\right) RT_c \left(1 - \left(\frac{p_2}{p_1}\right)^{\frac{k-1}{k}}\right)} \tag{20.3}$$

where

V_e is the exhaust velocity
k is the specific heat ratio (adiabatic index) of the exhaust gas
R is the specific gas constant of the exhaust gas
T_c is the temperature of propellant in the heat exchanger
P_2 is the pressure at nozzle exit
P_1 is the pressure of propellant in the heat exchanger

3. Satellite relay communication system: A relay satellite is needed to establish a communication link with the launch vehicle and provide actual telemetry data to the ground command station. This system should be designed to handle the high-speed transmission of large data volumes and provide as low as possible latency communication between the launch vehicle and the ground command station.

   The actual telemetry data, despite being delivered with latency, can still be helpful to provide information on the engine thrust, propellant mass flow rate (which helps to estimate vehicle mass), and heat exchanger pressure, which are needed to make final estimations of Delta-v. The telemetry data also will provide data to adjust analytic estimations.

$$\Delta v = I_{sp} \cdot g_0 \cdot \ln\left(\frac{m_0}{m_f}\right) \tag{20.4}$$

   where

   Δv: Delta-v, the change in velocity (m/s)
   $I_s p$: Specific impulse, a measure of how effectively a rocket uses propellant (seconds)
   g_0: Standard acceleration due to gravity (approximately 9.81 m/s^2)
   m_0: Initial mass of the rocket, including propellant (kg)
   m_f: Final mass of the rocket after expelling the propellant (kg)

4. Launch control computer: The telemetry data from both the indirect monitoring system and the satellite relay data link should be should be combined to evaluate the launch vehicle's performance and adjust the microwave beam tracking and

power density. The indirect monitoring system can serve as a baseline and provide the initial estimates for the launch vehicle's performance, while the satellite relay data link system can provide the real-time telemetry data to refine and adjust these estimates.

5. Beam adjustment: based on the data analysis, the microwave beam tracking and power density adjusted to maintain the desired trajectory, thrust to weight ratio, and energy absorption efficiency.

In the event of a failure or anomaly detected by the indirect monitoring system or transmitted via satellite relay data link, the ground station should be able to abort the mission and initiate contingency plans. The ground station should have the necessary control mechanisms in place to safely and effectively abort the launch, including the ability to adjust the microwave beam energy transfer, control the launch vehicle's attitude, and execute emergency procedures.

By combining indirect monitoring with a satellite relay data link system, it is possible to establish a robust and reliable communication and control system for the microwave beam launch technology. The indirect monitoring system can serve as the primary method for performance monitoring, while the satellite relay data link system can provide real-time data for refining and adjusting the analytical assumptions made by the indirect monitoring system. Additionally, the satellite relay communication system can provide a direct link to the launch vehicle for critical control commands and emergency procedures, enhancing the overall safety and reliability of the launch process.

The logic of the control loop for the combined indirect monitoring and satellite relay communication system can be outlined in the following steps

1. Initialization: Before the launch, the ground-based radar, infrared telescopes, and the satellite relay satellite must be set up and calibrated to ensure accurate data collection and transmission. The communication link between the launch vehicle and the ground station via the relay satellite must be established.
2. Launch phase monitoring: During the launch, the critical parameters must be continuously monitored using both the indirect monitoring system (ground-based radar and infrared telescope) and the satellite relay communication system. Data collected on the launch vehicle's velocity, altitude, attitude, engine heat signature, microwave beam alignment, main microwave power consumption.
3. Data integration and analysis: Data from the indirect monitoring system and the relay satellite data link must be merged at the ground command station. The combined data must be processed and analyzed to assess the launch vehicle's performance, trajectory, and microwave beam energy absorption.
4. Decision-making: Based on the integrated data analysis, it is determined whether any adjustments are needed to the microwave beam parameters, launch vehicle attitude, or propulsion system to maintain the desired trajectory, thrust to weight ratio, and energy absorption efficiency.

5. Control command generation: If adjustments are required, appropriate control commands are generated to modify the microwave beam parameters (e.g., intensity, phase, or focus) or to instruct the launch vehicle to adjust its attitude or propulsion system.
6. Command transmission: The control commands are transmitted to the launch vehicle using the established communication link through the relay satellite, with latency in the communication process accounted for.
7. Command execution: The launch vehicle receives the control commands and executes the necessary adjustments to its attitude or propulsion system, as well as providing feedback on the success of the executed commands.
8. Loop iteration: The control loop is continuously iterated throughout the launch process, with the ground station constantly monitoring the launch vehicle's performance, generating control commands, and adjusting the microwave beam parameters as needed to ensure optimal propulsion performance, trajectory control, and thrust to weight ratio.

Implementing this control loop logic, presumed to maintain a seamless flow of information between the ground station, the indirect monitoring system, and the launch vehicle. This enables real-time adjustments and ensures the success of the microwave beam launch while taking advantage of the benefits offered by the combined indirect monitoring and relay satellite communication approach.

Radar System requirements

For tracking the launch vehicle's altitude, velocity, and attitude, a high-resolution tracking radar system capable of providing precise and accurate measurements over long distances is needed. Some key features to look for in a suitable radar system include:

- High-frequency operation (e.g., X-band or higher) for improved resolution and accuracy.
- Phased-array or electronically steered antenna for rapid and precise target tracking.
- Adaptive beamforming and signal processing capabilities to minimize interference and noise.
- Weather-resistant design to ensure reliable operation under various environmental conditions.

Infrared Telescope requirements

For monitoring the engine heat signature, a high-performance infrared telescope system with adequate sensitivity, resolution, and range is required. Some essential features to consider in a suitable infrared telescope system include

- Mid-wave or long-wave infrared (MWIR or LWIR) detection capability for better sensitivity to the engine's heat signature.
- High-resolution sensor to accurately detect and track the launch vehicle's thermal emissions.

- Adequate cooling system (e.g., Stirling or pulse-tube cryocoolers) to ensure optimal sensor performance and sensitivity.
- Robust tracking system that can maintain a lock on the launch vehicle throughout its ascent.

Integration with advanced image processing and analysis software to extract relevant information from the collected data.

20.3.3 Economic Problems of Microwave Thermal Launch Systems

The application of hydrogen as a reaction mass causes technical complexity, consequently leading to cost growth. First of all, hydrogen has a low density (71 kg/m^3) which forces to include bulk tanks and use more heat shielding in the case of reusable vehicles. Even extra specific impulse and thus less mass of needed hydrogen does not allow avoiding the application of bulk tanks. Hydrogen requires a more expensive storage and distribution system with higher handling standards. Another problem with hydrogen is that microwaves cannot directly heat it, which causes certain limitations on the variety of propulsion system options.

The requirement for using two beam directors significantly limits the launch inclinations available for the launch system, as the vehicle must follow the line between two beam directors. Another problem is the two separate facilities' high capital expenditures for construction and maintenance costs.

Studies suggest using Gyrotrons as a ground-based microwave generator [10]. A Gyrotron is a maser capable of producing sub-terahertz electromagnetic waves at continuous wave power of up to 2.5 MW with a typical plug-to-beam efficiency of 50%. For the launch application, it is suggested to use 140 or 170 GHz class gyrotrons. However, to power a one metric-ton payload lifting launch vehicle, it is required to simultaneously use under 1000 gyrotrons at each ground station [1]. The economic problem is that gyrotrons have a significant market price of around 1–2 million USD, which means only gyrotrons would cost 2–4 billion USD for the 1-ton class launch vehicle powering.

At given conditions, the economic state of suggested configurations of the microwave launch systems is not competitive with conventional launch systems, which we will see in the «The outcome» section.

Proposed solution

As a solution to the hydrogen problem, methane can be a better alternative. Methane provides the closest to hydrogen specific impulse and, at the same time, has a significantly higher density (422 kg/m^3). In the case of reusable vehicles, because smaller tanks require less heat shielding, the methane-based launch vehicle wins in the payload mass ratio [1]. Another advantage is directly heating the methane reaction mass in the microwave transparent cavity. With advanced cooling solutions, this method promises much higher energies to be delivered than other methods. Methane also requires less demanding equipment standards.

The air launch method can be applied to overcome the requirement for two beam directors, thus reducing the ground-based beam director cost twice the initial assumption. The launch vehicle has to (1) take off from a runway, on the carrier aircraft or using an onboard jet engine, (2) raise an altitude of around 15 km, (3) approach the beam acquisition point at a distance of around 120 km from the beam director (Fig. 20.3). With this approach, the orbit inclination problem is also naturally solved.

Gyrotrons can be more economically efficient if we use 95 Ghz class gyrotrons because, in this case, it is possible to use cheaper non-cryo magnets, one of the significant cost drivers, and cheaper equipment, including less precision antennas.

Another step up is to build larger power supply units that can power up to 4 gyrotrons, as we found an effect economy of scale, which another drastic cost-reducing factor. A gyrotron cost might drop to $100 k.

The outcome

The following equation is determined to define the launch cost:

$$Lcp = \frac{Lec}{Pc} + \frac{\left(\frac{Lvc}{Lvf}\right) + \left(\frac{Ecq*Caw}{Alr}\right) + \left(\frac{Gbd*Aim}{Alr}\right)}{Pc} \tag{20.5}$$

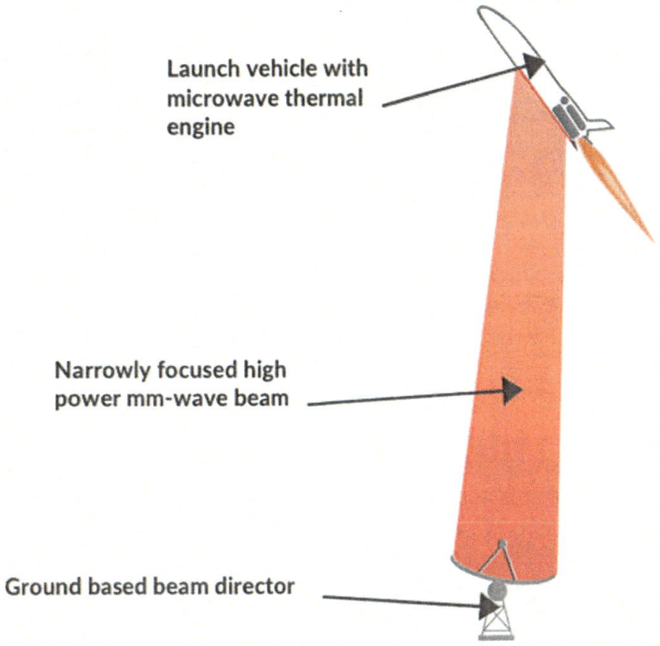

Fig. 20.3 Microwave beam launch conceptual illustration

where,

$$Lec = \frac{\left(\frac{Pc}{0.25}\right) * 300 * 1000}{3600} * 0.0726 \quad (20.6)$$

And,

$$Los = \left(\frac{Lvc}{Lvf}\right) + \left(\frac{Ecq * Caw}{Alr}\right) + \left(\frac{Gbd * Aim}{Alr}\right) \quad (20.7)$$

$\frac{Pc}{0.25}$ is the power required from wall plug at launch in MW, assumed that each kg of payload required one MW of power.

0.25 is the energy conversion efficiency of 25% from wall plug to microwave beam.
300 is the assumed launch time of 300 s.
1000 is for conversion from MW to KW.B
36,000 is for conversion from KJ to KW/h.
0.0726 is the cost of KW/h for industrial consumers in USD in the US for 23 Jan, 2023.

Nomenclature

Lcp	Launch cost per kg [$]
Los	Launch operations cost [$]
Lec	Launch electricity cost [$]
Lvc	Launch vehicle cost [$]
Lvf	Launch vehicle lifetime missions quantity
Ecq	Engaged crew quantity
Caw	Crew average wage [$]
Alr	Annual launch rate
Aim	Annual infrastructure maintenance cost [$]
Gbd	Ground beam director capital expenditures [$]
Pc	Payload capacity [kg]

Assumptions

- Annual infrastructure maintenance cost is assumed to be 5% of beam director capital expenditures.
- Launch vehicle manufacturing cost is set at $10 M in all scenarios because the calculation of actual cost is too out of the scope of this paper. The vehicle is set to have 300 missions lifetime.
- Each ground beam director is operated by a crew of 50 persons with a $130 k yearly wage.
- The annual launch rate is set to be 100 missions.

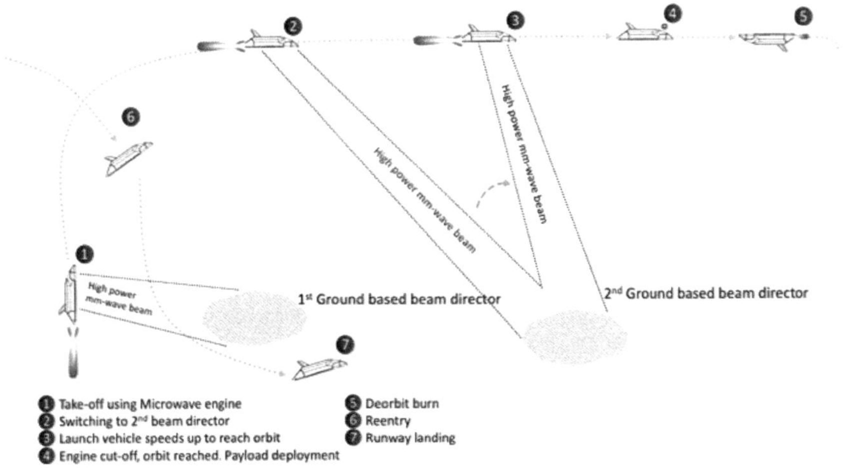

Fig. 20.4 Double beam director launch method

- Each ground station is set to have 1000 gyrotrons, antennas, aiming platforms, and 500 pcs of power supply units for suggested approach and 250 for improved approach.
- Energy is assumed to be taken from existing generating infrastructure.

Initially suggested approach costs for components of beam director

1. Gyrotrons \$1 M/pc.
2. Gyrotron power supply \$1.1 M per gyrotron.
3. The antenna and aiming platform cost \$130 k.

Offered optimization costs for components of beam director

1. Gyrotrons \$100 k/pc.
2. Gyrotron power supply unit \$275 k per gyrotron.
3. The antenna and aiming platform cost \$130 k.

During the research the authors consulted with manufacturers and confirmed the cost numbers for the initial approach. The numbers for the optimized version were set via consulting with developers, suppliers, and manufacturers.

The graph on Fig. 20.5 shows that the suggested technical and economical solutions lead to an order of magnitude improvement in the beamed energy launch cost. They are opening the opportunity to overcome conventional launchers on the free market. A launch system capable of lifting one metric ton might offer \$353/kg, whereas initial system design offers \$2331/kg. For comparison conventional launchers offer the lowest theoretical price of \$2350/kg (SpaceX 2023).

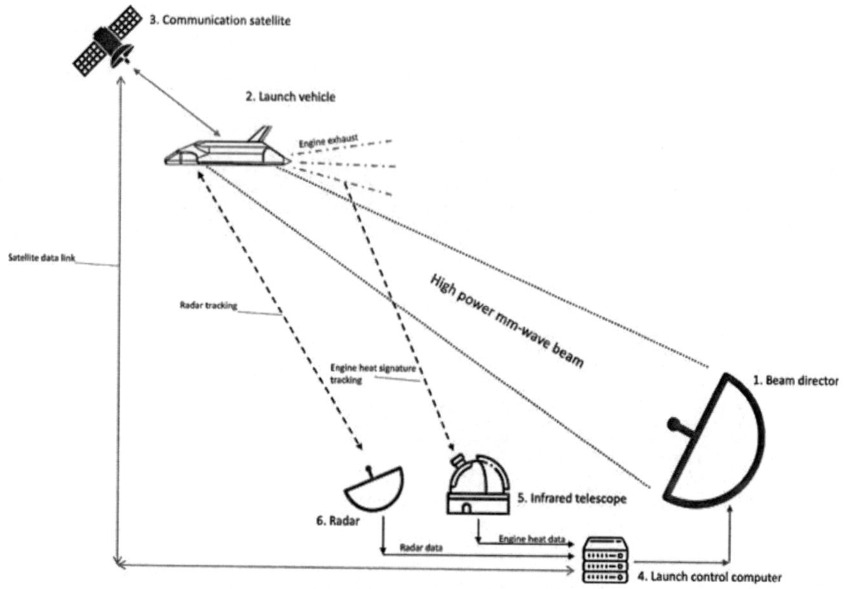

Fig. 20.5 Microwave beam launch control and monitoring system

20.4 Conclusions and Recommendations

The paper reviews the concept of beamed energy propulsion and focuses on microwave continuous wave-based propulsion solutions. Outlined the significant benefits of the technology, that could benefit the field of orbital transportation and thus benefit the whole space industry ecosystem. Eventually, the advancements in the space industry are presumed to deliver significant socio-economic benefits on Earth (Figs. 20.6 and 20.7).

The main restricting technical and economic problems of the concept were considered. The authors describe solutions that might bring significant optimization. The evaluation illustrates that these solutions might solve the command-and-control problem at launch process via establishing a robust and reliable communication system that consists of indirect monitoring methods and additional relay satellite communication. The proposed economic solutions provide significant cost reduction and open the way for realization in the near future. The authors suggest further studies of microwave thermal propulsion for launch application and testing the proposed methods.

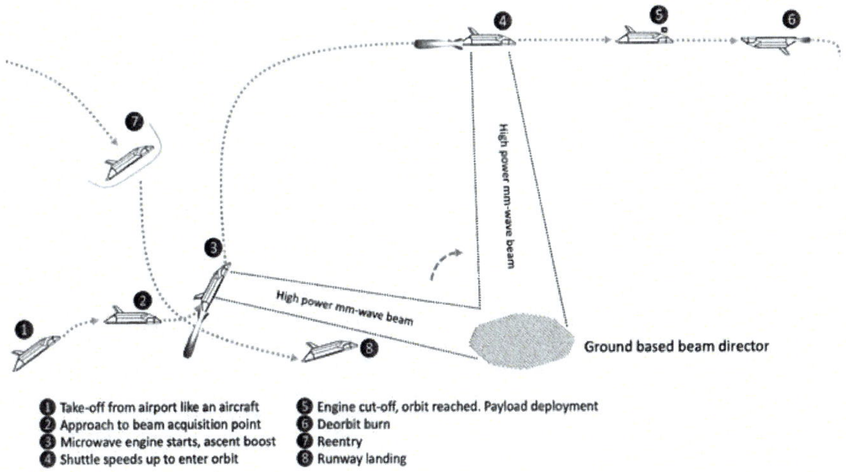

Fig. 20.6 Single beam director air-launch method

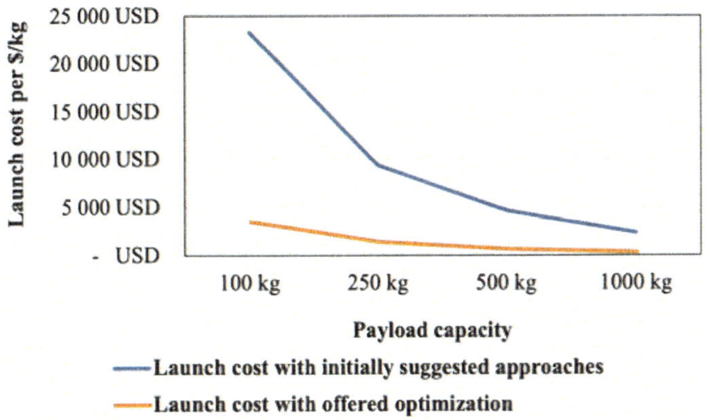

Fig. 20.7 The launch cost $/kg comparison of discussed approaches

References

1. Parkin, K., Lambot, T.: Microwave Thermal Propulsion. Report. NASA/TP-2017-219555. Carnegie-Mellon University Pittsburgh, PA (2017)
2. Paolo, B.: Overview of reusable space systems with a look to technology aspects. Acta Astronautica. **189**, 10–25 (2021). ISSN 0094-5765. https://doi.org/10.1016/j.actaastro.2021.07.039
3. Santiago, R.: Power dynamics in the age of space commercialization. Space Policy. **60**, 101472 (2022). ISSN 0265-9646. https://doi.org/10.1016/j.spacepol.2021.101472
4. Borowski, S.K., McCurdy, D.R., Packard, T.W.: Conventional and Bimodal Nuclear Thermal Rocket (NTR) Artificial Gravity Mars Transfer Vehicle Concepts, NASA/TM-2016-219393. Glenn Research Center Cleveland, OH (2016)

5. Tsiolkovsky, K.E. (1962) Izbrannye Trudy (Selected Works). In: Blagonravova, A.A. (ed.), Spaceship (Kosmicheskiy Korabl'). Izdatel'stvo Akademii Nauk SSSR, Moscow. p. 222
6. Masafumi, F., Toshikazu, Y., Yusuke, N., Kimiya, K., Yasuhisa, O., Ken, K., Koji, T., Keishi, S.: Thrust generation experiments on microwave rocket with a beam concentrator for long distance wireless power feeding, Acta Astronautica. **145**, 263–267 (2018). ISSN 0094-5765. https://doi.org/10.1016/j.actaastro.2018.01.057
7. Beamed-Energy Propulsion (BEP) Study, NASA/TM-2012-217014. Glenn Research Center Cleveland, OH (2012)
8. Kare, J., Parkin, K.: A comparison of laser and microwave approaches to CW beamed energy launch. In: AIP Conference Proceedings, p. 830 (2006). https://doi.org/10.1063/1.2203282
9. Parkin, K. L.G.: The Microwave Thermal Thruster and Its Application to the Launch Problem. California Institute of Technology (2006). https://doi.org/10.7907/T337-T709
10. Benford, J.N., Dickinson, R.: Space propulsion and power beaming using millimeter systems. In: Proceedings of SPIE 2557. Intense Microwave Pulses III (1995). https://doi.org/10.1117/12.218549
11. Kare, J.: Near-term laser launch capability: the heat exchanger thruster. AIP Conf. Proc. **664**, 442–453 (2003). https://doi.org/10.1063/1.1582132
12. Von der Porten, P.: Powered Explicit Guidance Modifications and Enhancements for Space Launch System Block-1 and Block-1B Vehicles, Conference Paper. AAS Paper 18-136 (2018)
13. Landis, M.A.: Overview of the fire control loop process for Aegis LEAP intercept. J. Johns Hopkins Appl. Tech. Digest. **22**, 436–446 (2001)
14. Rialland, V.: Infrared signature modelling of a rocket jet plume—comparison with flight measurements. J. Phys. Conf. Ser. **676**, 012020 (2016). https://doi.org/10.1088/1742-6596/676/1/012020
15. Andreas, N.S.: Space-based infrared system (SBIRS) system of systems. In: 1997 IEEE Aerospace Conference (n.d.). https://doi.org/10.1109/aero.1997.577525
16. Welch, B., Greenfeld, I.: Launch Vehicle Communications. Technical Memorandum (TM) NASA/TM-2005-213418 (2005)

Chapter 21
Models and Algorithms Script of Intellectual Dialog Systems

Bahodir Muminov

Abstract One of the main tasks of the concept of "Artificial intelligence", which is becoming a part of information technology and society, aimed at performing understandable thinking operations with the help of computer systems and their devices. It consists in generating a real process of interaction between human actions and computers. In this paper a new approach that generates appropriate questions during conversation is proposed.

Keywords Intellectual dialog systems · Fuzzy logic · Ontology

21.1 Introduction

Intellectual dialog systems (IDSs) may use to generate dialogs between human and computers. This requires AI-based software modules, architectures, information flow models, communication support methods, and algorithms to assist users in formulating answers/questions/queries of conversation [1].

In our research, we discovered that there are three different architectures of software modules that support the IDSs queries [2–7]. These include:

1. Generating proper questions to ***users***' utterances;
2. Selecting proper question from a set of ordered questions;
3. Hybrid approach.

In the process of fundamental analysis, presented in the architectures, the main tasks are identifying the intent of the user's utterance through linguistic, stylistic, morphological processing (examining the *style* of a text, segmentation of sentences, searching subject of sentences) of the text entered in the natural language.

B. Muminov (✉)
Tashkent State University of Economics, Karimov Street, 49, Tashkent, Uzbekistan 100066
e-mail: mbbahodir@gmail.com

D. Azimov (ed.), *Proceedings of the IUTAM Symposium on Optimal Guidance and Control for Autonomous Systems 2023*, IUTAM Bookseries 40,
https://doi.org/10.1007/978-3-031-39303-7_20

The technological analysis, presented in the architectures, include identification of W_i in the core of the query by processing words of context, that is prefix and suffix recognition of words, root search etc.

Holistic views of these architectures given in Figs. 21.1, 21.2 and 21.3.

This architecture takes a user's utterance as a question, and it has two requirements:

1. Building an ontology for the subject area;
2. Algorithm for determining the proximity of two objects (classes).

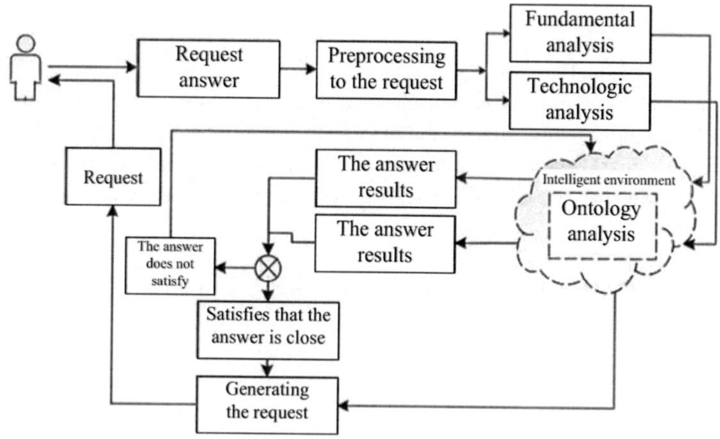

Fig. 21.1 An architecture of system that generates proper questions to the user's utterance

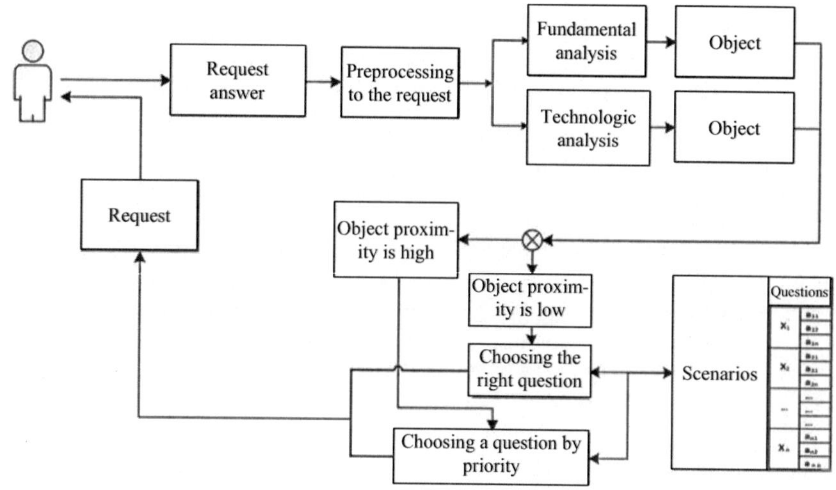

Fig. 21.2 Architecture of system that select proper question from a set of ordered questions

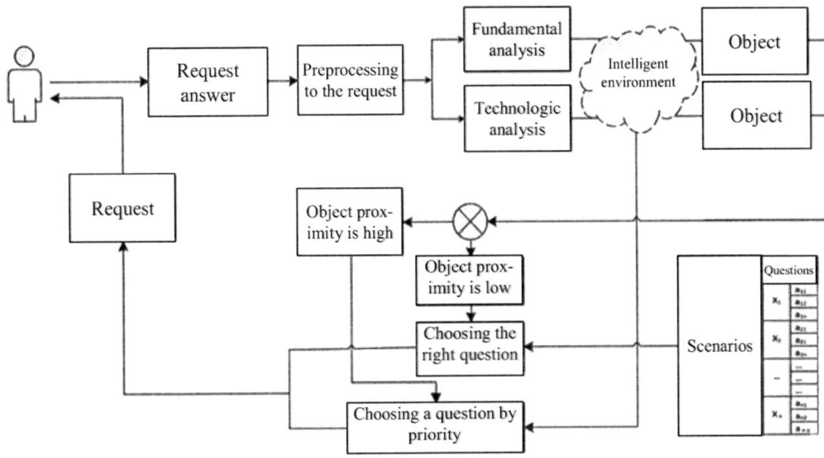

Fig. 21.3 Architecture of system that use hybrid approach

The intellectual environment is a classified subject area consisting of classes, sub-classes interrelationships, properties and phenomena, on which knowledge construction methods and ontology are built.

This architecture is a process that can implements any IDS dialog operation scenarios. Unlike the above architecture, it uses a knowledge and database, which are based on the proximity of questions and answers. Furthermore, through the analysis, the core elements of the question or answer will be determined and appropriate objects will be selected. Then, based on this object, the closest/the most relevant question or answer will be determined from the database. The proximity of objects is determined depending on their property range values and data types.

The last architecture is a combination of the two architectures above. In this approach, after the analysis, the object will be determined and their similarity will be found based on the ontology of the intellectual environment. If the proximity of the object is high ($s \rightarrow 0$), the corresponding answer to this object will be constructed based on the properties of the ontology and will be transmitted to the user in the form of a question. If the proximity of the object is low ($s \rightarrow 1$), the question corresponding to this object is selected from the database based on the knowledge base and presented to the user in the form of a question.

Based on the above architectures, the followings can be proposed as mathematical supports for building an intellectual environments and making desions in them:

1. Mathematical support of decision-making hierarchy based on formal logical knowledge;
2. Mathematical support based on logical semantic connected and production knowledge;
3. Mathematical support based on frame knowledge;
4. Mathematical support based on the construction of ontological knowledge;

It is necessary to build an intellectual environment for the formation of questions and to determine the relevant methods of intellectual processing based on ontological knowledge [8].

Representation of ontologies is important to ensure human-like conversation when creating questions. It is desirable to build ontologies for each object of architecture, to determine the mutual proximity of these objects and to combine objects with each other.

As the repetition of the process will be occurred, it is determined that an individual approach can be implemented in IDSs. That is, the main communication styles of IDSs are based on:

1. Question–answer;
2. Purposeful communication;
3. Aimless communication.

This requires the construction of ontologies for each SSD, the creation of the method of generating intellectual dialogue scenarios for organizing satisfactory conversation in IMTs, and also, this necessitate creation of methods for determining the proximity of their objects.

21.2 A Method of Constructing an Ontology of Specific Subject Domain Objects

Building an ontology for a specific subject domain (SSD) is determined on the basis of the following three variables (21.1).

$$O = \langle X, R, F \rangle \tag{21.1}$$

Here, $X = \{x_1, x_2, \ldots, x_n\}$, $n = |X|$—Number of classes given in SSD; $R = \{r_1, r_2, \ldots, r_k, \ldots, r_m\}$, $R: x_1 \times x_2 \times \ldots \times x_n$, $m = |R|$—the number of interactions of objects of a given class in SSD. These serve to create a law for the interaction of classes.

The method of constructing an ontology for SSD consists of two stages:

1. Conduct preliminary analysis for SSD and determine characteristics of class features;
2. Building an ontograph model for SSD. In this case, the Ontograph is a bigraph, its nodes are SSD classes (objects) and its edges are relations between them.

The first stage. In the construction of the ontology, two types of features are used for the SSD, and they differ from each other:

1. Property-value. It is a relationship between created classes and data types or character types;
2. Property-object. It is a relationship between two created classes.

There are many ways to delimit a relationship in an attribute definition. Some characteristics of features are introduced in order to use the most optimal methods for decision-making in feature-based intellectual processing. The proposed characteristics are used to introduce logical model conversions. They are as follows:

1. Transitivity. Transitivity is defined if for any a, b, c given in the set X, the relation property R(a, c) arise from the relation properties R(a, c) and R(b, c), that is: $\forall a, b, c \in X: R(a, b) \wedge R(b, c) \rightarrow R(a, c)$
2. Non-transitivity. If for any a, b, c given in the set X, the relation property ¬R(x, z) arise from the relation properties R(a, b) and R(b, c), the transitivity is undefined, i.e.: $\exists a, b, c \in X: R(a, b) \wedge R(b, c) \rightarrow \neg R(a, c)$
3. Antitransitivity. If for any a, b, c given in the set X, the relation property ¬R(a, c) arise from the relation properties R(a, b) and R(b, c), antitransitivity is defined, i.e.: $\forall a, b, c \in X: R(a, b) \wedge R(b, c) \rightarrow \neg R(a, c)$.

For a sub-feature of these features, the following is relevant:

- If the relation R is transitive, then the inverse relation R^{-1} is also transitive.
- If the relations R_1, R_2 are transitive, then the relation $R = R_1 \cap R_2$ is also transitive.
- The property of symmetry. If the relation R(a, b) = R(b, a) is appropriate for any a, b given in the set X, then the relation R has the property of symmetry.
- Functionality. If the property R is defined as an functionality, then for arbitrary a, b and c, from the relation R(a, b) and R(a, c), the equality b = c will be appropriate, i.e.
- $\forall a, b, c \in X: R(a, b) \wedge R(a, c) \rightarrow b = c$.
- Reversibility. If property P1 is defined as property P2, then for all x and y the statement $P_1(x, y)$ or $P_2(y, x)$ will be appropriate. It should be noted that the inversion property takes the name of the property as an argument. For example, for the relations A or B, A asserts B and B asserts A.
- Reverse functionality. If the property R is defined as an inverse functionality, then for arbitrary a, b and c, from the relation R(b, a) and R(c, a), the equality b = c will be appropriate, i.e. $\forall a, b, c \in X: R(b, a) \wedge R(c, a) \rightarrow b = c$.

In addition to defining additional property characteristics, it is possible to restrict properties for the current object in various ways. This is done on the basis of the capabilities of ontology creation languages.

There are languages and software tools for creating semantic network based ontologies. The most widely used of these is the Ontology Web Language (OWL).

The main concepts in the SSD should be structured as classes and have a "tree-like" structure with various taxonomic connections. Basic taxonomic constructors are defined by a general "subclass" relationship, that is, the relationship for the current class with its superior inheriting class. An attribute helps you work with common values of class members and special objects; and it has a binary relation.

21.2.1 The Second Stage

We introduce a fuzzy hypergraph-based data model to construct a bigraph model of SSD ontology.

Ontology is a very complex data structure that requires representation using a special mathematical apparatus [6]. We propose a binary approach to ontological knowledge processing based on the use of a fuzzy hypergraph data model for SSD. The constructed ontology is a fuzzy ontology (exact ontology is a special case of fuzzy ontology).

We assume that $X = \{x_i\}, i \in I, I = \{1, 2, \ldots, n\}$ is a finite set and $E = \left\{\breve{e}_j\right\}, j \in J, J = \{1, 2, \ldots, m\}$—X is a fuzzy subset of the set. The pair $\breve{H} = (X, E)$ is said to be a fuzzy directed hypergraph if $e_j \neq \emptyset, j \in J$ and $\bigcup_{j \in J} e_j = X$ are valid, and where $X = \{x_i\}$ elements are the nodes of the fuzzy hypergraph, and the set E consisting of elements $E = \left\{\breve{e}_j\right\}$ are the edges of the fuzzy hypergraph. From this, the degree of belongingness of the node x_i to the edge $\breve{e}_j$ is called the event of the node x_i and the edge $\breve{e}_j$ and is defined by $\mu_{e_j}(x_i)$.

We consider a group of objects that are related to each other by a certain type of relationship and form a stable class within a certain topic—a general fuzzy class.

If each of the objects is defined by some degree of belonging to the general class, then in the general fuzzy class this group of objects is united.

From this, a generalized fuzzy class can be viewed as an edge of a fuzzy hypergraph, the nodes included in the generalized edge form a generalized fuzzy class. However, the order of elements, the rules of relations between objects, and the direction of relations are important when constructing an ontology. Therefore, a fuzzy directed hypergraph of the first type is introduced into the fuzzy hypergraph data model.

A fuzzy directed hypergraph of the first type is defined as a set of pairs $\tilde{H} = (X, D)$, where $X = \{x_i\}, i \in I, I = \{1, 2, \ldots, n\}$ is a set of nodes and $D = \left\{\breve{d}_j\right\}, j \in J, J = \{1, 2, \ldots, m\}$ is a set of fuzzy directed edges. In this case, for each edge there is a fuzzy set which is defined as follows (2):

$$\breve{d}_j = \left(\frac{\mu_{d_j}(x_{i_1})}{x_{i_1}}, \frac{\mu_{d_j}(x_{i_2})}{x_{i_2}}, \ldots, \frac{\mu_{d_j}(x_{i_m})}{x_{i_m}} \right) \tag{21.2}$$

In such cases, some $x_y \in X$ nodes may be appeared repeatedly in the d_j set. In this case, depending on the location of x_y in the set, the value of the membership function $\mu_{d_j}(x_y)$ will be different.

Each d_j fuzzy edge of the hypergraph $\tilde{H}$ is expressed in the form of fuzzy directed graph as follows: $\tilde{G}(d_j) = \left(X, \tilde{U}_j\right)$. In this case, an arbitrary fuzzy ontology can be expressed in the form of a directed graph with fuzzy nodes (Fig. 21.4), i.e.

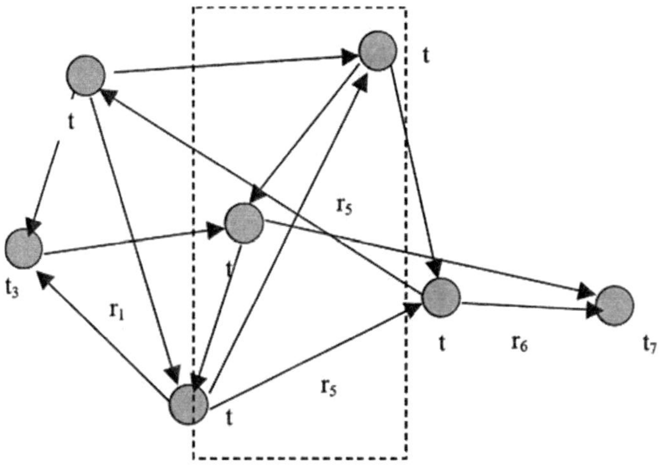

Fig. 21.4 A directed graph $\tilde{O}(H)$ with fuzzy nodes model for SSD intended ontology

$$O \to \tilde{X}(H) = \left(X, \tilde{U}\right) = \bigcup_{\tilde{d}_j \in D} \tilde{G}\left(d_j\right) \tag{21.3}$$

21.3 Algorithm for Constructing Intelligent Conversation Scenarios and an Algorithm to Help Form Queries Based on Them.

As discussed above-mentioned building systems of Intellectual dialog systems, they implements based on ontology-based question–answer techniques. Question–answer methods are mainly database-oriented and have two characteristics. The first method based on defining the roles in the dialog process, for example, the user is in the role of the requester, and the software module plays the role of the responder. The second answers-question formed based on key-value.

Such approaches are used in many complex software [9–12], for example, the DEDUCOM program that is a special case of Lisp—software modularity, using algorithms for building dialogue scenarios based on facts. The rules for deriving results based on Lisp expressions, as below:

$$R: ((X \to V) \wedge (V \to Y)) \Rightarrow f: (X \to Y) \tag{21.4}$$

As a result of the conducted study, intelligent dialog scenarios are a cyclic process, and if its elements are considered on the basis of nodes, then we present two of its transition schemes: a direct transition, a transition through a condition (Fig. 21.5).

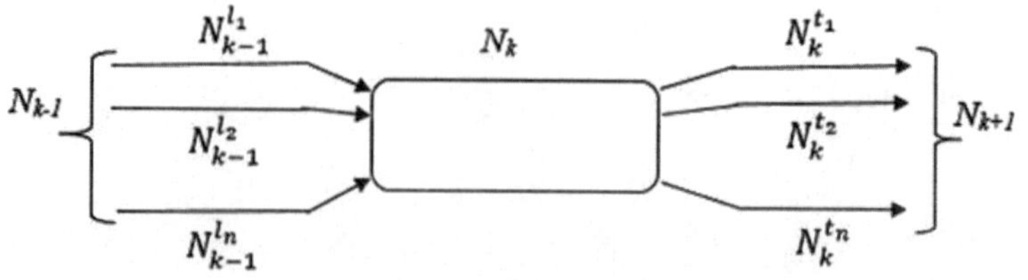

Fig. 21.5 Diagram of Intellectual dialog systems scenarios through nodes

Here, N—are nodes, i.e. N_{k-1}—incoming question, N_k—outgoing answer, formalized representation of questions corresponding to node l_i ($i = 1, \ldots, n$) and answers corresponding to node t_i ($i = 1, \ldots, n$) in the database.

The following rule is used for direct transition in the node transition scheme of Intellectual dialog systems scenarios:

$$NODE: \left(R: \left(Q\left(node: \left(N_{k-1} \to N_{k-1}^{l_j}\right)\right)\right)?A\left(N_k^{t_i}\right): A\left(N_k^{t_n}\right)\right) \to N_{k+1},$$
$$j = 1, .., n, i = 1, \ldots, n-1. \tag{21.5}$$

Here $A\left(N_k^{t_n}\right)$ *is the most universal answer to the given question. Therefore, if there are no suitable answers to the question, this most universal answer is used. n-the number of options suitable for the question.*

In the node traversal scheme of Intellectual dialog systems *scenarios, the traversing condition is based on the following rule:*

$$NODE: \left(R: \left(\left(\exists Q\left(node: \left(N_{k-1} \to N_{k-1}^{l_j}\right)\right)\right) \wedge \left(\forall Q\left(node: \left(N_c \to N_c^{l_j}\right)\right)\right)\right)? \right.$$
$$\left. A\left(N_k^{t_i}\right): A\left(N_k^{t_n}\right)\right)$$
$$\to N_{k+1}, j = 1, .., n,$$
$$c = 1, \ldots, m \ll k-1, \quad i = 1, \ldots, n-1. \tag{21.6}$$

Here $A\left(N_k^{t_n}\right)$—is the most universal answer to the given question and n is the number of options corresponding to the question.

It can be seen from the transition schemes above that in Intellectual dialog systems scenarios, the reference to previous questions determines the existence of their memory of questions and the purpose of the communication, creates the desired request for the user.

In order to achieve the desired query, the following conditions must be satisfied for scenarios used in Intellectual dialog systems:

The goal of the dialogue scenario process can be achieved in limited transactions, and with the beginning of problem solving, all the questions and answers necessary for the formation of these transactions are defined;

Changing the set of expected responses always produces a new set that differs from the current state.

Based on these conditions, Intellectual dialog systems scenarios can be constructed based on Milli's automaton, and its construction method can be written based on the following five variables:

$$D = (Q, A, T, f, g) \tag{21.7}$$

where, D—represents the dialog scenario.

$Q = \{q_i\}; i = 1, \ldots, k$—A set of identifiers (names) of questions of an active conversation scenario satisfying condition (1) to form an optional transaction. $A = \{a_i\}; i = 1, \ldots, l$—is a set of responses satisfied by condition (1) to form an optional transaction in the dialogue scenario method. $T = \{t_i\}; i = 1, \ldots, m$—denotes the set of stable states of the automaton in the dialogue scenario method and corresponds to the set of names of the dialogue record. $f: (T \times A) \rightarrow T$—is the transition function and it is the transition from the previous scenario to the new state scenario (the new transaction of the scenario). $g: (T \times A) \rightarrow Q$- Indicates the output function that determines the name of the next question, the name of the current transaction, and the current response of the reactive communication node.

The transition and output functions correspond to the classical definition of a Milli's automaton. It is common to use a transaction function when modeling a memory access method [13, 14]. Therefore, the function φ is used instead of functions f and g.

$$\varphi: (T \times A) \rightarrow (T \times Q) \tag{21.8}$$

From this, the transaction function determines the pair. The name of the next question and the name of the transaction identify the name of the current answer and the name of the current transaction.

In the first transaction, the active dialogue scenario produces the question Q_1, and the following answers can be obtained:

- A_1 is an understandable answer indicating a request for a copy of the current question. After completing the request, there is no transition to the next transaction.
- A_3 indicates the conversation duration, Q_3 question is formed after the transaction t_3.
- A_4 indicates the conversation duration, it is transferred to transaction t_4 and the question Q_4 is formed.
- A_5 indicates the conversation duration, it is transferred to transaction t_2 and the question Q_2 is formed.
- t_2 transaction, an active dialog scenario—it generates a corresponding question Q_2 based on the reformulation of question Q_1. In transaction t_2, it is expected to receive answers to questions in transaction t_1. Also, the response A_2 serves to create a request to return to the transaction t_1 and repeat the question Q_1.

An automatic model of an intellectual dialog system scenario with the feature of remembering answers is mathematically expressed as follows:

$$Q = \{Q_1, Q_2, Q_3, Q_4\} \tag{21.9}$$

$$A = \{A_1, A_2, A_3, A_4, A_5\} \tag{21.10}$$

$$T = \{t_1, t_2, t_3, t_4\} \tag{21.11}$$

$$\left.\begin{array}{l} \varphi : (t_1, A_1) \to (t_1, Q_1) \\ \varphi : (t_1, A_3) \to (t_3, Q_3) \\ \varphi : (t_1, A_4) \to (t_4, Q_4) \\ \varphi : (t_1, A_5) \to (t_2, Q_2) \\ \varphi : (t_2, A_1) \to (t_2, Q_2) \\ \varphi : (t_2, A_2) \to (t_1, Q_1) \\ \varphi : (t_2, A_3) \to (t_3, Q_3) \\ \varphi : (t_2, A_4) \to (t_4, Q_4) \end{array}\right\}. \tag{21.12}$$

The transaction function represented by the function φ (), on the one hand, describes the logic of entering the set of questions, and in this sense it can be considered as a mathematical model of the method of entering the set of questions, and on the other hand, as the logic of the method of the dialog scenario of solving the problem.

In the algorithm of the dialogue scenario, $A_i? = 1$—is an comprehensible answer; $A_i? = 0$—is an incomprehensible answer; Next(t_i)—is the next transaction; Q_K—is a question with an unclear answer; Q_S is a Q_S question on which a t_K—transaction should be formed on the basis of a Q_K; $Q_i? = Q_S$ This means that the current question has been repeated, determining Q_1—is part of the algorithm (Fig. 21.6).

The proposed dialog scenario also contains matching answers named or indexed to the questions and a set of matching questions to the incoming question. It is clear from the above algorithm that all questions except the first incoming question are generated by the transactions of the dialogue scenario. This raises the question of how to understand the initial incoming question. For this, the following steps must be completed:

1. Having an intellectual environment based on ontology;
2. The management structure of the data and knowledge base corresponding to the ontology has been developed;
3. Class properties and events, their values as objects are in the base.

There should be a way to split the request into objects. Fundamental and technological analyzes of communication system architecture are provided.

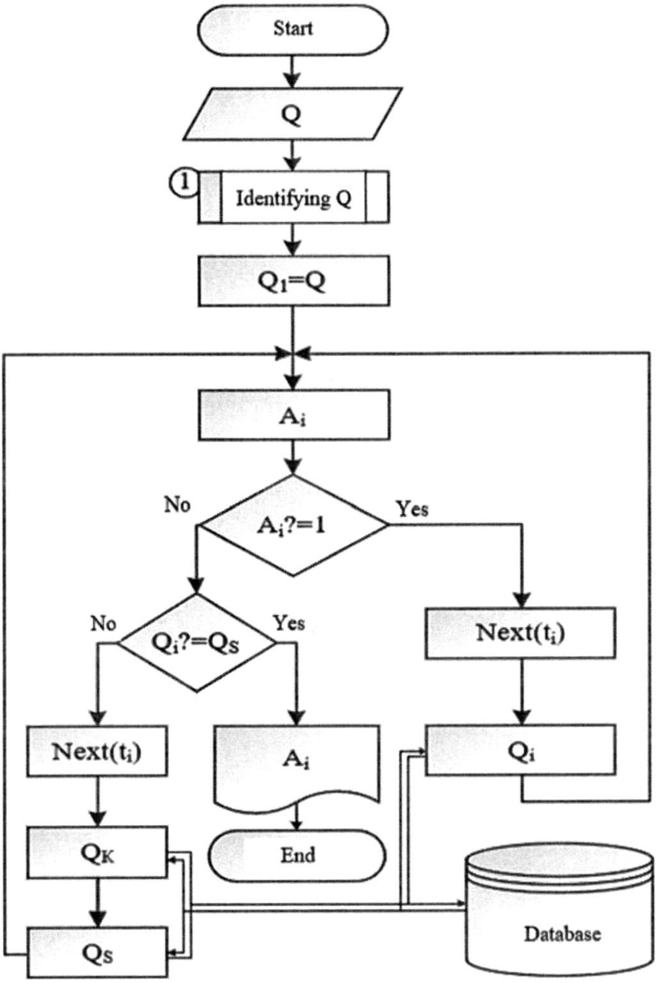

Fig. 21.6 The algorithm of the dialog scenario described in expression (21.12)

21.4 Conclusion

The main results that help to properly formulate further research on models and algorithms of scenarios of intelligent communication systems were formed:

- Using IMTs, we proposed architectures of software modules that help to form queries;
- the core of the request was identified through various methods in the fundamental and technological analysis presented in the architectures;
- methods were proposed as mathematical support for decision-making;

- It was found that the individual approach of repetition of architectures process can be implemented in IMTs and the main communication styles are question–answer, targeted communication and non-targeted communication;
- the method of ontology construction for the subject area was analyzed;
- For the CIS intended ontology, $\tilde{O}(H)$ fuzzy nodes directed graph model was created;
- a mathematical representation of an automaton model with the characteristic of remembering answers of the intellectual dialogue scenario was developed;
- the algorithm of the dialogue scenario was developed;
- the proximity of objects in the intellectual environment was determined.

References

1. Brabra, H., Báez, M, Benatallah, B., Gaaloul, W., Bouguelia, S., Zamanirad, S.: Dialogue management in conversational systems: a review of approaches, challenges, and opportunities. IEEE Trans. Cogn. Dev. Syst. **3**(14), 783–798 (2022). https://doi.org/10.1109/TCDS.2021.3086565
2. McGuinness, D.L.: Ontologies come of age. In: Fensel, D., Hendler, J. et al. (eds.) Spinning the Semantic Web: Bringing the World Wide Web to Its Full Potential. MIT Press (2003)
3. Mertins, K., Heisig, P., Vorbeck, J. (eds.): Knowledge Management: concepts and best practices, 2nd edn, 383–290. Springer Verlag, Berlin (2003)
4. Mizoguchi, R., Kozaki, K., Sano, T., Kitamura, Y.: Construction and deployment of a plant ontology. In: Proceedings of the 12th European Workshop on Knowledge Acquisition, Modeling and Management, 113–128 (2000)
5. Oberle, D., Spyns, P.: The knowledge portal on to web. In: Staab, S., Studer, R. (eds.) Handbook on Ontologies, 499–551. Springer, chapter IV (2004)
6. Muminov, B.B.: The mathematic models of searching data in the centers of information resource. IJRET Int. J. Res. Eng. Technol. **05**(10), 43–46 (2016)
7. Rumizen, M.K.: Knowledge Management, p. 128. Publishing House Astrel (2004). (in Russian)
8. Abdul, Q., Chuan, L., Dorian, M., Fatma, O., Jeffrey, K., Robert, J.M., Vasilis, E.: An ontology-based conversation system for knowledge bases. In: The 2020 ACM SIGMOD International Conference on Management of Data (SIGMOD'20). Association for Computing Machinery, New York, NY, USA, 361–376 (2020). https://doi.org/10.1145/3318464.3386139
9. Muminov, B.B.: The calculating rating of electronic resources. Euro. Sci. Rev. Sci. J. **7–8**, 20–21 (2016)
10. Muminov, B.B., Bekmurodov, U.B.: FSV-technology: intellectual search tools. IIUM Eng. J. **22**(1), 118–128 (2021)
11. Muminov, B.B., Bekmurodov, U.B.: Software database design and management method. IJHCS Int. J. Human Comput. Stud. **2**(3), 318–320 (2021)
12. Muminov, B.B.: Logical semantic search model in data search modules, actual problems of applied mathematics and information technology—al-Khorezmi 2016. In: Third International Conference Held on Bukhara, 94–97, 9–10 Nov 2016
13. Muminov, B.B.: The model of logical semantic searching. In The Uzbek-Japan Symposium on Ecotechnologies, Innovation for Sustainability Harmonizing Science Technology and Economic Development with Human and Natural Environment. Tashkent, 72–77, 2016
14. Muminov, B.B.: Algorithms and methods of indexing, compression in intellectual search of data. Probl. Comput. Appl. Mathe. **1**(7), 92–105 (2017)

Part VI
Nonlinear Control Systems and Machine Learning

Chapter 22
Data-Driven Reinforcement Learning for Mission Engineering and Combat Simulation

Althea Henslee, Indu Shukla, Haley Dozier, Brandon Hansen, Thomas Arnold, Jo Jabour, Brianna Thompson, Griffin Turner, Jules White, and Ian Dettwiller

Abstract Historically, operational analysis has largely utilized human operators and subject matter experts (SMEs) with the military knowledge and expertise necessary to make realistic decisions about combat engagements or missions. Recent operational challenges have prompted a renewed interest in machine-assisted mission engineering tools that can advance decision-making and innovation in defense. These tools have the ability to simulate high-fidelity combat engagements to analyze operational concepts, platforms, systems, and capabilities in multiple domains to develop optimal solutions to the intricate and difficult issues within the field of mission engineering. The ability of reinforcement learning (RL) to master highly complex win-loss games has highlighted its potential within the field of Mission Engineering. Agents trained through RL can interpret the environment and learn to navigate the game through experience, a success that, in theory, should translate well to combat simulations. However, in practice, using combat simulation tools in combination with RL methods poses particular challenges due to the difficulty of agent interaction with simulators not built for RL engagement. However, offline—or data-driven—RL overcomes these challenges by removing the agent's dependence on interaction with the environment, and instead using historical data for training. This study proposes training an agent with synthetic combat simulation data by using offline, or data-driven, RL.

Keywords Reinforcement learning · Mission engineering · Combat simulation

A. Henslee (✉) · I. Shukla · H. Dozier · B. Hansen · T. Arnold · J. Jabour · B. Thompson · G. Turner · I. Dettwiller
US Army Engineer Research and Development Center, Vicksburg, MS 39180, USA
e-mail: althea.c.henslee@erdc.dren.mil

J. White
Vanderbilt University, Nashville, TN 37235, USA

D. Azimov (ed.), *Proceedings of the IUTAM Symposium on Optimal Guidance and Control for Autonomous Systems 2023*, IUTAM Bookseries 40,
https://doi.org/10.1007/978-3-031-39303-7_21

22.1 Introduction

Recent operational challenges have prompted a renewed interest in mission-engineering tools that can advance decision-making and innovation in defense. These tools have the ability to simulate high-fidelity combat engagements to analyze operational concepts, platforms, systems, and capabilities in multiple domains to develop optimal solutions to the intricate and difficult issues within the field of mission engineering. Often these combat simulation tools include a variety of applications and plugins that can be used for systems engineering as well as engagement, mission, or campaign-level simulation [17, 18].

The evaluation of simulated mission performance using tools, such as the Advanced Framework for Simulation, Integration, and Modeling (AFSIM) owned by AFRL, is often difficult due to the complexity and dynamic nature of military missions and systems as well as the need for multiple simulation outcomes to analyze parameter impact [19]. Over the years, this process has been improved through the use of automation and the deployment of parameter sweeps, optimization, and/or machine learning algorithms. These algorithms have the ability to run and evaluate multiple simulations using various input parameters automatically and can provide decision-makers unbiased and optimal ways of managing combat units, improving strategic planning or resource allocation, and finding a tactical advantage in both offensive and defensive engagements.

In recent years, a machine learning methodology called reinforcement learning (RL) has been investigated for AI-enabled decision making and planning within mission-engineering tools. This is due to RL's ability to master highly complex win-loss video games such as StarCraft II, Chess, Shogi, and Go through the use of a software "agent" that can interpret the environment and learn to navigate the game through experience [20]. In theory, the success shown by RL's application to games should translate well to combat simulations. "Winning" strategies for both video games and mission engineering simulations require an understanding of the balance between multiple objectives for an individual "game piece" performance as well as the performance of the entire system of pieces working together.

In practice however, the application of RL to a mission engineering tool like AFSIM has presented a significant challenge due to the inability of the software agents deployed through state-of-the-art, open-source RL tools to interact with the simulator. One solution to this challenge is to develop a low fidelity surrogate model of the simulation outside of AFSIM that an RL software agent can interact with [21]. However, using an external surrogate model does not utilize the high fidelity physics engine within combat simulation tools such as AFSIM.

Offline, or data-driven, Reinforcement Learning, also referred to as batch reinforcement learning [22], is a promising sub-field of RL, aiming at overcoming the challenges to the widespread applicability of RL. Agents developed through offline RL methods discover policy through large batch, historical datasets rather than through individual interactions with an environment. Unlike traditional online RL, offline RL methods allow the agent to leverage large datasets, therefore allowing for the use of RL even when interacting with the environment is not possible. Further, offline RL methods allow for utilizing both simulation and real-world data. This combination makes data-driven RL methods particularly useful to ME problems.

The approach outlined here will utilize a data-driven, or “offline”, approach where an RL agent learns a successful policy from large batch, pre-collected data of historical interactions within a simulation environment. Unlike previous approaches, this method can work in conjunction with “black box” combat simulation tools and utilize their high fidelity physics engine for engagement simulation without the need for dynamic agent interaction within the environment.

22.2 Background

22.2.1 Mission Engineering

Due to their complex nature, military operations require an interconnected system of systems to successfully execute the mission. Research over the last decade has recognized this interdependence, and indicates the necessity for a systematic approach to mission planning and execution which takes into account this system of system structure. With the publication of the Mission Engineering Guide in 2015, the Department of Defense formalized the sub-field of Mission Engineering to encompass this system of systems approach, including the planning, analysis, and execution of a mission through the application of systems engineering principles [1].

Current standard practices in the field of ME depend heavily on operational analysis done by human operators and subject matter experts (SMEs) that have the military knowledge and expertise to be able to make realistic decisions about combat engagements or missions. This dependence on human operators poses particular challenges. Not only does this reliance result in additional time expense, but human operators are also predisposed to unintentionally introduce bias.

Often SMEs are used to design mission critical scenarios for simulation to test out new platforms or engagement strategies and generally perform operational analysis. These scenarios can be difficult to setup, operate, or change, as they require a lot of human interaction or input, and analysis is often limited by computational power of the machine running the simulations.

22.2.2 Reinforcement Learning for Mission Engineering

By augmenting human SME input with Machine Learning techniques, Machine-assisted Mission Engineering can be employed to inform decision makers on unbiased and optimal ways of managing combat units, improving strategic planning or resource allocation, and finding a tactical advantage in both offensive and defensive engagements. Utilizing Machine Learning for operational analysis can provide the Department of Defense unconventional strategies for Mission Engineering and have a significant impact through the reduction of cost, improvement of the development and use of weapons systems, reduction of loss of platforms in an engagement, identification of unique decisions and strategies that increase military effectiveness and response, and most importantly, the reduction of the loss of life.

However, the fact that RL algorithms provide a fundamentally online learning paradigm presents a substantial obstacle to their widespread adoption. In many settings, online interaction is impractical because data collection is expensive and dangerous. Further, the online interaction required of traditional RL methods pose specific challenges to the use of RL with combat simulation software that act as black boxes. As an example, many combat simulation tools in their current form does not easily allow for online learning, as there is no opportunity for the agent to interact with its environment.

Offline, or data-driven, RL aims to learn policies entirely from a large batch of previously collected data without the dynamic environmental interaction [8]. This approach has potential to make tremendous progress in a number of real-world decision-making problems where active data collection is expensive. In offline RL, the agent must learn the best policy from the provided static dataset and construct a policy that attains the largest possible cumulative award. However, offline RL is still a challenging approach because the agent no longer has the ability to interact with the environment to collect additional transitions and improve exploration, and therefore requiring careful data curation to ensure the dataset is representative of the environment to be simulated [7, 9].

22.2.3 Offline Reinforcement Algorithms

Offline RL is a learning-based control method that learns exclusively from a dataset containing previously collected interactions. While many other learning-based control methods, such as imitation learning, simply "copy" optimal behavior in the dataset that lead to desired outcomes, offline methods strive to learn a policy that can generalize from experience and lead to desired outcomes using datasets that do not necessarily contain optimal behaviors.

Behavior Cloning (BC) Offline RL models can learn from data resembling demonstrations. One possible solution to that is through Behavior Cloning (BC). In its simplest form, BC can be framed as imitation learning, where a dataset consisting of states and actions are obtained from an expert demonstrator, often times a human expert. The policy then attempts to match the actions from the behavior data via a supervised loss.

The BC method can be very effective when used in combination with high-quality data. However, there are many situations where BC is not an appropriate choice, including when there is a large proportion of random behavior in the dataset; or when the learned policy generates a trajectory distribution that differs significantly from the dataset under consideration [10].

Twin Delayed Deep Deterministic Policy Gradients algorithm (TD3 The TD3 algorithm is a Deep Deterministic Policy Gradients (DDPG) extension. DDPG agents may overestimate value functions, resulting in suboptimal policies. A TD3 agent learns two Q-value functions $Q1$ and $Q2$, using mean square Bellman error minimization [11] i.e., the minimum value function estimate during policy updates. In a TD3 setting, the policy and goals are updated by agents less frequently than by Q functions. As a result, when updating the policy, the agent adds noise to the target action, which makes the policy less likely to exploit actions with high Q-value estimates.

Soft Actor-Critic algorithm (SAC) Soft Actor-Critic (SAC) is an actor-critic deep RL algorithm that is not policy-driven (i.e., off-policy). Entropy regularization is a key element of SAC. The policy is trained to optimize a trade-off between anticipated return and entropy, which is a measure of the policy's randomness. This is related to the exploration-exploitation trade-off: increased entropy leads to greater exploration, which might speed up learning later on. It can also keep the policy from getting to a poor local optimum too soon [12].

Conservative Q-Learning Conservative Q-Learning (CQL) is Q-Learning, another actor-critic algorithm. CQL learns the Q-functions through determining the lower bounds of the Q-value functions for previously unseen actions in a state [13]. Through maximizing behavior with Q-functions for out of distribution state inputs, deep RL algorithms may overestimate values for previously unencountered situations. CQL takes the lower bound estimation of the Q-value functions and create that lower-bound property [5].

22.2.4 Combat Simulation Software

Three combat simulation environments stand out in comprehensiveness and detail: AFSIM, OneSAF, and Command Professional Edition (Command PE).

AFSIM The Advanced Framework for Simulation, Integration, and Modeling or AFSIM is a suite of tools revolving around, as its name implies, simulation and modeling for military purposes. This US-owned tool is suitable for battle simulation from one-on-one engagements up to mission-level planning. The behavior of AFSIM platforms and command structures is defined using its scripting environment with additional command line-based support.

OneSAF. OneSAF or One Semi-Automated Forces provides both fully and semi-automated forces simulation for Army readiness. This US Army product is open source and is meant to be the first truly composable, full-spectrum system for managing thee entire constructive simulation life cycle. The platform is primarily used for experimentation and training purposes [14].

Command PE The wargaming software Command PE is an advanced military simulation software suite hosting interactive and systems analysis of hardware, systems, tactics and CONOPS, and logistics [15, 16].

22.3 Data and Methodology

22.3.1 Scenario

Although in practice, many degrees of freedom are required in mission planning, this work focuses on the development of a proof of concept for an offline approach to reinforcement learning with a simulation environment, and therefore mission specifications are restricted to motion planning in a single dimension. However, the scenario depicted is in line with other RL-based AFSIM research.

The scenario described here is similar to the one-dimensional, red versus blue, air defense scenario proposed by Li Zhang et al. in "Air Dominance Through Machine Learning: A Preliminary Exploration of Artificial Intelligence—Assisted Mission Planning" [21], although efforts have been made to more realistically depict jammer behavior.

The composition of this scenario includes four platforms: a blue team consisting of two jammer aircraft and a strike fighter, and a red team consisting of a surface-to-air-missile. Each aircraft is constrained to move, or fly, along a constant line of Latitude from east to west while the SAM remains still at a designated location along that line of Latitude. The speed and altitude of both aircraft remain constant. Additionally, the firing range of both the fighter aircraft and SAM are fixed, along with the jamming distance of the jammer aircraft.

For this study, the scenario is as follows: the agent is represented as a fighter jet following two jammer aircrafts in order to avoid detection by a SAM long enough to deliver its payload. Once the payload has been released, the agent changes course to reverse direction and leave the area. The goal for each team is relatively simple: to survive the scenario while simultaneously trying to destroy the designated platform on the opposing side. Specifically, the SAM targets the fighter aircraft while the fighter aircraft targets the SAM.

Strike Fighter The fighter follows a randomly-chosen route containing three points which are set at the beginning of the run: an ingress point, an egress point and a terminal point. The ingress point designates the initial position of the platform and remains unchanged between episodes of the simulation. The platform moves towards the egress point, and upon reaching it, changes its flight path towards the terminal point, marking the final direction of flight, which it will continue towards until interrupted. This Strike Fighter, upon being in detecting range to the SAM, will launch an Air-to-Surface missile at the SAM with the intent to destroy it. A potentially desired behavior from this platform is that the routing would be configured to bring

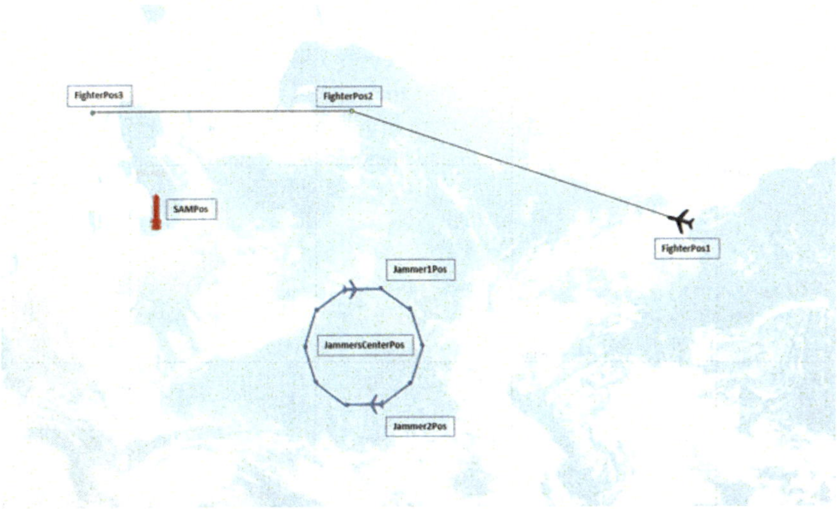

Fig. 22.1 The combat scenario features a fighter jet following two jammer aircrafts in order to avoid detection by a SAM long enough to deliver its payload

the Fighter close enough to the SAM to fire its missile and be able to perform any defensive maneuvers to defeat the kinematic range of the SAM missile.

In this scenario, the reinforcement learning agent will control the flight path of the fighter aircraft. The desired learned behavior of the agent is to fly behind the jammers until it can fire at the SAM, turn around, and head back to its base. Additionally, the fighter agent will need to learn at what distance from the jammers that the jammers will effectively ensure safe passage (Fig. 22.1).

Jammer 1 and 2 Both jammer platform agents within the scenario play similar functions and make a formation conceptualized as a single entity in the simulation. Each jammer flies in a circular formation opposite of the other jammer in a clockwise orientation at identical speeds and with identical waypoints in their routes. Each jammer is equipped with a jamming weapon that blocks radar detection devices. The purpose of this formation is to maintain as constant a jammer coverage as possible (on the SAM) in order to assist the Strike Fighter in approaching the SAM without detection.

Surface-to-Air Missile (SAM) The SAM is a stationary platform agent on the red team. When the Strike Fighter approaches the SAM, it will launch a missile at the Strike Fighter. The goal of the simulation at large is for the Strike Fighter to disable the SAM without being detected and fired at themselves. Even though the fighter is constantly mobile, when the SAM fires at it, it is an almost guaranteed hit. Although the SAM location is fixed during run-time, the location varies in each round and is randomly chosen.

22.3.2 Data

At the core of data-driven RL techniques is the ability to produce high quality policies by learning, without exploration, from historically-collected data without any additional data collection or interaction with the environment. A data-driven RL model requires an experience dataset D of state-action tuples, (s_i, a_i, r_i, s_i^p), where each tuple performs action a_i from state s_i to transition to state s^p while assigning rewards to r_i. The dataset is made up of many episodes or trajectories of the type $(s_1, a_1, r_1, s_2, \ldots, s_H)$, where H denotes the length of the trajectory.

A dataset must contain enough information for the learning algorithm to fully comprehend the dynamical system that underlies the MDP [2]. There are different methods for data collection, including random dataset [5], human expert, mixed policy, noisy and replay [3, 4].

The dataset used here was created using a simulation, which collected a total of 65 parameters, represented by five categories. These categories include hits, positions, engagement, damage, and waypoints. The hits parameters include the number of platforms destroyed on the blue and red side in the simulation. If zero is returned, no platforms were destroyed, but this does not necessarily imply no platforms sustained damaged. The positions parameters include the positions for each of the four platforms at three points, including the ingress, egress, and terminal positions. The engagement parameters include various information about platform engagement, such as the time and position a platform fires its weapon, in addition to the number of times that weapon is fired. The parameters include all information on damage sustained by each platform, including the number of times a platform is hit, the platform's location at time of hit, and the amount of damage caused; the lowest value signifying no damage, and the highest value signifying complete destruction. The final parameters, waypoints, denotes the location of the jammers at the time the fighter reaches its first and second waypoints.

Data Preparation The simulation parameters were combined into a Pandas dataframe, and each run was divided into three time steps representing the ingress, egress, and terminal positions. Rewards were determined using the "hits" parameters, and applied to the final time step. Continuous action values were then scaled using sklearn MinMaxScaler.

In order to apply the offline RL Python package d3rlpy, data was formatted using MDPDataset provided by d3rlpy. The dataset was first broken into observations, actions, rewards, and terminals, and then MDPDataset was applied.

Finally, the formatted dataset was split into testing (20%) and training data (80%).

22.3.3 Environment Design

Although data-driven reinforcement learning allows an agent to be trained without access to interact with a training environment, a RL environment is developed for more accurate testing. RL environments must be carefully constructed to both accurately replicate real world problem spaces and provide an interpretable representation of them to an Artificially Intelligent software agent. To design an environment appropriate for Reinforcement Learning, the process to be "learned" must first be defined as a Markov Decision Process, or MDP. MDPs are mathematical frameworks that act as a discrete-time stochastic control process under uncertainty. MDP's are primarily made up of states, $s \epsilon S$, actions, $a \epsilon A$, policies or transition probabilities, $P = P(s_{t+1}, r|s_t, a_t)$, and a reward function, $R(s_t, a_t, s_{t+1})$. Therefore, to design a successful RL environment, each component of the MDP must be well-defined. Designing these components must be created with meticulous detail to what the agent can perceive or "observe" within the environment, how the agent can interact or take "action" within the environment and the "rewards" given to the agent for making good decisions. The actions available to a reinforcement learning agent are explicitly designated as the action space, while the observation space is the space available for the agent to operate within.

Action States For this environment, the agent acts as the Strike Fighter and may move within a field defined by specified geo-located points. Therefore, the agent's actions are recorded as continuous values.

Observation States The observation space encompasses the simulated battlefield, as defined by user-specified geo-located points. While this battlefield is set in the simulation, the environment's observation space must match the space defined in simulation. This simulated battlefield then acts as the basis for the observation space, representing all the possible locations an agent can go.

Each observation shows the battlefield at that time step, including the agent's current location, along with the locations of the SAM and both jammers.

Rewards The agent is rewarded with +1 if the fighter's missile hits the SAM and returns to base, indicated by a 1–0 encoding. A positive reward of 0.25 is also given if neither the fighter nor SAM makes a successful hit. If the SAM hits the fighter pilot without the fighter pilot successfully returning fire, the agent is negatively rewarded with − 1, indicated by a 0–1 encoding. A negative reward of − 0.25 is given if both the fighter pilot and SAM make successful hits. Rewards were only given at the final time step.

22.3.4 Model Selection

To measure offline RL performance, an online Soft Actor-Critic (SAC) algorithm [24, 25] is compared to 4 offline methodologies: Behavior Cloning (BC) [23], Offline SAC, Twin Delayed Deep Deterministic Policy Gradient (TD3) [26], and Conservative Q-Learning (CQL) [27]. SAC, TD3, and CQL can all be considered model-free, off-policy reinforcement learning algorithms, while BC strives to imitate actions within a given dataset using supervised learning as opposed to traditional reinforcement learning.

Behavior Cloning Behavior Cloning is a simple form of imitation learning, which uses reinforcement learning to learn policy by evaluating an expert dataset [23].

Soft Actor-Critic Soft Actor Critic is a Deep Deterministic Policy Gradient (DDPG) based off-policy RL algorithm, seeking to maximize policy entropy [24, 25].

Twin Delayed Deep Deterministic Policy Gradient Twin Delayed Deep Deterministic Policy Gradient is also a DDPG-based method that improved upon the DDPG algorithm throgh the implementation of twin Q functions and adding noise to avoid overfitting [26].

Conservative Q-Learning Conservative Q-Learning is a Q-learning RL method that seeks to address the failure of many other off-policy RL methods by lower-bounding the Q-fuction's true value to prevent the overestimation of Q-values by traditional methods [27].

22.4 Results and Discussion

All models trained on 15,000 steps, gathered from 5000 episodes with 3 steps per episode. Models were then tested on 50 randomized games to obtain results (Fig. 22.2).

As seen in the graph, the SAC algorithm was best able to capture these results. This outcome is as expected, as BC is known as a better choice for online methods, and Q-learning was not optimal for this scenario. It is worth noting that behavior cloning most likely always avoids getting a reward of 1 because it uses a supervised learning algorithm, and it is probably easy for a supervised learning algorithm to recognize that smaller numbers (positions) for the fighter equates to a small reward. It may have trouble recognizing an appropriate aircraft ingress distance.

22.5 Conclusions and Future Work

Offline reinforcement learning methods have the potential to offer new avenues of expansion for RL use with "black box" simulation software within the Department of Defense. Though the environment studied here represents a simplistic engagement scenario, this serves as a proof of concept, highlighting the applications of a data-driven approach to more advanced combat simulations.

There are many opportunities for future work in this area, including addressing more complex scenarios for combat, including the development of multi-task agents, and creating explainable AI solutions.

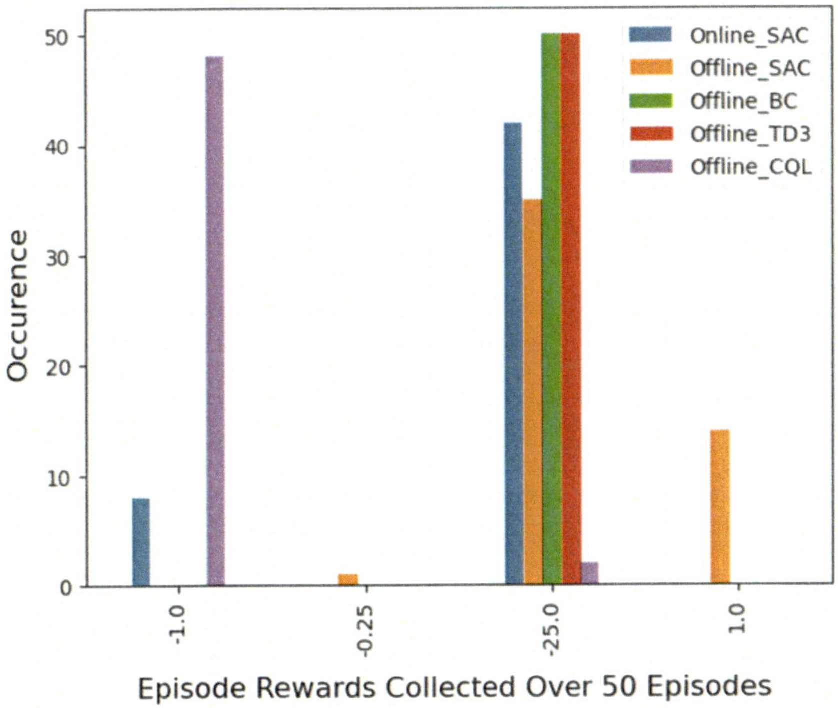

Fig. 22.2 Model performance for an online SAC algorithm, offline SAC, offline BC, offline TD3, and offline CQL. Each algorithm was trained using 5000 episodes and then tested on 50 randomly generated scenarios

22.5.1 *Complexity*

More complex combat scenarios, including the addition of multiple agents, is critical to the successful deployment of ORL methods in the Department of Defense. The scenario studied here is simplistic by design, in order to show early proof of concepts. However, the methodology used can be transferred to other applications, including scenarios involving other platforms and more complex behaviors, and scenarios modeled through other simulation software.

22.5.2 *Simulation Software*

Although this study currently focuses on implementation using data gathered through a simulation that is typically done in AFSIM, the same methodology could be applied to other simulation software used through out the Department of Defense, including OneSAF and Command PE. As described earlier, though the use of AFSIM is widespread, there are other software used for combat simulation in other DOD groups. This wide range of applicability ensures the contributions of this effort are far-reaching, with impacts across the DOD, regardless of the simulation software most appropriate for the targeted problem.

22.5.3 *Explainable AI for Mission Engineering*

A major obstacle to the widespread adoption of AI solutions to Mission Engineering problems is the lack of explainability for complex models. Studies have proven that decision makers are less likely to trust AI methods when there is little or no insight into model output [29]. An AI product's accessibility by non-experts is critical to the successful deployment of the product. Furthermore, the accessibility of a product relies on its explainability. With Mission Engineering, this gap is widened by a lack of direct communication between model developers and decision makers.

Beyond the issue of trust, there are also ethical challenges to ME decisions that are based on unexplainable, "black box" AI techniques. ME decision makers are often not subject matter experts in the field of AI. At the same time, AI developers are often not subject matter experts in the fields of ME, whether this is logistics, combat scenarios, etc. Without an acceptable level of explainability, it is possible that model inaccuracies will be lost in this knowledge gap. The Toronto Declaration, a vision statement on AI ethics authored by representatives of human rights nonprofits along with ML ethicists, underscores the importance of this gap, arguing that decision makers need to become fluent in ML, while developers of ML need to become fluent in policy [30]. Without model transparency, it will likely be impossible for decision makers who are not also ML experts to reach an acceptable level of fluency.

Transparency of design is also a critical component of ethical AI for Mission Engineering. A common finding in many AI ethics vision statements is the understanding that inherent in any AI design are human values [30]. In RL, the inclusion of the developer's values is a particularly strong concern. Rewards engineering plays a important role in the agent's behaviors, and these rewards are based on the developer's understanding of the value of a specific outcome.

References

1. Mission Engineering Guide. Guide prepared by the Office of the Under Secretary of Defense for Research and Engineering (2020)
2. Levine, S., Kumar, A., Tucker, G., Fu, J.: Offline reinforcement learning: tutorial, review, and perspectives on open problems. arXiv (2020)
3. Schweighofer, K., Hofmarcher, M., Dinu, M., Renz, P., Bitto-Nemling, A., Patil, V., Hochreiter, S.: Understanding the effects of dataset characteristics on offline reinforcement learning. arXiv (2021)

4. Nair, A., Dalal, M., Gupta, A., Levine, S.: Accelerating online reinforcement learning with offline datasets. arXiv (2020)
5. Kumar, A., Zhou, A., Tucker, G., Levine, S.: Conservative Q-learning for offline reinforcement learning. CoRR (2020)
6. Mission Engineering Guide. Guide prepared by the Office of the Under Secretary of Defense for Research and Engineering (2019)
7. Kumar, A., Fu, J., Tucker, G., Levine, S.: Stabilizing off-policy q-learning via bootstrapping error reduction. In: 33rd Conference on Neural Information Processing Systems (2019)
8. Kumar, A., Zhou, A., Tucker, G., Levine, S.: Conservative q-learning for offline reinforcement learning. In: Proceedings of the 34th Conference on Neural Information Proceeding Systems (2020)
9. Wu, Y., Zhai, S., Srivastava, N., Susskind, J., Zhang, J., Salakhutdinov, R., Goh, H.: Uncertainty weighted actor-critic for offline reinforcement learning. In: Proceedings of the 38th International Conference on Machine Learning (2021)
10. Ross, S., Gordon, G., Bagnell, D.: A reduction of imitation learning and structured prediction to no-regret online learning. In: Proceedings of the Fourteenth International Conference on Artificial Intelligence and Statistics (2011)
11. Fujimoto, S., Hoof, H., Meger, D.: Addressing function approximation error in actor-critic methods. In: International Conference on Machine Learning, PMLR, pp. 1587–1596 (2018)
12. Haarnoja, T., Zhou, A., Abbeel, P., Levine, S.: Soft actor-critic: off-policy maximum entropy deep reinforcement learning with a stochastic actor. In: International Conference on Machine Learning, PMLR, pp. 1861–1870 (2018)
13. Kumar, A., Singh, A.: Offline reinforcement learning: how conservative algorithms can enable new applications. In: The Berkeley Artificial Intelligence Research Blog (2020)
14. One Semi-Automated Forces (OneSAF). Leidos (2019)
15. Staff, MCWL Wargaming Division: Wargaming with Command Professional Edition. Marine Corps Gazette 18–20 (2021)
16. Velazquez, L.E.: Automation of wargaming capabilities. Marine Corps Gazette 43–44 (2021)
17. Clive, P.D., Johnson, J.A., Moss, M.J., Zeh, J.M., Birkmire, B.M., Hodson, D.D.: Advanced framework for simulation, integration and modeling. In: Proceedings of the International Conference on Scientific Computing, pp. 73–77 (2015)
18. West, T.D., Birkmire, B.M.: AFSIM: the air force research laboratory's approach to making M&S ubiquitous in the weapon system concept development process. J. Cyber Sec. Inform. Syst. 50–55
19. Solyeman, S., Khosla, D.: Multi-agent mission planning with reinforcement learning. In: Proceedings of the AAAI Symposium on the 2nd Workshop on Deep Models and Artificial Intelligence for Defense Applications: Potentials, Theories, Practices, Tools, and Risks (2020)
20. Silver, D., Hubert, T., Schrittwieser, J., Antonoglou, I., Lai, M., Guez, A., Lanctot, M., Sifre, L., Kumaran, D., Graepel, T., Lillicrap, T., Simonyan, K., Hassabis, D.: A general reinforcement learning algorithm that masters chess, shogi, and go through self-play. Science **362**, 1140–1144 (2018)
21. Zhang, L.A., Xu, J., Gold, D., Hagen, J., Kochhar, A., Lohn, A., Osoba, O.: Air Dominance Through Machine Learning: A Preliminary Exploration of Artificial Intelligence Assisted Mission Planning. RAND Corporation, Santa Monica, CA (2020)
22. Thomas, G., Luo, Y., Ma, T.: Safe reinforcement learning by imagining the near future. In: Advances in Neural Information Processing Systems, vol. 34, pp. 13859–13869 (2021)
23. Michie, D., Bain, M., Hayes-Miches, J.: Cognitive models from subcognitive skills. IEE Control Eng. Ser. **44**, 71–99 (1990)
24. Haarnoja, T., Zhou, A., Abbeel, P., Levine, S.: Soft actor-critic: off-policy maximum entropy deep reinforcement learning with a stochastic actor. CoRR, vol. abs/1801.01290 (2018)
25. Haarnoja, T., Zhou, A., Hartikainen, K., Tucker, G., Ha, S., Tan, J., Kumar, V., Zhu, H., Gupta, A., Abbeel, P., Levine, S.: Soft actor-critic algorithms and applications. CoRR, vol. abs/1812.05905 (2018)

26. Fujimoto, S., van Hoof, H., Meger, D.: Addressing function approximation error in actor-critic methods. CoRR, vol. abs/1802.09477 (2018)
27. Kumar, A., Zhou, A., Tucker, G., Levine, S.: Conservative Q-learning for offline reinforcement learning. CoRR, vol. abs/2006.04779 (2020)
28. Silver, D., Lever, G., Heess, N., Degris, T., Wierstra, D., Riedmiller, M.: Deterministic policy gradient algorithms. In: Proceedings of the 31st International Conference on Machine Learning, pp. 387–395. PMLR (2014)
29. Puiutta, E., Veith, E.M.S.P.: Explainable reinforcement learning: a survey. In: Machine Learning and Knowledge Extraction, vol. 12279, pp. 77–95 (2020)
30. Greene, D., Hoffmann, A.L., Stark, L.: Better, Nicer, Clearer, Fairer: a critical assessment of the movement for ethical artificial intelligence and machine learning. In: Proceedings of the 52nd Hawaii International Conference on System Science, pp. 2122–2131 (2019)

Chapter 23
Initial Development of Cooperative Control and Localization of Multiple Spacecraft Using a Multi-Agent Mission Operations System

Trevor Sorensen, Eric Pilger, Miguel Nunes, James Lewis, Scott Ginoza, Chris Battista, Lillian Marie Shibata, and Zhuoyuan Song

Abstract Multi-satellite swarms are becoming very popular due to their low costs and short development time. Instead of large and costly monolithic satellites, small satellite swarms can be flown as distributed sensing applications for atmospheric sampling, distributed antennas, synthetic apertures among other exciting applications, delivering an even greater mission capability. This paper reports on the results of a NASA STTR Phase-I project that contributes to the development and demonstration of a mission operations system for robust, coordinated operation of mobile agent swarms in dynamic environments with a high degree of autonomy. Through a collaboration with the University of Hawai'i at Mānoa, Interstel Technologies' Comprehensive Open-architecture Solution for Mission Operations Systems (iCOSMOS™) is being enhanced to coordinate and control swarms of space vehicles and other assets. The proposed iCOSMOS-Swarm™ will enable motion planning for large numbers of agents in densely crowded areas and robust position estimation with built-in cooperative localization. The major tasks for Phase-I included (1) the development of a scalable multi-agent coordination module to coordinate large agent swarms, a multi-nodal software architecture for diverse (heterogeneous) assets, and hierarchical cooperative localization module for robust inter-agent positioning, (2) enhanced system performance with improved data handling and nodal message passing and dynamic system configuration for node addition and removal, and (3) significantly enhanced simulation capabilities to eventually support up to at least 100 simultaneous nodes and end-to-end simulation of five satellite nodes in real time or up to at least 1000x real time. The results for Phase-I include the design of the iCOSMOS-Swarm™ product and the verification of the methodology using simulation results from a baseline benchmark mission with one microsat and four CubeSats

T. Sorensen (✉) · E. Pilger · M. Nunes · J. Lewis · S. Ginoza
Interstel Technologies, Inc., Honolulu, USA
e-mail: info@interstel.tech

C. Battista · L. M. Shibata · Z. Song
University of Hawai'i at Mānoa, Honolulu, HI, USA
e-mail: zsong@hawaii.edu

D. Azimov (ed.), *Proceedings of the IUTAM Symposium on Optimal Guidance and Control for Autonomous Systems 2023*, IUTAM Bookseries 40,
https://doi.org/10.1007/978-3-031-39303-7_22

to collect dynamic, multi-dimensional data sets over a wildfire outbreak or similar event through the use of multiple detectors, spread out in time, space, and spectrum.

Keywords Mission operations · Spacecraft software · System of systems · Nodal architecture · Distributed control · Cooperative localization

23.1 Introduction

The Hawaii Space Flight Laboratory (HSFL) at the University of Hawai'i at Mānoa (UHM) developed the Comprehensive Open-architecture for Space Mission Operations Support (COSMOS) with the support of a three-year NASA grant with NASA Ames Research Center as the program monitor. The original purpose of this project was to develop a comprehensive suite of software and hardware that could be used to support the mission operations of multiple small satellites, especially for universities [1]. The emphasis was to make it very easy to adapt a new vehicle to COSMOS or to introduce COSMOS to a new mission operations center. COSMOS utilizes a "plug and play" framework that allows legacy or other external tools to be used within its system [2]. The university helped initiate a technology start-up company to further develop and commercialize COSMOS. The result was the formation of Interstel Technologies, Inc. in December 2014. The commercial version of COSMOS is called Interstel's Comprehensive Open-architecture Solution for Mission Operations Systems (iCOSMOS™).

In 2020, Interstel with UHM won a NASA STTR Phase-I contract to develop a version of iCOSMOS, called iCOSMOS-Swarm™, designed to support the development and operations of unmanned vehicle swarms such as small spacecraft, but also unmanned aerial vehicles (UAVs), autonomous underwater vehicles (AUVs), mobile robots, etc. In the STTR Phase-I project, we designed the iCOSMOS-Swarm™ to simulate a swarm-chain of Earth-orbiting CubeSats that demonstrated our innovations in swarm cooperation and control that will be directly applicable to future NASA and commercial applications. During this phase, we developed the software to the stage where we can do simulations to prove our concept. iCOSMOS™ is currently at Technology Readiness Level (TRL) 6. iCOSMOS-Swarm™ was at TRL-3 at the end of this Phase-I project.

23.2 iCOSMOS™ Description

23.2.1 Functional Architecture

iCOSMOS™ is an interacting framework of software and hardware elements. Originally designed for spacecraft, it is meant to address all elements and phases of the

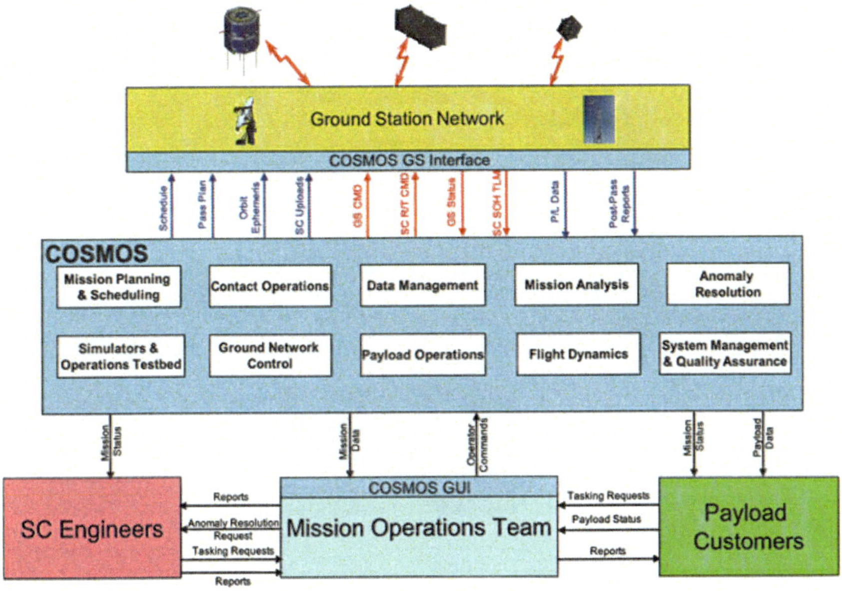

Fig. 23.1 COSMOS/iCOSMOS functional architecture for spacecraft

mission life cycle: design, development, testing, implementation, and operations. It provides elements for the creation of fully interactive simulators and testbeds, and both flight and operations software (Fig. 23.1). A guiding principle of the iCOSMOS™ suite is that it will be easy to port to different platforms, locations, and missions. To achieve this goal, it has been developed with an open architecture (meaning both open source and providing a framework to integrate external tools), a modular approach to the system elements (vehicles, ground stations, operations centers, etc.), and a friendly and assisting user interface. The iCOSMOS™ design has evolved to facilitate the ease of monitoring and control of many simultaneous modules (Nodes) including spacecraft, ground stations, UAVs, autonomous delivery vehicles, etc. A high degree of automation helps to accomplish this goal.

In several technical papers, we have explained how COSMOS can be used for satellite remote sensing missions [3], rover missions [4], and the operations of multiple satellites [5]. iCOSMOS™ is a unified software framework not only for mission control operations tools, but also for ground stations and flight software. To aid in making COSMOS easily scalable and adaptable to these various situations, we have based iCOSMOS™ on a limited number of key design elements, put together in a layered approach. These key elements are based as much as possible on existing protocols and approaches, encapsulated in a number of libraries and program elements. The foundation of COSMOS consists of a set of libraries supporting the various functionalities available in the suite. This includes mathematical functions, orbital and coordinate calculations, protocol support, and hardware and simulation support for the Operations Test Bed (OTB).

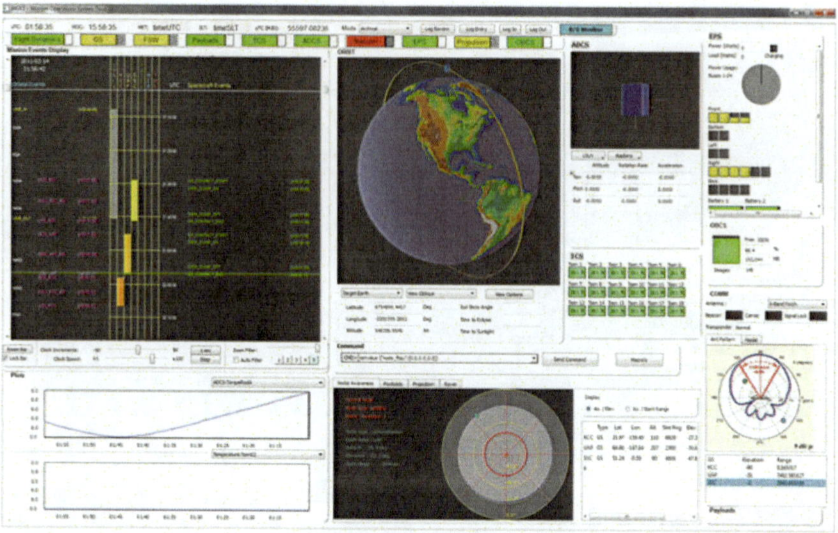

Fig. 23.2 MOST main display for the HawaiiSat-1 mission

23.2.2 *iCOSMOS™ Primary Tools*

The iCOSMOS™ Tools are the data receiving and commanding applications for the mission, providing system visualization and interfaces to the users. The major iCOSMOS™ tools that are relevant to the current project are as follows.

Mission Operations Support Tool (MOST): Originally designed as a standalone graphics program, MOST provides detailed command and control of a single satellite. MOST is the primary element of iCOSMOS™ and is the visualization tool designed specifically for supporting real-time operations. However, MOST can also be used for supporting the following major operations functions: spacecraft and payload monitor and control, mission planning, simulations and rehearsals, trending and analysis, and anomaly resolution. The main concept behind MOST (and all of the iCOSMOS™ Tools) is that panels of the interface are given over to specific areas of concern. Each panel supplies summary information concerning its area. Where appropriate, it also supplies access to a more highly detailed panel specific to that area (i.e., data mining). Figure 23.2 shows the MOST main display used for the *HawaiiSat-1* mission.

Mission Planning and Scheduling Tool (MPST): MPST converts mission status and tasking inputs into executable command loads or sequences, schedules, and timelines (Fig. 23.3). The MPST provides a top-level interface to the mission planner (human) and consists of the following major functions: Scheduler, Optimizer, Timeliner, and Command Script Generator.

COSMOS Executive Operator (CEO): CEO provides situational awareness of multiple vehicles or simulated vehicles simultaneously (Fig. 23.4). Its initial design

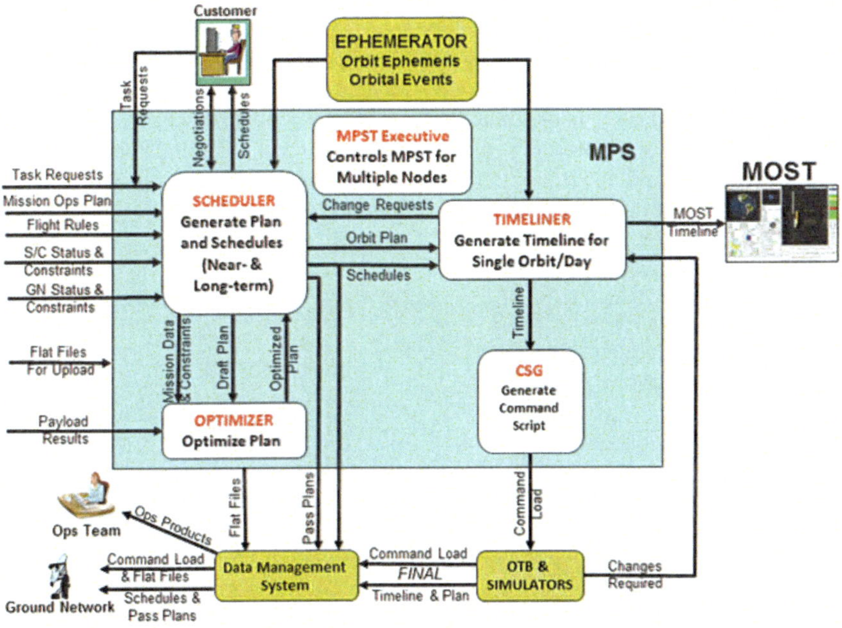

Fig. 23.3 MPST function flow block diagram

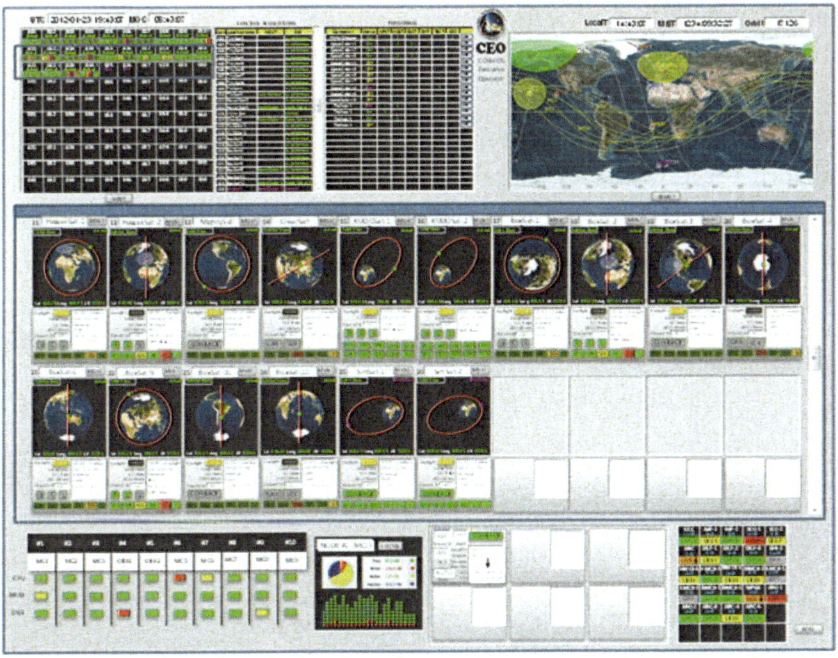

Fig. 23.4 CEO main display set up for monitoring up to 100 satellites

required that CEO can handle 100 satellites and 35 ground stations, but there is no reason it cannot be modified to handle more if needed. It can also be modified to monitor other types of nodes (e.g., UAVs). CEO collects information from MOST Engines or other agents running with the Data Management System. CEO also provides situational awareness of the Ground Segment, including the status and operations of all the ground stations in the network. It can launch the Ground Segment Control Tool for more detailed information or commanding of the Ground Segment nodes. The CEO can also launch any of the iCOSMOS™ tools, such as the MPST, MOST, and analysis tools.

23.3 iCOSMOS™ Design

The iCOSMOS™ software has adopted a paradigm that empowers a generic and modular approach for software reusability. When these rules are implemented the software becomes "iCOSMOS-aware" and part of the iCOSMOS™ ecosystem. iCOSMOS™ has a software agent-based paradigm, which can effectively enable autonomous and scalable swarm operations. The multi-agent system software of iCOSMOS-Swarm™ can operate swarms of small spacecraft by augmenting the operational autonomy of the distributed assets to help eliminate the need for complex ground operations. To achieve autonomy, satellite swarms require high-level commanding for tracking events. The distributed agent software can autonomously coordinate and control each spacecraft. One potential application is coordinated image-taking events, such that the swarm will coordinate the attitude of the satellites and track a specific target on the ground, distribute the attitude information of each element to the other satellites for automatic corrections, and coordinate the pointing campaigns. Some of the immediate benefits of a multi-agent system are scalability, augmented autonomy, reduced processing load, improved tolerance to subsystem faults, and modularity.

The iCOSMOS-Swarm™ is defined as a system with multiple *Agents* (with various degrees of intelligence depending on the application) that interact with each other, by sharing telemetry and command messages between each other, with limited understanding of the complete set of roles and responsibilities but in the pursuit of a common mission goal. We present a single scenario that demonstrates all these elements, tied together toward one common purpose. A benchmark mission that encapsulates all the elements, is technically feasible, and is open to full simulation. In this particular case the mission goal is autonomy of a swarm mission for coordinated tracking using high-level commands and breaking those commands into specialized commands that are executed by individual satellites. The software agents also have drivers that operate the hardware. The multi-agent system addresses critical aspects of swarm technologies such as *Knowledge*, *Communication*, and *Control*. Each satellite in the swarm produces the 'Knowledge' information such that it reports the position and attitude to the other members of the swarm through the structured telemetry. This information is 'Communicated' to the swarm using the Agent message passing mechanism: the broadcast and request messages. The Agent produces a 'state-of-health' message that is regularly broadcast to every other satellite. The 'request' messages are specific pieces of information that can be requested at run time by any other Agent. This may be useful to Agents that need to create a decision matrix and require more information than the one available in the regular broadcast messages. The multi-agent system can also handle the timing synchronization between the members of the swarm if no GPS signal is available (e.g., extraterrestrial missions). This helps to maintain the data consistency and configuration of the swarm to enable precise attitude formation. The aspect of 'Control' is handled by the distributed attitude control Agents that coordinate the configuration and reconfiguration of the swarm to track mutual targets of interest. The multi-agent system is also responsible for the planning and scheduling of tasks for efficient use of the swarm.

Modularity and autonomy are critical aspects of the iCOSMOS™ paradigm, incorporated at a number of levels. At the lowest levels, the commanding and querying of hardware elements is generalized into named variables, accessible through a common *Namespace*. These hardware elements are then incorporated into multiple *Agent* and *Client* programs, which communicate with each other using a common language, based on this shared Namespace. Agents are unique in their ability to accept requests from other programs, while both Agents and Clients share the ability to collect information and generate requests. At the next level of modularity assemblages of Agents and Clients, and their associated hardware, are collected into *Nodes* (e.g., spacecraft, ground station, etc.). Each Node is designed to be autonomous and self-sufficient to some degree, but will also be tightly or loosely coupled to other Nodes. Each Node typically supports a standard set of Agents as well as additional special featured Agents. Standard Agents perform such functions as gathering state-of-health information and supporting command queues. Each Node will have its own set of common information elements, such as bus voltage, position vector, etc., which is stored in a common Namespace. At the highest level, groups of nodes are assembled into *Realms*, within which they share knowledge about each other, allowing for quick communication back and forth.

23.4 Baseline Mission

For the baseline mission, we use a chain of Earth-orbiting satellites as our swarm (Fig. 23.5). To demonstrate the applicability of the iCOSMOS-Swarm™, we consider a sample application to use a swarm of five small satellites for wildfire tracking. This application, although focused on an Earth-observing technology demonstration mission, demonstrates the usability to potential applications for swarms such as asteroid tracking or mapping, solar flare events, and even to support Artemis missions to the Moon, such as providing a Lunar Positioning System using crater recognition methods, or characterizing the RF quiescence of the lunar far side.

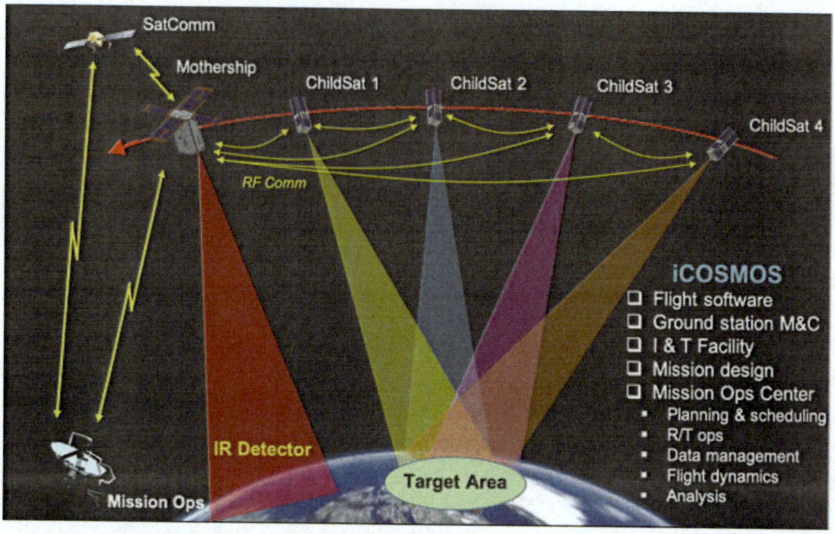

Fig. 23.5 Baseline Mission using one Mothership and four ChildSats

A progressing wildfire is a highly volatile and dynamic situation. Even with the possibility of multiple visits per day that will be provided with the growing number of micro satellites, single snapshots of data will be lacking vital information. The swarm technology we are proposing provides the opportunity to collect dynamic, multi-dimensional data sets through the use of multiple detectors, spread out in time, space and spectrum. As an example of the way this "data fusion" could be used to provide an enhanced data set over the wildfire event.

23.5 Cooperative Control and Localization

Two new modules were introduced to iCOSMOS-Swarm™ to expand its capabilities in handling multi-agent systems for a wide range of applications. A multi-agent coordination (MAC) module enables top-level mission planning in distributed coordination of large swarms of spacecraft nodes. A highly scalable and fully distributed multi-agent swarming scheme was implemented as the backbone of MAC. This coordination scheme allows swarms of homogeneous or heterogeneous spacecraft to form and maintain cohesive formations, avoid collisions among agents and with foreign objects, conduct self-organized way-point navigation, and restructure constellations in reaction to various mission objectives. A hierarchical cooperative localization (HCL) module is essential in space missions with limited or no GPS support, such as the collection of pathfinder data in lunar orbit using satellite constellations to map the spatial extent of the RF quiescent zone on the lunar farside. HCL improves the localization accuracy of agents without ground navigation support (e.g., due to signal

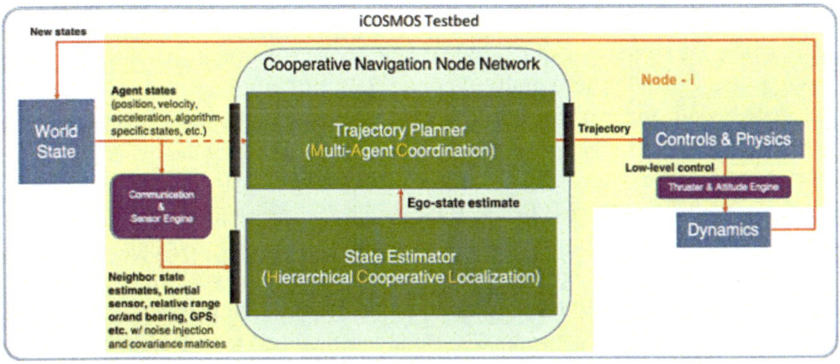

Fig. 23.6 Nodal architecture for integrating the HCL and MAC modules into iCOSMOS-Swarm™

blockage) by leveraging well-localized neighboring agents or better-equipped relay agents (e.g., the Mothership) as secondary navigation references. HCL performs decentralized information fusion on each node to combine inertial navigation sensor measurements, inter-agent range and bearing measurements, and state estimates of neighboring nodes. With the existing capabilities of iCOSMOS™, the addition of MAC and HCL modules adds capacities to support swarms of various nodal types including satellites, UAVs, rovers, etc. Figure 23.6 illustrates the interface mechanism of the MAC and HCL modules with the relevant components of iCOSMOS™.

23.5.1 Multi-agent Coordination (MAC) Module

Mobile sensor networks consisting of collaborative agents are of great significance as an engineering solution to various large-scale monitoring and surveillance applications. The ability to control and coordinate multiple collaborating nodes in a distributed fashion is a crucial step to a robust space sensing system. Recent development in the telecommunication satellite constellations from the private space industry, including SpaceX's Starlink, OneWeb, and Amazon's Kuiper, further underpins the significance of such capabilities. Many techniques have been introduced in the past for the control of artificial swarm include flocking control, formation control, and hierarchical cooperation control. The fundamental requirements for any swarming algorithms include cohesion maintenance, collision avoidance, and sometimes heading or velocity consensus.

A distributed multi-agent coordination approach that naturally combines swarming control with swarm guidance was implemented for the MAC module. MAC uses smoothed particle hydrodynamics (SPH) [6] as its algorithmic backbone to enable cohesion, separation, and alignment among the controlled agents, three fundamental elements for flocking behaviors [7]. A decentralized version of MAC was implemented and demonstrated for Newtonian particle dynamics. This SPH-based multi-agent control draws inspiration from the physics intuitions provided by fluid dynamics and generates decentralized control laws for multi-agent systems to naturally satisfy collision avoidance, cohesion, and motion alignment. This provides a convenient paradigm for multi-agent system control. MAC generates "virtual forces" corresponding to the dynamics of a desired virtual hydrodynamic system to control the behavior of the satellites. Because the multi-agent system is controlled using SPH, the resulting flock can resemble different hydrodynamic systems depending on the parameter settings. For instance, the SPH formulation was simplified for an initial mission demonstration by using an implementation of the ideal gas law for the virtual hydrodynamics; the final system can incorporate other hydrodynamic properties, such as viscosity and other background forces.

In our implementation for Phase-I, the calculated virtual forces were used directly to generate control for satellite movements. The final MAC implementation in iCOSMOS-Swarm™ can integrate these SPH forces to either generate navigation waypoints or provide accelerations the system should deliver in addition to other mission-specific actuation. These waypoints or accelerations are fed into the existing satellite control algorithms in iCOSMOS-Swarm™. The MAC module considers constraints on satellite dynamics, actuation bounds, tolerance over formation errors, etc., ensuring that the resulting control policies are compatible with the spacecraft and mission.

MAC enables guidance of the agent or swarm by defining virtual attracting (low-density) SPH particles, which create low-pressure regions attracting all the assigned agents toward them [8]. This allows for leader-follower mission configurations by assigning a virtual attracting SPH particle at the leader's location. Controlling heterogeneous groups with multiple leaders is also possible through MAC by defining the visibility of each leader to a subgroup of followers. MAC realizes collision avoidance naturally since SPH enforces the conservation of mass such that two SPH particles cannot occupy the same location in space at the same time, which will be the case when inter-agent collisions occur [9]. Intuitively, two SPH particles will naturally repel each other when they come too close to one another. Collision avoidance with respect to external obstacles is also possible by assigning virtual repelling (high-density) SPH particles to these obstacles.

A simulation environment was developed for testing the MAC module. This simulation environment is capable of simulating multi-agent systems under Newtonian particle dynamics, using a numerical force integration approach. Multiple test cases were implemented and tested including a case with time-varying hydrodynamic parameters, which represents possible scenarios of dynamic formation adjustment, and the cases with varying numbers of satellites.

The formation control performance of MAC is demonstrated through the benchmark mission where one Mothership leads four ChildSats. The Mothership leads the string-of-pearl formation following the ISS orbit. In the first case (Fig. 23.7a), the four ChildSats that initially stayed close to each other spaced out to 100 m spacing. Each was assigned a virtual SPH attractor located at its goal location on the target formation. Figure 23.7a shows the changes of separation distances between each

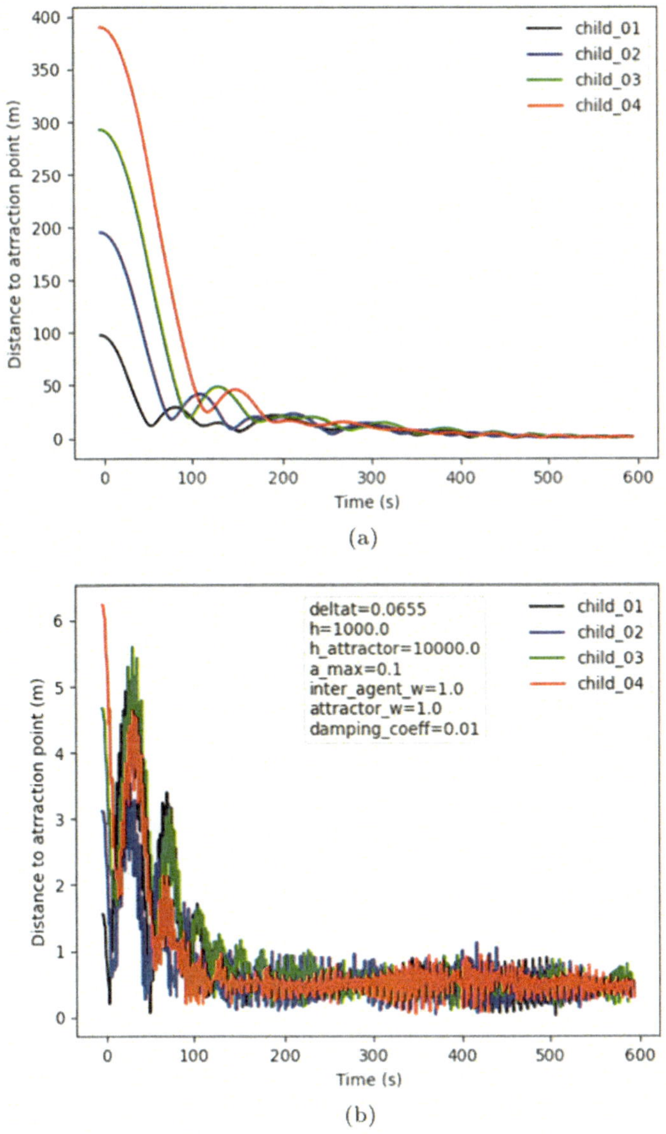

Fig. 23.7 Separation distance between each ChildSat and its goal location on the target formation in the case of **a** spacing out and **b** station keeping

ChildSat and its attractor over time. The ChildSats are spaced out about every 500 m apart along the orbit. In the second case (Fig. 23.7b), we considered a station-keeping scenario where each ChildSat tries to stay at its location on the orbiting formation. Figure 23.7b shows the distance between each ChildSat and its attractor used for MAC guidance. The ChildSats started out a small distance from their attraction points with ChildSat-1 being closest and ChildSat-4 being the farthest. Oscillation patterns can be observed at the initial phase as they close in on their goals, indicating that the MAC parameters need to be tuned to minimize both the fuel consumption and and the control error.

Parameter tuning is often necessary for multi-agent control systems in order to achieve the desired performance. We conducted a series of comparative testing to explore the MAC parameter space in order to identify the optimal sets of parameters for station-keeping in the string-of-pearl formation. Two metrics are used to quantify the performance: (1) the average final distance from the ChildSats to their attractor, and (2) the cumulative impulse of all ChildSats throughout the simulation period. We investigated the optimal SPH smoothing lengths for both the inter-agent interactions (h) and guidance towards the attraction points (h_attractor) (see Fig. 23.8). This parameter tuning method can be used to identify the optimal MAC parameters for different formation changes and mission types.

23.5.2 Hierarchical Cooperative Localization (HCL) Module

Precise knowledge of location is a critical prerequisite for accurate control and coordination. For space missions far away from the Earth, consistent navigation support such as the ground navigation tracking becomes less accessible, if at all. For instance, in mapping the lunar farside with a swarm of small satellites, the satellites will lose navigation support from the Earth's GPS system due to the shielding effect of the Moon. To counteract this issue, the HCL module enables iCOSMOS-Swarm™ to improve the localization accuracy of nodal agents through information fusion and inter-agent measurements. The HCL module is applicable to both hierarchical (leader-follower) and anarchic (agent-agent) swarm structures, where agents with high-localization accuracy are utilized as secondary navigation references by other agents, and inter-agent measurements and state estimate exchange are further utilized in a decentralized and ad-hoc fashion for localization improvement.

Cooperative localization has been previously studied for different types of mobile agents and applications. With a few exceptions, one common assumption is a connected vehicle network throughout the task. The HCL module was based on the cooperative localization solutions introduced by the authors previously for multi-agent systems in situations where GPS assess is limited or intermittent at best [10, 11]. A mother-child cooperative localization hierarchy for applications where one or more Motherships with high localization accuracy can act as moving landmarks to assist the localization of many ChildSats through opportunistic information exchange and relative measurements. Cooperative localization improves the navigation accuracy of a team of collaborating mobile agents by allowing information exchange and relative measurements in a centralized or distributed fashion. It has been demonstrated that the combination of moving landmarks and cooperative localization can efficiently mitigate the localization error increase among agents. The number of Motherships also plays a positive role in the localization performance of the ChildSats.

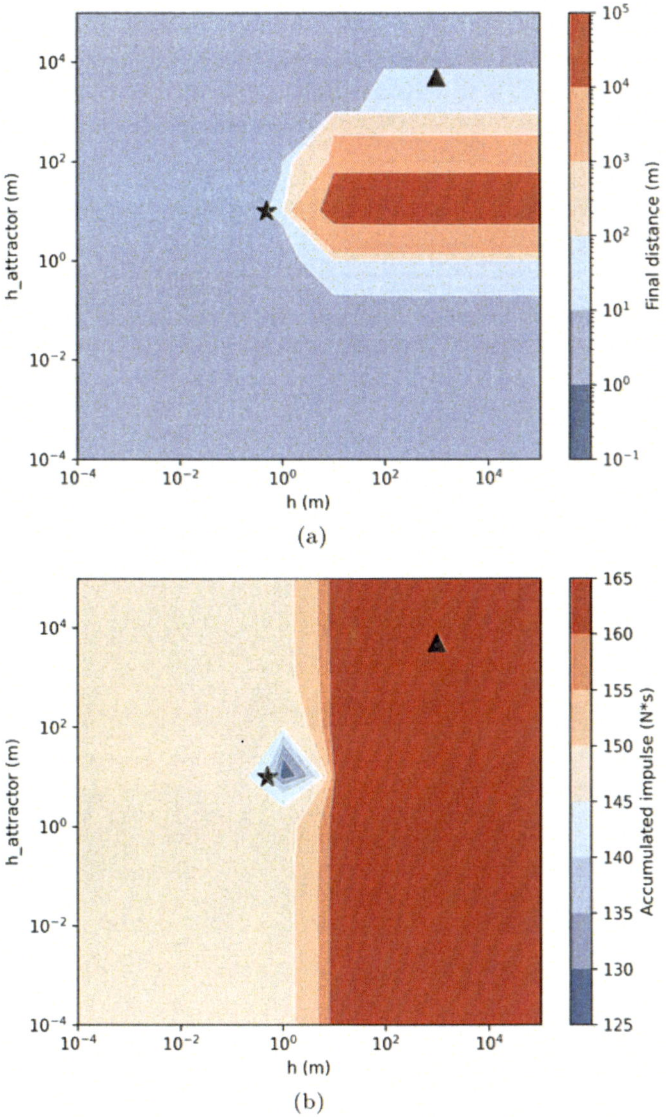

Fig. 23.8 Optimal SPH smoothing lengths for inter-ChildSat interactions (h) and guidance attractors (h_attractor) in terms of **a** the average final distance and **b** the cumulative impulse. The triangle represents the default parameter value and the star represents an improved value set

The HCL module acts as a decentralized state estimator that provides the most up-to-date state estimation of each satellite by fusing all the available navigation sensor measurements. The backbone of HCL is a Kalman filter. It performs dead-reckoning state estimation based on noisy velocity measurements as its basis. Currently, HCL supports direct position fixes provided by an external navigation and tracking system

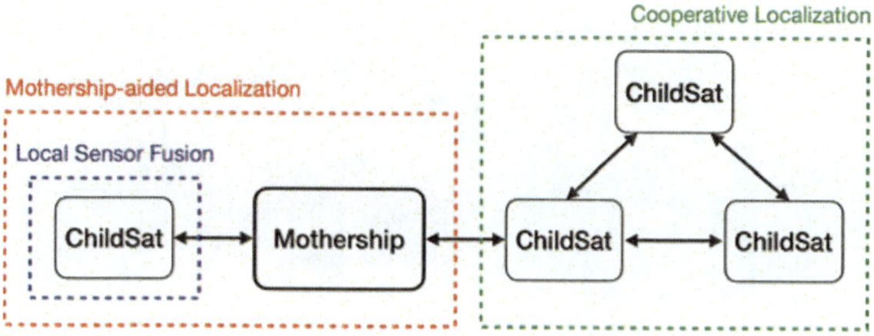

Fig. 23.9 HCL module provides robust and uninterrupted position estimates to all agents in the swarm through three different modes: local sensor fusion, Mothership-aided localization, and cooperative localization

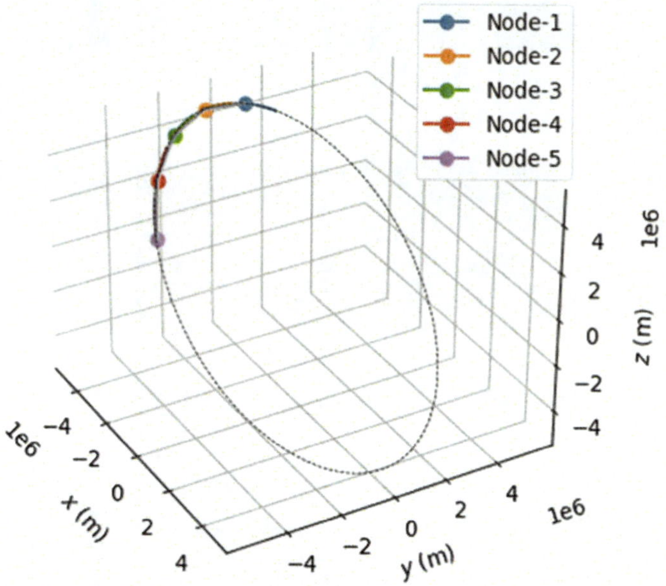

Fig. 23.10 Baseline mission scenario with five representative satellites following the ISS orbit

(e.g., Global Navigation Satellite System (GNSS), ground-tracking, star-tracking, etc.) as well as cooperative localization by measuring relative positioning information with respect to neighboring agents and receiving these neighbors' state estimates. The HCL module combines all the measurements from various navigation sensor modalities that are available, including positioning information provided by neighboring agents (ChildSats), a leader agent (Mothership), or a ground-tracking system, to provide an improved position estimate. Additional sensor modalities can be easily added. Figure 23.9 explains the different types of operation modes supported by the HCL module.

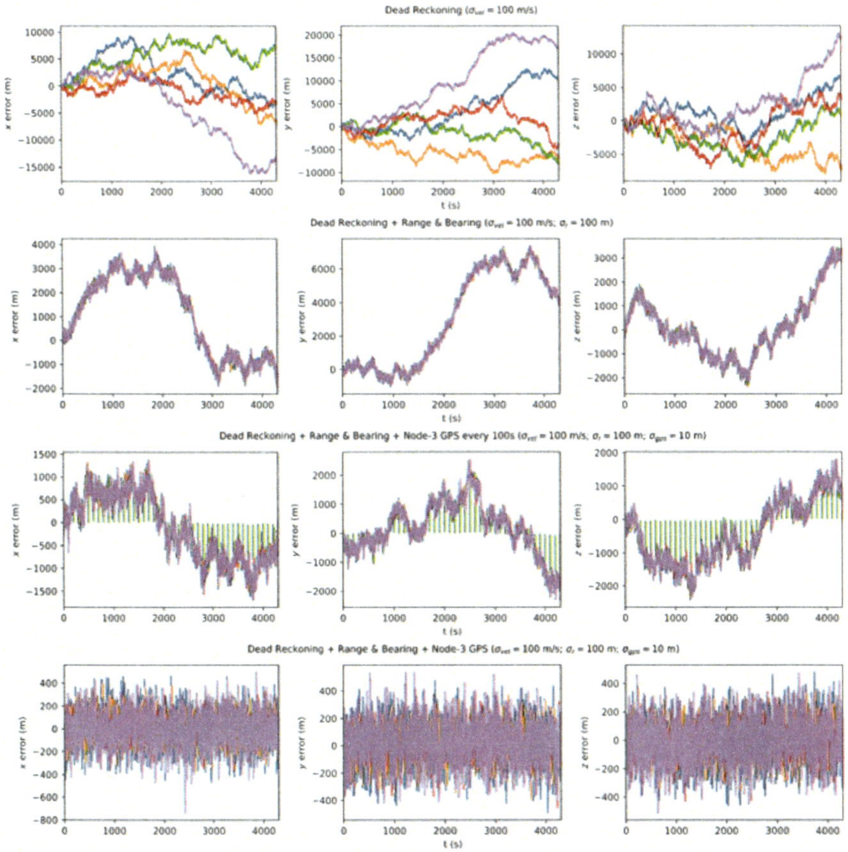

Fig. 23.11 Demonstration of position estimation using HCL in a 5-agent case on the ISS orbit in Earth-Centered Inertial (ECI) coordinates. Standard deviations and frequency of specific sensor measurements are indicated. Case 1: All agents estimate their positions using only noisy velocity measurements. Case 2: Cooperative localization using HCL is added based on the inter-satellite communication/measurement topology shown by the edges in Fig. 23.10. Case 3: Only Agent-3 can receive GPS position fixes at 1 Hz. Case-4: Only Agent-3 can receive GPS position fixes at 0.01 Hz

An implementation of the HCL system is shown in Fig. 23.10 to demonstrate the performance of the approach. This implementation used a group of five satellites, corresponding to one Mothership and four ChildSats from the benchmark mission. The satellites followed the orbit of the International Space Station (ISS), separated by a time delay of five minutes. Here we assume that inter-satellite communication and measurements are only possible with the closest neighboring agents, i.e., 1–2, 2–3, 3–4, and 4–5. The performance of the HCL algorithm for the base demonstration test case under different sensor availability is shown in Fig. 23.11. HCL establishes a coupling relationship among the positioning errors of all the agents that participate in collaborative localization through inter-agent position measurements and state estimate exchange. This coupled state estimation network allows position fixes to any agent to benefit the positioning of all the other coupled agents.

23.6 Conclusion

This paper presented the initial develop of iCOSMOS-Swarm™ under the NASA STTR Phase-I collaboration between Interstel Technologies, Inc. and UHM. The design of two new modules unique to iCOSMOS-Swarm™ were introduced including the multi-agent coordination module based on smoothed particle hydrodynamics and the hierarchical cooperative localization module based on opportunistic sensor fusion. Preliminary results from multiple simulations for the five-satellite baseline mission were presented.

Interstel and UHM won a two-year Phase-II contract starting in March 2022 to further the develop of iCOSMOS-Swarm™. Among several other major improvements to the iCOSMOS™ Core, the MAC and HCL modules will be expanded. The MAC module will be expanded with optimal trajectory planning capability to generate energy-optimal trajectories for dynamic virtual attractors to improve formation change efficiency. Similarly, the HCL module will be expanded with proper handling of anisotropic and non-stationary sensor noise distributions, attitude estimation, and automatic sensor fault detection, isolation, and recovery. Finally, Phase-II will involve hardware-in-the-loop simulations of using engineering models of representative spacecraft avionics and sensors to assess and validate the simulations and performance of iCOSMOS-Swarm™.

References

1. Sorensen, T.C., Pilger, E.J., Wood, M.S., Nunes, M.A., Yost, B.D.: A university-developed Comprehensive Open-architecture Space Mission Operations System (COSMOS) to operate multiple space vehicles. In SpaceOps: Conference, p. 2013. Stockholm, Sweden (2012, Mar)
2. Sorensen, T., Pilger, E., Yost, B., Nunes, M., Differding, J.: Plug and play mission operations. In: 2012 IEEE Aerospace Conference, pp. 1–13 (2012)
3. Sorensen, T.C., Pilger, E.J., Wood, M.S., Nunes, M.A., Yoneshige, L.K.: Mission design and operations of a constellation of small satellites for remote sensing. SPIE Proc. **8739**, 873906 (2013)
4. Sorensen, T., Pilger, E., Wood, M., Nunes, M., Garbeil, H., Wessel, E., Kruzelecky, R., Jamroz, W.: Adapting an open-architecture mission operations system for a Lunar rover mission. In: 63rd International Astronautical Congress, Naples, Italy (January 2012)
5. Sorensen, T.C., Pilger, E.J., Wood, M.S., Nunes, M.A., Yost, B.D.: Development of a comprehensive mission operations system designed to operate multiple small satellites. In: AIAA/USU Conference on Small Satellites, January 2011
6. Monaghan, J.J.: Smoothed particle hydrodynamics. Rep. Prog. Phys. **68**(8), 1703 (2005)

7. Reynolds, C.W.: Flocks, herds and schools: a distributed behavioral model. In: Proceedings of the 4th Annual Conference on Computer Graphics and Interactive Techniques (SIGGRAPH), vol. 21, pp. 25–34, New York, NY, USA. ACM (1987)
8. Song, Z., Lipinski, D., Mohseni, K.: Multi-vehicle cooperation and nearly fuel-optimal flock guidance in strong background flows. Ocean Eng. **141**, 388–404 (2017)
9. Silic, M.B., Song, Z., Mohseni, K.: Anisotropic flocking control of distributed multi-agent systems using fluid abstraction. In: AIAA Information Systems-AIAA Infotech @ Aerospace, Jan 2018
10. Song, Z., Mohseni, K.: Hierarchical underwater localization in dominating background flow fields. In: Proceedings of the IEEE/RSJ International Conference on Intelligent Robots and Systems (IROS), pp. 3356–3361, Tokyo, Japan (2013)
11. Song, Z., Mohseni, K.: A distributed localization hierarchy for an AUV swarm. In: Proceedings of the American Control Conference (ACC), pp. 4721–4726 (June 2014)

Chapter 24
Application of Cybenko's Theorem and Algebraic Geometry in Solving Modified E-Guidance Equations

Matthew Leonard and Dilmurat Azimov

Abstract Current guidance methods are typically derived from control solution parameters. These include proportional guidance, Zero-Effort Miss (ZEM) guidance, and optimal guidance. However, a method where guidance parameters are tuned independent of a control solution has not been investigated to date. Additionally, the development of a strict methodology for tuning guidance parameters has not been investigated to date. Having a more rigorous approach to tuning guidance parameters is advantageous, as it provides qualitative and quantitative metrics to understand the performance of the guidance parameters. Additionally, it allows for more sophisticated techniques such as targeting, retargeting, and coordinated vehicle operation to occur. These two novelties, Cybenko's theorem and algebraic geometry, allow for better performance to be achieved for a certain class of problems. These novelties attempt to add more rigour to the generation of e-guidance solutions.

Keywords Tuning · Guidance · UAS · Algebraic geometry

24.1 Introduction

Typical methods for determining guidance solutions involve defining an analytical model of the vehicle being guided, and having a control engineer in the loop to tune the solution. Then, after a series of iterative verification tests, the appropriate guidance coefficients and parameter equations are determined to be within an acceptable performance range. This process for tuning guidance solutions is time consuming, and is often left as guess work [9]. However, a systematic approach to determining the

M. Leonard · D. Azimov (✉)
Department of Mechanical Engineering, University of Hawai'i at Manoa, 2540 Dole Street, Honolulu, HI 96822, USA
e-mail: azimov@hawaii.edu

M. Leonard
e-mail: mbleonar@hawaii.edu

D. Azimov (ed.), *Proceedings of the IUTAM Symposium on Optimal Guidance and Control for Autonomous Systems 2023*, IUTAM Bookseries 40,
https://doi.org/10.1007/978-3-031-39303-7_23

parameters for a guidance solution has not been demonstrated to date [7]. This is as guidance solutions are typically formulated after the generation of a control solution, rather than before the control solution is derived. It is proposed to leverage mathematical techniques, such as Cybenko's theorem and algebraic geometry, to provide a more effective formalism to generate a desired guidance solution. This is to allow for more robust performance, and allow for more unconventional maneuvers to become guided. These efforts are intended to provide a more systematic method for tuning guidance parameters. It is proposed to utilize algebraic geometry and Cybenko's theorem to generate a set of guidance solution parameters for a Unmanned Aerial System (UAS) with vertical take-off capabilities. This is considered an open problem [6], and it will be shown that traditional Explicit Guidance (E-Guidance) methods cannot satisfy initial and final take-off conditions. E-Guidance has been demonstrated to work for horizontal translational motion waypoint guidance, but has not been formulated for a UAS take-off from ground problem [7]. Only via the direct application of Cybenko's theorem, combined with algebro-geometric solutions, can these conditions be satisfied. The result is a modified E-guidance solution, where the higher order terms are bounded and regulated, while the lowest order term becomes unbounded.

24.2 Background

E-guidance has been utilized in vehicle guidance for over 60 years [1]. In his seminal document, Cherry proposes the development of guidance laws which are dependent on reaching a desired final state. A guidance law generated using Cherry's E-guidance method will result in a guidance law where guidance commands are produced relative to a single desired final state. Typically, this is in terms of a desired final position, but it can be tailored to provide single solution outputs to angular rates and acceleration terms [9–11]. However, there is no method for providing guidance commands for multiple desired states with a single set of parameter values. Additionally, most methods for determining the guidance solution parameters are derived analytically, and do not use approximation methods. This is as guidance is integrated into Guidance, Navigation, and Control (GNC), where navigation provides a current state and a discrete time, and guidance outputs a command which is generated from the a-priori analytical solution. The utilization of approximation methods, specifically Cybenko's theorem, allows for more robust guidance performance. This can be seen in allowing the direct application of algebraic geometry to allow for more intuitive maneuvers, negating the need for a null pointer in the GNC software set-up, and providing a platform for which real-time targeting and retargeting is able to be executed. Additionally, the use of approximation methods means that certain performance factors can be met, such as minimizing mechanically invalid guidance commands from being generated [7]. This may include guidance commands commanding the vehicle to go into the ground, guidance commands which are unstable, or guidance commands which violate a mechanical condition (such as angle of attack constraints). These additional constraints can be understood in the context of algebraic geometry. The application of Cybenko's theorem and algebraic geometry towards solving guidance problems is formulated to provide a reference trajectory to a vertical waypoint for a VTOL capable UAS. This specific guidance problem has not been shown to be able to be solved explicitly, due to the nature of E-guidance not allowing for coupled higher order constraints. It is then shown that Cybenko's theorem, coupled with algebraic geometry, can provide an admissible solution to these additional constraints posed by the problem.

24.3 Problem Preliminaries

In UAS platforms, GNC functions typically are oriented around maneuvering towards specified waypoints. These waypoints are typically specified by a mission planner, although targeting capabilities may allow for waypoints to be selected by the vehicle during mission operations [7]. The use of waypoints for UAS platforms is advantageous as it provides the mission planner artifacts which can be used to assess UAS performance, as well as waypoints provide a tangible reference for vehicles to use for navigation. The ability to utilize waypoints leverages technologies such as GPS and Terrain Relative Navigation (TRN), which provide reference points which inform the vehicle about its state. Additionally, waypoints provide boundary conditions from which trajectories can be generated. From this, guidance and control solutions can be generated to satisfy this waypoint condition. It is advantageous to provide analytic solutions with a definitive end-point boundary condition, such as a waypoint, as it gives context for targeting to be considered in the GNC scheme. An outline of the kinematics for a single Degree Of Freedom (DOF) is provided below. That is, what are the kinematic terms acting on a UAS when terms below the snap vector are neglected [1, 8] (Fig. 24.1):

$$Waypoint(t) = -f(r(t)) \tag{24.1}$$

$$r_0 = r(t_0) \tag{24.2}$$

$$v(t) = v * t \tag{24.3}$$

$$a(t) = C * t - g * t \tag{24.4}$$

$$r(t) = r_0 + v(t) + a(t) \tag{24.5}$$

$$\psi = \cos(Waypoint(t)/r(t)) \tag{24.6}$$

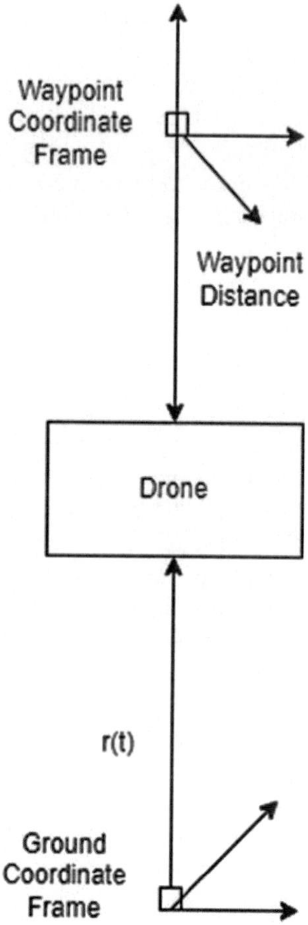

Fig. 24.1 Coordinate system of problem

This problem formulation defines the waypoint as being ahead of the vehicle, and thus is tracked from behind. This can be seen as the distance to go, which may serve to inform guidance commands about the magnitudes of the commands provided. The problem outlined is for a single DOF kinematic reference solution which can satisfy certain conditions. Specifically, there are boundary conditions on the initial position, initial velocity, initial acceleration, final velocity, and final acceleration. These terms are constrained, as it is desired to produce a reference solution which will guide the vehicle to the waypoint and meet certain final conditions. An application of waypoint guidance to UAS platforms is for a VTOL maneuver. In the Vertical Take-Off maneuver, there are strict initial conditions, as the vehicle is starting from rest. Also, if you wish to provide a reference solution for attaining a certain altitude and not overshooting the waypoint, additional constraints are placed on the reference solution. The equation takes the form of E-guidance, and is specified as solving for a polynomial function. An example for thrust acceleration is provided below:

$$a(t) = c_1 p(t)_1 + c_2 p(t)_2 \tag{24.7}$$

It is proposed to leverage algebraic geometry to provide better geometries for the polynomial equations that must be solved for. In order to utilize algebraic geometry, Cybenko's theorem needs to be applied. Additionally, to meet these boundary conditions, it is proposed to utilize Cybenko's theorem and algebraic geometry to solve for an admissible reference solution.

24.4 Problem Statement

It is desired to generate a reference solution to a Vertical Take-Off maneuver where initial and final boundary conditions are satisfied. Additionally, the solution formulation shall allow for the adjustment of tuning coefficients. This allows for the final state to be adjusted to fit the mission planner's objectives. Specifically, the authors wish to develop a tuneable guidance law which can satisfy higher order boundary conditions. The polynomials must be able to be mapped into an E-guidance formulation. This formulation is as follows:

$$c_0 + c_1\tau + c_2\tau^2 \tag{24.8}$$

This is the classical E-guidance formulation, and is found in the form of a second-order polynomial. c_0 is a position term, c_1 is a velocity term, c_2 is an acceleration term, and τ is a time evolution term. These values are analogous to the physical units of position, velocity, and acceleration. The goal of applying Cybenko's theorem to solve for either the coefficients, or the characteristic analytical equations which define the guidance law, is to provide a more robust solution generation process. This is achieved via structured programming, where there are multiple "hidden" layers, which are used to tune the characteristic equations. This tuning is to achieve higher order boundary conditions, which are placed on the initial position, initial velocity, initial acceleration, final velocity, and final acceleration. These conditions are described as modified Nagumo conditions. This is as it is a modification of the classical Nagumo necessary condition for higher-order boundary conditions. The classical statement of Nagumo's necessary condition says that for a characteristic equation to have bounded higher order terms, the lower order becomes unbounded. In the modification of Nagumo's condition, we postulate that it is possible to have a polynomial and its derivative satisfy topological bounds on the same metric distance, while having the lowest order term become unbounded, but scaleable. The conditions are as follows:

$$\text{Initial conditions}: r_0 = \partial r_0/\partial t = \partial v_0/\partial t = 0 \tag{24.9}$$

$$\text{Final conditions} : (\partial r_f/\partial t) = (\partial v_f/\partial t) = 0 \tag{24.10}$$

The initial conditions are informed by physical constraints for the UAS. This is as the UAS has to take off from an initial point on the ground, and can have no directional velocity or acceleration components. Simply put, this is for a UAS taking off from a take-off platform. The final conditions are posed in order to control the UAS once it has reached the waypoint. This is intended to provide a set of reference solutions which will not overshoot or undershoot the waypoint by having unbounded velocity or acceleration terms. This is to have a reference solution which can provide guidance commands which will command the vehicle to a stop at or near a waypoint. The analysis of waypoint convergence is not analyzed here.

This is done via application of Cybenko's theorem to algebraic geometry to generate these solutions. A direct statement of Cybenko's theorem is provided in the next section.

24.5 Cybenko's Theorem

The explicit use of Cybenko's theorem to tune guidance parameters has not been demonstrated to the authors' knowledge. The use of other techniques to meet the boundary conditions stated in the previous section has not been demonstrated [6, 7]. A certain technique, which leverages Cybenko's theorem is utilizing a set of algebraic varieties which are injectively mapped into each other. This is known as a superposition, and can be used to generate coupled parameter sets for higher order functions. This is necessary to satisfy the higher order boundary conditions, which are not able to be met with alternate methods. The direct mathematical equivalent of mapping various sets into each other is Cybenko's universal approximation theorem. Cybenko's theorem states that a univariate function can approximate the superposition of multiple other functions. An explicit statement of the theorem is provided below:

Let σ be any continuous discriminatory function. Then finite sums of the form

$$G(x) = \sum_{j=1}^{N} \alpha_j \sigma\left(y_j^T x + \theta_j\right) \tag{24.11}$$

Are dense in $C(I_n)$. In other words, given any $f \in C(I_n)$ and $\varepsilon > 0$, , there is a sum, $G(x)$, of the above form, for which

$$|G(x) - f(x)| < \varepsilon \text{ for all } x \in I_n \tag{24.12}$$

This is pulled directly from Cybenko's seminal 1989 paper, and is proposed to tune E-guidance parameters. However, modifications need to be made to generalize it to algebraic varieties. A generalized version of the universal approximation theorem is presented below:

Let there be an abstract variety, CURVE, which has a restricted domain (x, y). There exists n-many affine varieties which can be superimposed together to generate an approximate solution $n \circ CURVE$ which satisfies the curve evaluation metric. This curve represents an analytic polynomial, with n many hidden coefficient layers in it. The metric condition is defined as the evaluation of a measure M on n, which is approximately equivalent to a CURVE measure Q.

$$M/Q \approx 1 \tag{24.13}$$

If this condition is satisfied, then n can be injectively mapped into CURVE. This provides a formalism to utilize multiple varieties to tune E-guidance parameters.

24.6 Problem Set-Up

In order to allow for the application of algebraic varieties to tune E-guidance parameters, a network structure needs to be configured. This is to allow for each layer to have tuning weights applied only to the next layer. This allows for a more systematic approach to tuning E-guidance parameters. The network structure is outlined in Fig. 24.2.

This network structure leverages structured programming to allow the GNC engineer more flexibility in tuning the guidance parameters. See Fig. 24.2 for the network structure this is as the behavior of the guidance parameters can be tracked at each layer, and adjusted appropriately. This provides a methodology for which guidance parameters can be dynamically tuned, to fit different performance criteria. Dynamic, real-time tuning is not addressed in this paper. The purpose of Cybenko's theorem is to provide a set of layers, which after a set of operations, renormalizes the units to become consistent with physical measurements. The set of varieties is presented below:

$$\mathrm{a(t) = f(v, t, r, h)} \tag{24.14}$$

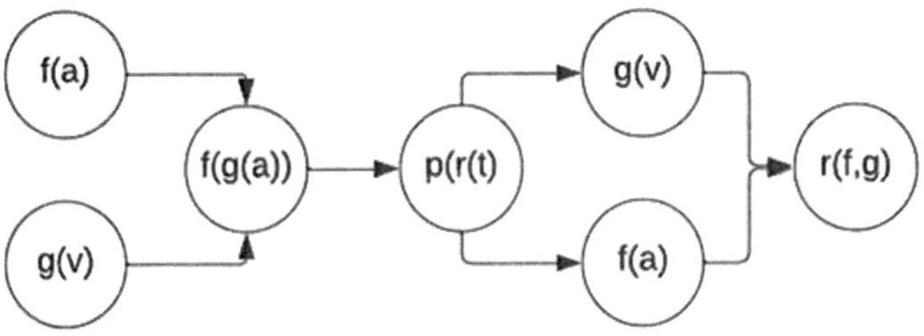

Fig. 24.2 Flowblock diagram of network tuning setup

$$\mathrm{v(t) = g(a, c)} \tag{24.15}$$

$$\mathrm{r(t) = q(v, a)} \tag{24.16}$$

These varieties are functions of multiple inputs, and do not require consistent units. Rather, they represent a unique state that must be reached at a given point along the index. The intermediate equations, which represent "neurons" in the network are outlined below.

$$f(a^{init}) = \left((v^2)/2\right) * t + r + h \tag{24.17}$$

$$g(v^{init}) = \left((-2f(a^{init}))/2\right) - c \tag{24.18}$$

$$\mathrm{f}(a_2^{init}) = -\left(\left((g(v^{init}))/2\right) * t\right) + r + h \tag{24.19}$$

$$p(r(t)) = r\left(g\left(v^{init}\left(f\left(a_2^{init}(t)\right)\right)\right)\right) = \left(\left(\left(v^{init}\right)^2\right)/2\right) + a_2^{init} \tag{24.20}$$

$$g(v_2^{init}) = -\left(((a_2^{init})^2)/2\right)/r(t) \tag{24.21}$$

$$\mathrm{f}(a_3^{init}) = -\left((v_2^{init})^2/2\right) * t + (h - r(t)) \tag{24.22}$$

$$r^{reference}(f(), g()) = \left(((v_2^{init})^2)/2\right) + a_3^{init} \tag{24.23}$$

These equations act as independent varieties which are mapped into each other. These parameters are intended to appropriately scale initial conditions into a form where there are negligibly small shaping coefficients which shape the curves to meet the desired boundary conditions. These equations are intermediate equations which were fed into each other to generate variable tuning parameters. The intermediate tuning plots are presented in Figs. 24.3, 24.4, 24.5, 24.6, 24.7, 24.8, 24.9 and 24.10.

In these intermediate plots, it is shown that by feeding in previous values into a new analytical expression, you get changes in the curve behavior which can be tuned. This is achieved via a structured programming set-up, where each layer's behavior can be tracked. Additionally, it provides the structure to map one variety into another variety. By solving for one portion of the E-guidance solution, then feeding the values into the lower order term, solution coupling is achieved. This is advantageous as it means that the higher order term can be tuned independently of the lower order term, but still meet the same boundary condition on the same measure. The proof for why Cybenko's theorem is applied is presented in the next section.

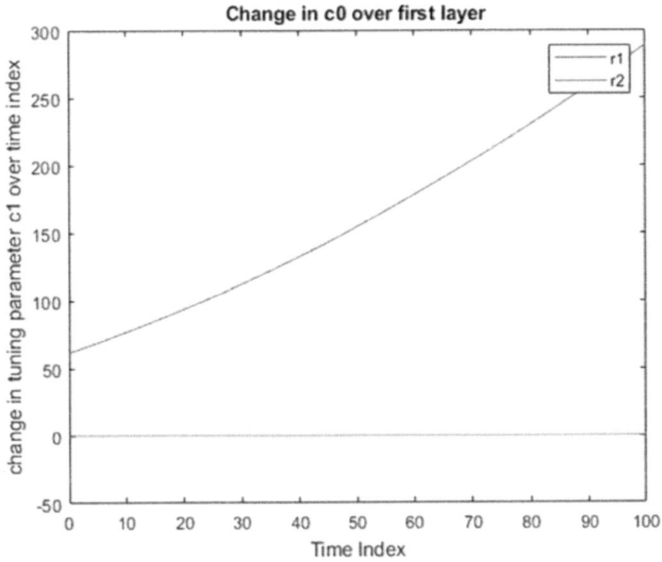

Fig. 24.3 Tune 1 in position term

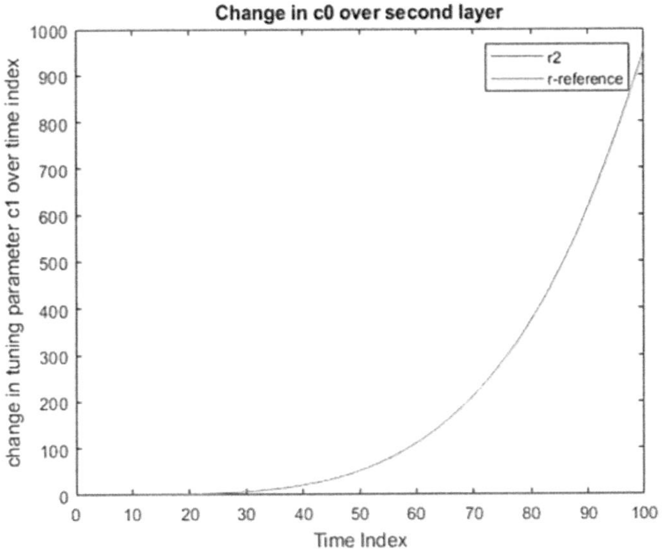

Fig. 24.4 Tune 2 in position term

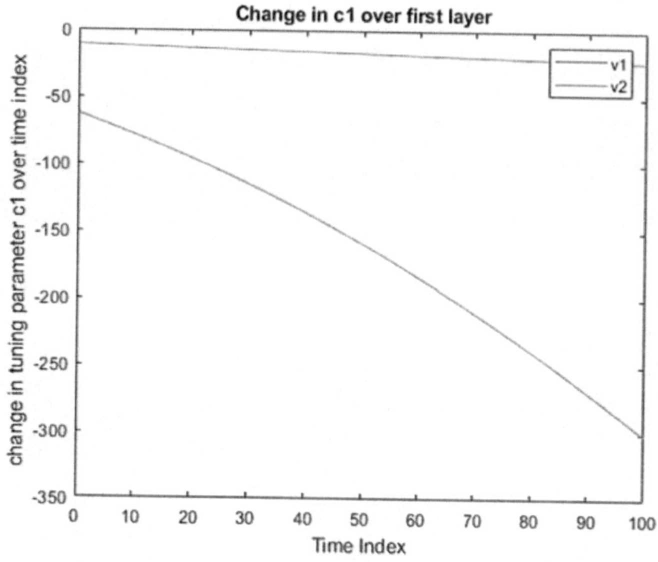

Fig. 24.5 Tune 1 in velocity term

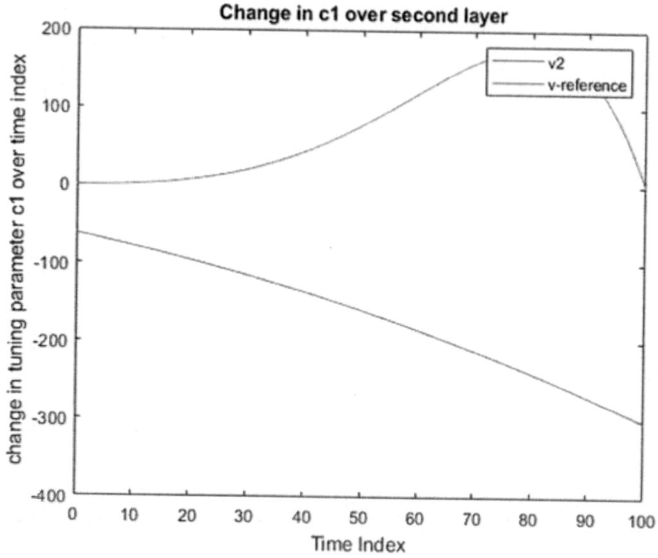

Fig. 24.6 Tune 2 in velocity term

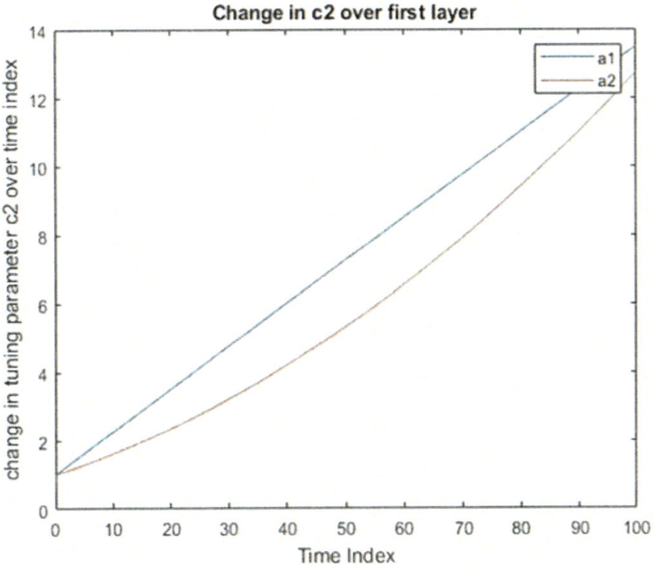

Fig. 24.7 Tune 1 in acceleration term

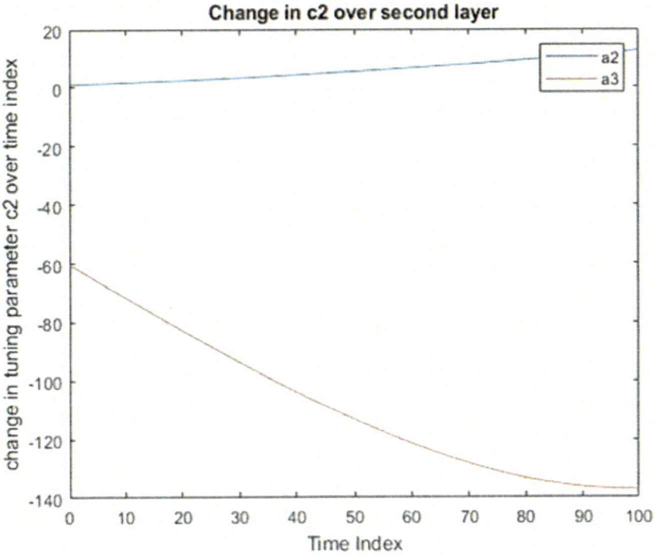

Fig. 24.8 Tune 2 in acceleration term

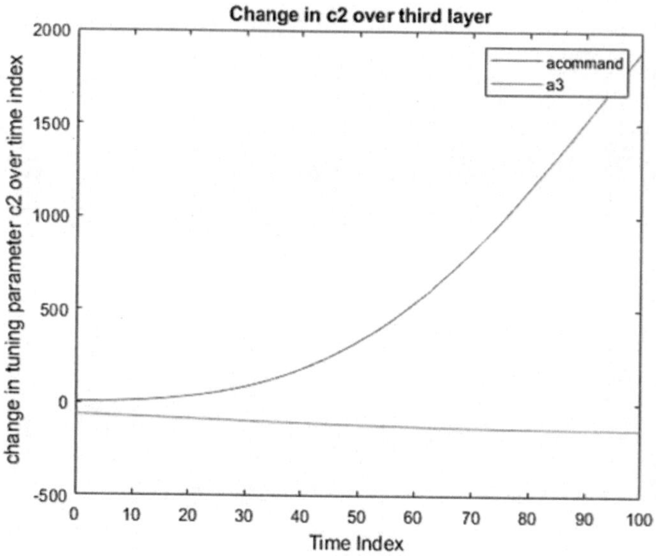

Fig. 24.9 Tune 3 in acceleration term

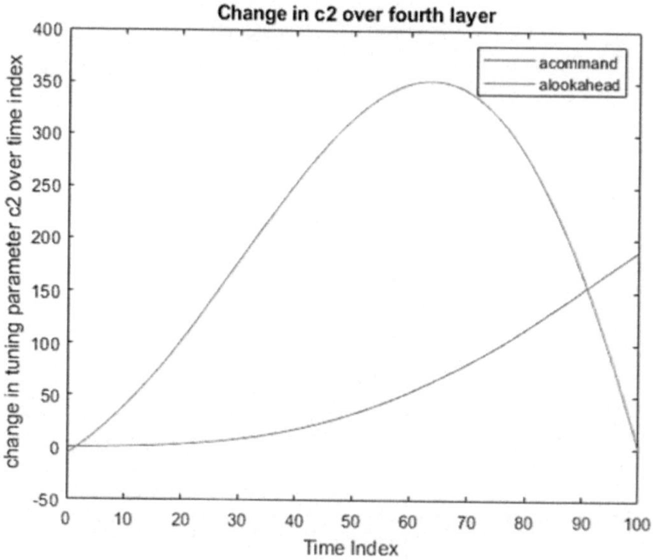

Fig. 24.10 Tune 4 in acceleration term

24.7 Difference Between Analytical Guidance and Numerico-Analytic Guidance

There exists a precise reason why higher order boundary conditions are difficult to solve in a traditional context. This has to do with the nature of basic calculus, and how derivatives are classically understood. Derivatives are classically understood as an arbitrarily small slope of a tangent line between two points. The closer the two points get, the more precise the derivative becomes. However, when evaluating higher order terms, a similar logic is followed. Specifically, a higher order derivative is the slope of a tangent line between two arbitrarily spaced points. However, these points rely on information from the lower order term. This means that it takes two lower order terms to satisfy the tangent condition for the derivative. Once you get to a high order term, such as a second order derivative, you are relying on information from both the first order derivative, and the original expression. Thus, if you are looking to take the derivative of a first order term, you are relying on information from the first order term and the zeroth "base" form. This means that the boundary for the lower order term is going to be an arbitrarily small difference than the boundary for the first order term. Because of this, it implies that no non-trivial solutions exist to satisfy these boundary conditions. This is known as the stepwise evaluation of functions and their derivatives [1]. There is a more rigorous way to describe the necessity of this modification. This method describes that when generating values analytically, you would not be able to achieve the desired behavior in the boundary conditions. Let x1 and x2 be conditions, lying in a set X. X is a closed set. There exists n-many variations along an interval between x1 and x2. If x1 after an arbitrarily small variation equals x2, then x1 is in the neighborhood of x2. There exist n many subsets in which the arbitrarily small variations lie. The larger the variations get, the further x1 and x2 are spaced apart. This then implies that a variational condition exists for having x1 lie in the neighborhood of x2. However, if the variation can be partitioned into smaller, open subsets along the variational interval. Thus, if you evaluate x1 and x2 as open sets, x1 does not lie in the neighborhood of x2 after a sufficiently large variation. This implies that any monotonic function cannot be solved for these boundary conditions when evaluated along a closed interval. This is as the initial and final boundary conditions are at the same magnitude along the interval, and no nontrivial variational distance between the two will satisfy the modified Nagumo conditions. Additionally, no function and its derivative can satisfy the same boundary conditions for an arbitrarily small variation via the stepwise evaluation of a function and its derivative. The only exception is for a trivial "zero" case. Cybenko's theorem allows for the violation of the stepwise principle by immediate coupling of lower order values which are fed into higher order terms. This is done by superimposing functional "values" onto an analytical expression which reclassifies the values. A variety of a certain value, such as from Eq. 24.14, is injectively mapped into an equation, like Eq. 24.16. This not only allows for a range of tuning parameters to be applied to the E-guidance formulation, but gives more degrees of freedom to tune the equation. This is as rather than having static coefficients and functions, the E-guidance equation has variable coefficients which change over the course of the index. It is necessitated that these parameters be along the same index, otherwise there will be discontinuities and errors in the tuning process. The application of algebraic geometry to solving guidance problems has not been shown in the literature yet. This is as most guidance methods are derived from a control solution, which acts as a reference for which solutions can be imputed into the controller. However, via application of Cybenko's theorem, we can leverage the power of algebraic geometry to shape reference trajectories which may not have otherwise been possible. Algebraic geometry is focally oriented around understanding the behavior of algebraic "zeroes" within a solution set. Specifically, algebraic geometry provides a geometric framework for understanding geometric instances of algebraic solutions. For the problem we are considering, we look to leverage Cybenko's theorem to generate two separate algebraic varieties which are topologically related. This topological relation comes in the form of a set of boundary condition values. These boundary condition values are necessitated for way-point guidance, which imposes higher order conditions on all possible guidance solutions. This seemingly violates the notion of explicit guidance providing an analytical solution to meet a single final condition, as there are multiple conditions imposed on the guidance formulation. This is verified from an example simulation outlined in Figs. 24.11 and 24.12.

The left plot shows what guidance solution results would be if you relied purely on analytical integration of the kinematic parameters, starting from an acceleration term. The right plot shows the coupled guidance solution after feeding the solutions through the network. An important note of this is that the thrust acceleration and the thrust velocity are commanded to zero at the exact same time instance along the

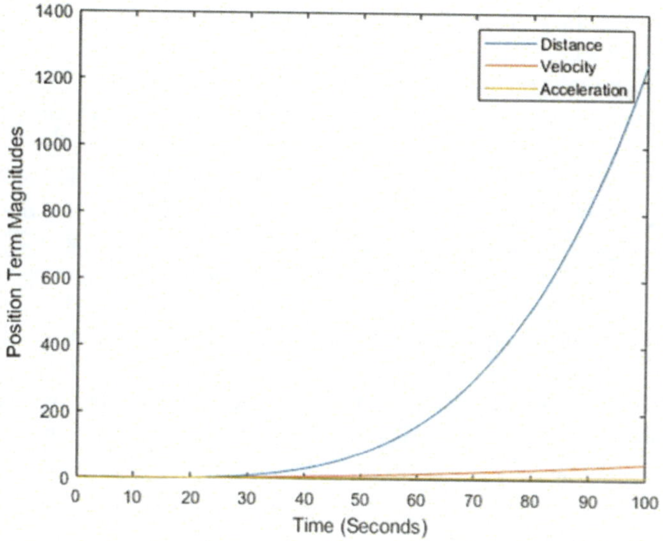

Fig. 24.11 Plots Without Feedthrough

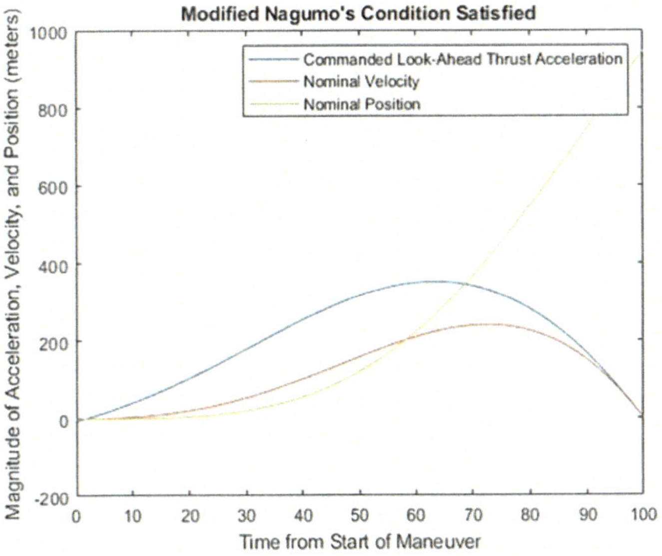

Fig. 24.12 Plots With Feedthrough

maneuver. This is important as it prevents overshooting of the waypoint. Another note which makes this result interesting is that there is no negative commanded thrust acceleration or thrust velocity. This is not only a novel result which could not be achieved by Kawamura, but has important implications towards optimal control. This is as a necessary condition for optimal control is a non-negativity condition, in which control state results cannot be negative [3]. The same analytical expressions were used for both plots, except the plot on the right utilizes Cybenko's theorem to feed through tuning values to generate the desired results. These are as follows:

$$a^{command} = \left((a_3^{init})^2 * t^2\right) - a_3^{init} \tag{24.24}$$

$$a_{command}^{lookahead} = \left(\left(a_3^{init}\right) * t\right) - \left(a_3^{init} * t\right) - a^{command} \tag{24.25}$$

$$v^{guided} = a_{command}^{lookahead} * t \tag{24.26}$$

$$r_{reference}^{modified} = \left(v^{guided} * t\right) + (a^{command} * t^2)/2 + h \tag{24.27}$$

This noticeable difference in output performance reflects the power of Cybenko's theorem to tune guidance equation parameters. This is as a number of layers are tuned to achieve certain initial and endpoint conditions.

24.8 Conclusion

The application of Cybenko's theorem to guidance allows for better tuning to occur. It can be used to both tune the guidance coefficients, as well as generate parametric functions which serve as a reference trajectory. This form of tuning is novel, as no work has presented a method for rigorously tuning guidance parameters. By combining algebraic geometry with Cybenko's theorem, a better formalism can be achieved for GNC tuning.

References

1. Cherry, G.: A general, explicit, optimizing guidance law for rocket propelled spaceflight (1964)
2. Cybenko, G.: Approximation by superpositions of a sigmoidal function. In: Mathematics of Controls, Signals, and Systems Journal. Springer (1989)
3. Dmitruk, A., Kaganovich, A.: The Hybrid Maximum Principle is a consequence of Pontryagin Maximum Principle. Lomonosov Moscow State University
4. Hartshorne, R.: Algebraic Geometry. Springer (1977)
5. Hull, D.: Fundamentals of Airplane Flight Mechanics. Springer (2007)
6. Kawamura, E.: Integrated Targeting, Guidance, Navigation, and Control for Unmanned Aerial Vehicles, December 2020
7. Kawamura, E., Azimov, D.: Extremal control and modified explicit guidance for autonomous unmanned aerial vehicles. J. Autonomous Veh. Syst. (2022)
8. Klumpp, A.: Apollo Lunar Descent Guidance (1971)
9. Simplicio et al.: Parametrized Laws for Guidance and Control of Planetary Landers (2017)
10. Siouris, G.: Missile Guidance and Control Systems. Springer (2004)
11. Tillerson, M., Inalhan, G., How, J.: Coordination and control of distributed spacecraft systems using convex optimization techniques. Int. J. Robust Nonlinear Control (2002)

Chapter 25
A Novel Dynamic and Aerodynamic Intelligent Morphing System (DA-IMS) for the Stability of an Autonomous Utility Truck with the Boom Equipment

Parth Y. Patel, Roy Koomullil, and Vladimir Vantsevich

Abstract Utility trucks with boom equipment, also known as bucket trucks, are the first responders in severe environments, such as natural disasters or extreme weather conditions. These bucket trucks can provide critical support for emergency responders to restore essential services quickly; hence, the utility trucks' stability while driving in hazardous weather conditions is a crucial aspect of safety of the rescue crew. This paper represents a novel comprehensive study of the DA-IMS system, which utilizes the active aerodynamics and truck dynamics to manage the moment of inertia, tire-road forces, and the aero-dynamic multi-phase forces, e.g., air wind, meteorology, and several multi-phase hazardous kinds of weather and roadway conditions, thus, improving its stability and safety under dangerous critical weather conditions. The system employs the stability criteria, analyzes the data, and subsequently adjusts the utility truck's shape to enhance its driving conditions on the road. By continuously adapting to changing driving conditions, such as sudden maneuvers and crosswinds, the DA-IMS system can mitigate the risk of accidents and enhance the overall safety of the utility truck. This study offers a novel solution for improving the safety and stability of an autonomous utility truck in extreme weather conditions.

Keywords Autonomous utility truck · Stability · Intelligent morphing system · Boom equipment · Morphing device · Multi-body dynamics · Aerodynamics

---

P. Y. Patel (✉) · R. Koomullil · V. Vantsevich
University of Alabama at Birmingham, Birmingham, AL 35294, USA
e-mail: parth144@uab.edu

R. Koomullil
e-mail: rkoomul@uab.edu

V. Vantsevich
e-mail: vantsevi@uab.edu

D. Azimov (ed.), *Proceedings of the IUTAM Symposium on Optimal Guidance and Control for Autonomous Systems 2023*, IUTAM Bookseries 40,
https://doi.org/10.1007/978-3-031-39303-7_24

25.1 Introduction

Severe weather conditions are natural occurrences that can seriously affect human life, property, and infrastructure. A new pattern of more frequent and intense extreme weather events has emerged, including hurricanes, floods, blizzards, tornadoes, hailstorms, and ice storms. These extreme events have led to growing concerns about their impact on public safety, the economy, and the environment [1]. Such events can cause catastrophic flooding and, most likely, intensify hurricanes and cause severe damage to both residential and public properties. Therefore, the safety of the rescue and repair team is of paramount importance when they attend to emergencies due to these extreme weather conditions.

Utility trucks (also known as boom trucks) are the first responders in extreme climate and weather situations, from cutting trees to restoring traffic, rescuing living beings from destroyed properties, repairing electric posts, and restoring power. A stable utility truck will be beneficial under this kind of situation, making the utility truck remain on the ground without skidding or even driving on the road to attend to such emergencies. According to the US Census Bureau, there were approximately 15 million trucks currently in operation across the country operated by 2.83 million drivers, 28.2% of whom drive various utility trucks [2]. Such trucks with morphing capabilities of the boom equipment can increase the possibility of road accidents in many ways and, additionally, create dangerous situations on the roads, and on off-road conditions, while moving, and performing tasks.

It has been established that the airflow impacts, (i) vehicle productivity, and (ii) energy efficiency. The influence of truck systems and bluff body geometry on the aero drag was studied in [3–6], including air resistance produced by the wheels of the trucks [7], the cabin design [8, 9], the shape of the truck bottom, the rear surface of the trailer [4], the height of a tractor/semitrailer, and the gap between them [4, 7, 10]. Approximately 25% of the air resistance comes from the air-cabin interaction, 30% is added by the aero drag beneath the trailer, and the gap between the cabin and the trailer makes 20%, while 25% comes from the air vortex behind the trailer [11]. Passive head/side deflectors, frontal air dam [9, 12–14], spoilers [15–17], half-balloon-shaped airfoils, trailer skirts to prevent the air flow from entering the space under the trailer, and rear-end taper [4, 13, 18]. Truck platooning reduces the aero-drag [19] while hydrophobic coating decreases the aero-drag/fuel consumption [20]. It has been found that the tire rolling resistance power loss can potentially go up to 30–40%, while air resistance power loss makes 35–55% of the power needed to move a truck [21]. A significant effort to analyze morphing vehicles has also been taken in the areas of multidisciplinary design optimization methods, design and control of airfoil and fuselage shapes, and aeroelasticity. Morphing was defined as a real-time adaption problem to enable a multi-point optimized performance [22]. Considerable research has been performed on the aeroelasticity, and load redistribution effects [23], adaptive skeletal structures [24, 25], adaptive aeroelastic wings [26, 27], and new methods for morphing designs [28, 29].

It is important to emphasize that an additional vehicle–road force domain makes a principal difference between morphing aircraft and ground vehicles. Therefore, the morphing capabilities of ground vehicles need to be investigated together with the vehicle–road forces to hold vehicles safe on the roads and improve productivity/efficiency. Towards those directions, a novel technical paradigm of model-based design as a two-domain interactive problem, comprising of (i) the truck aerodynamics and (ii) the truck-road dynamics, are presented in this paper for the stability of an autonomous utility truck. This interactive nature allows for determining the optimal morphological characteristics of the DA-IMS and counteractive flow characteristics of the active add-on devices. Furthermore, the severe weather condition is modeled as an input in a two-domain interactive model-based design. The mathematical formulation provides a base to generate fields of aerodynamic forces around the utility truck's structure to determine the tire-road forces sufficient to hold the utility truck within a *safe zone*.

25.2 Utility Truck—Aerodynamics and—Road Dynamics (Two-Domain) Interactive Models

The formulation of the utility truck-aero and -road dynamics results in determining forces that should be applied to a vehicle to keep required/given kinematic parameters of the truck in motion. Figure 25.1 shows a free-body diagram that was used for developing a mathematical model of a utility truck with the boom equipment on a support surface that can be modelled as both deformable and non-deformable. Some fundamentals of the models were first introduced in [29–32]. This paper presents the further development in which an impact of the severe weather conditions is modeled as an input in a two-domain interactive model-based design to investigate the stability of an autonomous utility truck with the boom equipment.

25.2.1 *Aerodynamic Forces and Moments*

When the utility truck is moving under severe weather conditions with speed v_x and the speed of the absolute wind is u and is aligned to the direction of a truck movement at an angle α, then the speed of the relative wind that acts on the truck is V and is aligned to the direction of the truck movement by angle β. The angles α and β are taken positive in the compass direction as shown in Fig. 25.2.

With the help of Fig. 25.2, the vector equation $v_x + V = u$ indicates that the relative wind speed V and its angle β are given by the following equations when u and α are given [36, 37].

$$V^2 = (v_x + u\cos\alpha)^2 + u^2\sin^2\alpha \tag{25.1}$$

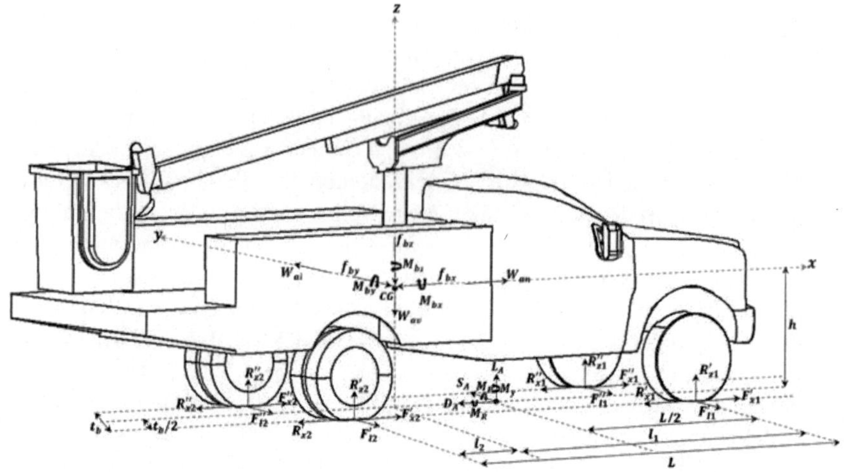

Fig. 25.1 Dimensions, reaction forces, wind-induced resultant forces, and moments on the utility truck with the boom equipment

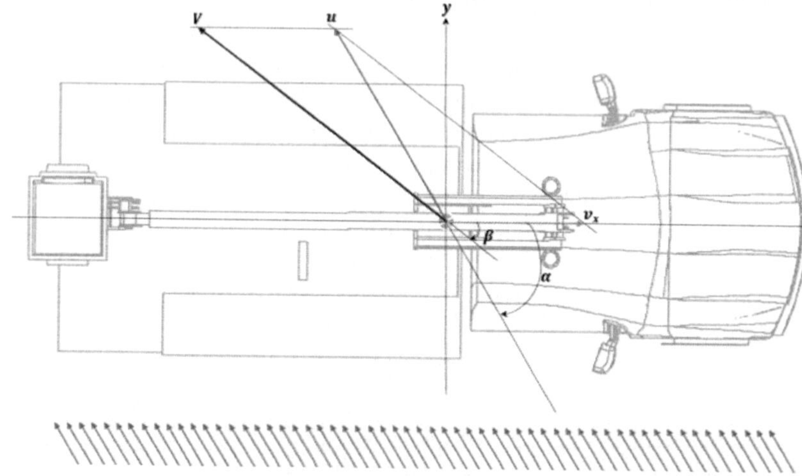

Fig. 25.2 Truck velocity, v_x, absolute wind velocity, u, and relative wind velocity, V

$$\beta = \arctan \frac{u \sin \alpha}{v_x + u \cos \alpha} \tag{25.2}$$

The interaction of a truck and air results in shear stress and normal pressure on the truck's surface, which produces the aerodynamic forces and moments, which are given by the following equations [38].

$$D_A = \frac{1}{2}C_D\rho AV^2 \quad L_A = \frac{1}{2}C_L\rho AV^2 \quad S_A = \frac{1}{2}C_S\rho AV^2 \tag{25.3}$$

$$M_P = \frac{1}{2}C_{PM}\rho AV^2L \quad M_R = \frac{1}{2}C_{RM}\rho AV^2L \quad M_Y = \frac{1}{2}C_{YM}\rho AV^2L \tag{25.4}$$

where D_A is the drag force, L_A is the lift force, S_A is the side force, M_P is the pitching moment, M_R is the rolling moment, M_Y is the yawing moment, C_D is the drag coefficient, C_L is the lift coefficient, C_s is the side force coefficient, C_{PM} is the pitching moment coefficient, C_{RM} is the rolling moment coefficient, C_{YM} is the yawing moment coefficient, ρ is the density of the air, A is the frontal reference area of the truck, V is the total velocity of vehicle speed and wind speed, and L is the reference length.

In this analysis, the aerodynamic forces and moments are applied at a SAE aerodynamic reference point [39] as a wheelbase mid-point along the intersection of the vehicle plane of symmetry and at the ground, as shown in Fig. 25.1. Computational fluid dynamics (CFD) numerical models, validation, experimental and numerical results, and flow field visualization of the utility truck in severe weather conditions were presented in [34, 35].

25.2.2 *Utility Truck's Road-Dynamics*

In this analysis, the boom equipment, which morphs its geometry and orientation relative to the utility truck's frame, is considered as a 5-degree-of-freedom (DOF) robotic manipulator, as shown in Fig. 25.3a. The static and dynamic modelling of the morphing boom equipment was developed and implemented symbolically in compact form using the recursive Newton–Euler dynamical formulation. Also, a novel approach, a method, and mathematical models were presented to investigate the dynamic coupling of the 5-DOF with the utility truck, as described in [30, 32]. Therefore, the motion of the boom equipment can be represented by forces (f_{bx}, f_{by}, f_{bz}) and moments (M_{bx}, M_{by}, M_{bz}) at the truck's center of gravity, as shown in Fig. 25.3b.

The free-body diagram in Fig. 25.1 leads to the following equation of the longitudinal motion of the utility truck under crosswind conditions:

$$F_{x\sum} = \sum_{i=1}^{2} F_{xi}^{\prime(\prime\prime)} = \frac{W_{av}a_x\delta_r}{g} \pm W_{an}sin\theta_n + \sum_{i=1}^{2} R_{xi}^{\prime(\prime\prime)} + D_A + f_{Bx} \tag{25.5}$$

here, W_{av} is the truck's weight; a_x is the truck's longitudinal acceleration that presents the required/desired motion; δ_r is the factor that reflects the influence on the truck rotating mass; D_A is the aerodynamic drag force, θ_n is the road micro profile. The steering system of the truck is decoupled from the driveline system and the hybrid-electric power transmitting unit (HE-PTU) concept [40] is used to alter the split of torque between the drive axles to better fit an optimal power distribution. Thus,

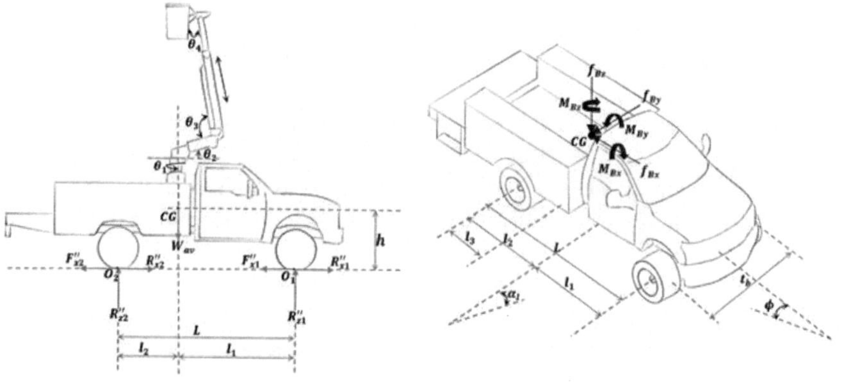

(a) (b)

Fig. 25.3 **a** 5-DOF Boom equipment representation, **b** representation of the boom equipment as three forces and moments

Eq. (25.5) determines the total circumferential force of the truck, $F_{x\sum}$ in the time domain. This force is equal to the total resistance to motion including inertia force due to the required motion.

Following is a general algorithm for determining the total circumferential force of the truck $F_{x\sum}$:

1. Define severe weather conditions.
2. Calculate the inertia force $\frac{W_{av}a_x\delta_r}{g}$. Find the speed v_x of the truck by integrating the acceleration and calculate the aerodynamic forces and moments using CFD approach.
3. Simulate a macro profile of the surface of motion and determine the longitudinal component of the weight. e.g., $\pm W_{an}\sin\theta_n$.
4. Calculate the static normal reactions of the wheels, $R_{zsi}^{\prime(\prime\prime)}$, $i = 1, 2$ using static equilibrium condition as described in Sect. 24.2.2.1.
5. Simulate a micro profile of the surface of motion and determine the dynamic normal reactions of the wheels, $R_{zi}^{\prime(\prime\prime)}$, $i = 1, 2$.
6. Using the normal reactions $R_{zi}^{\prime(\prime\prime)}$ (found in item 5) and truck speed v_x (found in item 1), determine the rolling resistance forces $R_{xi}^{\prime(\prime\prime)}$, $i = 1, 2$.
7. Substituting the findings from item 1–6 in Eq. (25.5), finally determine the total circumferential force of the truck $F_{x\sum}$.

25.2.2.1 Static Normal Reactions of the Wheels

The following assumptions have been made to calculate the static normal reactions:

- The utility truck is subject to a uniform crosswind that moves in a steady, straight line.

- The utility truck is treated as a rigid body, and thus, the equilibrium conditions of the forces and moments with respect to truck center of gravity yield six equilibrium equations.
- The truck system is statically indeterminable. For this reason, normal stiffness of the tires and suspension are used.
- Sudden wind gust is ignored.

After considering the equilibrium equations, a constraint equation, and the aerodynamic forces and moments with the applied wind direction, the solution of the system produces the following static normal reaction forces

$$R''_{zs1} = \frac{k_1}{k_1+k_2}\left[\frac{h}{t_b}(W_{al}+f_{By}) + \frac{1}{t_b}(M_R - M_{Bx})\right] + \frac{W_{av} - f_{Bz} - L_A}{2} - \frac{1}{2(l_1+l_2)}\left[-h(W_{an}+f_{Bx}) - l_1(f_{Bz}+L_A-W_{av}) - M_{By} - M_p + L_A\left(l_1 - \frac{L}{2}\right)\right] \tag{25.6}$$

$$R'_{zs1} = \frac{-k_1}{k_1+k_2}\left[\frac{h}{t_b}(W_{al}+f_{By}) + \frac{1}{t_b}(M_R - M_{Bx})\right] + \frac{W_{av} - f_{Bz} - L_A}{2} - \frac{1}{2(l_1+l_2)}\left[-h(W_{an}+f_{Bx}) - l_1(f_{Bz}+L_A-W_{av}) - M_{By} - M_p + L_A\left(l_1 - \frac{L}{2}\right)\right] \tag{25.7}$$

$$R'_{zs2} = \frac{-k_2}{k_1+k_2}\left[\frac{h}{t_b}(W_{al}+f_{By}) + \frac{1}{t_b}(M_R - M_{Bx})\right] + \frac{1}{2(l_1+l_2)}\left[-h(W_{an}+f_{Bx}) - l_1(f_{Bz}+L_A-W_{av}) - M_{By} - M_p + L_A\left(l_1 - \frac{L}{2}\right)\right] \tag{25.8}$$

$$R''_{zs2} = \frac{k_2}{k_1+k_2}\left[\frac{h}{t_b}(W_{al}+f_{By}) + \frac{1}{t_b}(M_R - M_{Bx})\right] + \frac{1}{2(l_1+l_2)}\left[-h(W_{an}+f_{Bx}) - l_1(f_{Bz}+L_A-W_{av}) - M_{By} - M_p + L_A\left(l_1 - \frac{L}{2}\right)\right] \tag{25.9}$$

here, $k_i, i = 1, 2$ is the normal stiffness of the tires and suspension; W_{al}, W_{an}, and W_{av} are the utility truck gross weight lateral, longitudinal and vertical component, respectively, and $'$ and $''$ superscript indicates right and left wheel(s).

25.2.2.2 Dynamic Normal Reactions of the Wheels

Rolling resistance forces $R_{xi}^{'('')}$ can be calculated only when the values of the normal wheel reactions $R_{zi}^{'('')}$ are known. These are determined from reactions $R_{zsi}^{'('')}$ obtained for the cases of motion over a straight inclined/horizontal surface (suspension's action included) with allowance for dynamic components induced by the tires and micro-topography:

$$R_{zi}^{'('')} = R_{zsi}^{'('')} + C_{ti}^{'('')}\left(\xi_i^{'('')} - q_i^{'('')}\right) + K_{ti}^{'('')}\left(\dot{\xi}_i^{'('')} - \dot{q}_i^{'('')}\right) \tag{25.10}$$

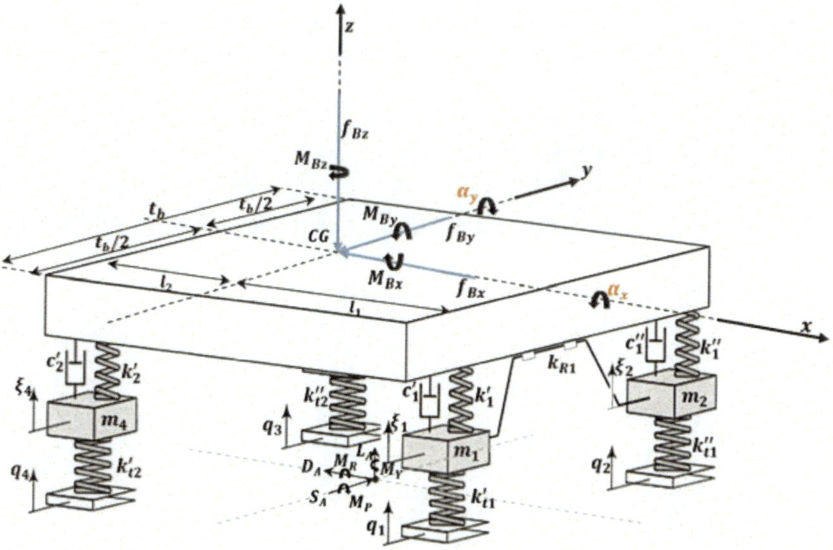

Fig. 25.4 Computational layout of the sprung and unsprung masses in the transverse plane

where $C_{ti}^{\prime(\prime\prime)}$ is the normal stiffness of the tire, $K_{ti}^{\prime(\prime\prime)}$ is the tire's damping factor, $q_i^{\prime(\prime\prime)}$ is the height of the surface micro-profile, and $\xi_i^{\prime(\prime\prime)}$ is the translational displacement of the unsprung mass normal to the surface of motion, $m_i^{\prime(\prime\prime)}$, $i = 1, 2$. Computational layout of the sprung and unsprung masses in the transverse plane is shown in Fig. 25.4.

The values of $\xi_i^{\prime(\prime\prime)}$ and $q_i^{\prime(\prime\prime)}$ and their derivatives needed for performing the calculations from formula (25.10) are found by compiling equations of oscillations of the sprung and unsprung masses of the truck using second-order Lagrange equations. The generalized coordinates for describing the oscillations of the sprung mass are the following: translation z of the center of sprung mass normal to the support surface and α_x and α_y, the angles of rotation about the truck's longitudinal and transverse axes. The generalized coordinates for the unsprung masses $m_i^{\prime(\prime\prime)}$ consisted for their motions normal to the surface $\xi_i^{\prime(\prime\prime)}$. The equations of oscillations are written as

$$\begin{aligned} M\ddot{z} &+ C'_{p1}\left(\dot{z} - \dot{\xi}'_1 + 0.5t_b\dot{\alpha}_x - l_1\dot{\alpha}_y\right) + C''_{p1}\left(\dot{z} - \dot{\xi}''_2 - 0.5t_b\dot{\alpha}_x - l_1\dot{\alpha}_y\right) \\ &+ C''_{p2}\left(\dot{z} - \dot{\xi}''_3 - 0.5t_b\dot{\alpha}_x + l_2\dot{\alpha}_y\right) + C'_{p2}\left(\dot{z} - \dot{\xi}'_4 + 0.5t_b\dot{\alpha}_x + l_2\dot{\alpha}_y\right) \\ &+ K'_{p1}\left(z - \xi'_1 + 0.5t_b\alpha_x - l_1\alpha_y\right) + K''_{p1}\left(z - \xi''_2 - 0.5t_b\alpha_x - l_1\alpha_y\right) \\ &+ K''_{p2}\left(z - \xi''_3 - 0.5t_b\alpha_x + l_2\alpha_y\right) + K'_{p2}\left(z - \xi'_4 + 0.5t_b\alpha_x + l_2\alpha_y\right) \\ &- F_{Bz} + L_A = 0 \end{aligned} \tag{25.11}$$

$$\begin{aligned} I_x\ddot{\alpha}_x &+ 0.5t_b\{C'_{p1}\left(\dot{z} - \dot{\xi}'_1 + 0.5t_b\dot{\alpha}_x - l_1\dot{\alpha}_y\right) - C''_{p1}\left(\dot{z} - \dot{\xi}''_2 - 0.5t_b\dot{\alpha}_x - l_1\dot{\alpha}_y\right) \\ &- C''_{p2}\left(\dot{z} - \dot{\xi}''_3 - 0.5t_b\dot{\alpha}_x + l_2\dot{\alpha}_y\right) + C'_{p2}\left(\dot{z} - \dot{\xi}'_4 + 0.5t_b\dot{\alpha}_x + l_2\dot{\alpha}_y\right) \end{aligned}$$

$$
\begin{aligned}
&+ K'_{p1}(z - \xi'_1 + 0.5t_b\alpha_x - l_1\alpha_y) - K''_{p1}(z - \xi''_2 - 0.5t_b\alpha_x - l_1\alpha_y) \\
&- K''_{p2}(z - \xi''_3 - 0.5t_b\alpha_x + l_2\alpha_y) + K'_{p2}(z - \xi'_4 + 0.5t_b\alpha_x + l_2\alpha_y)\} \\
&+ K_{Rf}\left(\alpha_x - \frac{\xi'_1 - \xi''_2}{t_b}\right) - M_{Bx} - M_R = 0
\end{aligned} \tag{25.12}
$$

$$
\begin{aligned}
&I_y\ddot{\alpha}_y - l_1C'_{p1}(\dot{z} - \dot{\xi}'_1 + 0.5t_b\dot{\alpha}_x - l_1\dot{\alpha}_y) - l_1C''_{p1}(\dot{z} - \dot{\xi}''_2 - 0.5t_b\dot{\alpha}_x - l_1\dot{\alpha}_y) \\
&+ l_2C''_{p2}(\dot{z} - \dot{\xi}''_3 - 0.5t_b\dot{\alpha}_x + l_2\dot{\alpha}_y) + l_2C'_{p2}(\dot{z} - \dot{\xi}'_4 + 0.5t_b\dot{\alpha}_x + l_2\dot{\alpha}_y) \\
&- l_1K'_{p1}(z - \xi'_1 + 0.5t_b\alpha_x - l_1\alpha_y) - l_1K''_{p1}(z - \xi''_2 - 0.5t_b\alpha_x - l_1\alpha_y) \\
&+ l_2K''_{p2}(z - \xi''_3 - 0.5t_b\alpha_x + l_2\alpha_y) + l_2K'_{p2}(z - \xi'_4 + 0.5t_b\alpha_x + l_2\alpha_y) \\
&- M_{By} + M_P = 0
\end{aligned} \tag{25.13}
$$

$$
\begin{aligned}
&m'_1\ddot{\xi}'_1 - C'_{p1}(\dot{z} - \dot{\xi}'_1 + 0.5t_b\dot{\alpha}_x - l_1\dot{\alpha}_y) - K'_{p1}(z - \xi'_1 + 0.5t_b\alpha_x - l_1\alpha_y) \\
&- \frac{K_{Rf}}{t_b}\left(\alpha_x - \frac{\xi'_1 - \xi''_2}{t_b}\right) + K'_{t1}(\xi'_1 - q'_1) - F_{Bz} + L_A = 0
\end{aligned} \tag{24.14}
$$

$$
\begin{aligned}
&m''_2\ddot{\xi}''_2 - C''_{p1}(\dot{z} - \dot{\xi}''_2 - 0.5t_b\dot{\alpha}_x - l_1\dot{\alpha}_y) - K''_{p1}(z - \xi''_2 - 0.5t_b\alpha_x - l_1\alpha_y) \\
&+ \frac{K_{Rf}}{t_b}\left(\alpha_x - \frac{\xi_1\prime - \xi''_2}{t_b}\right) + K''_{t1}(\xi''_2 - q''_2) - F_{Bz} + L_A = 0
\end{aligned} \tag{24.15}
$$

$$
\begin{aligned}
&m''_3\ddot{\xi}''_3 - C''_{p2}(\dot{z} - \dot{\xi}''_3 - 0.5t_b\dot{\alpha}_x + l_2\dot{\alpha}_y) - K''_{p2}(z - \xi''_3 - 0.5t_b\alpha_x + l_2\alpha_y) \\
&+ K''_{t2}(\xi''_3 - q''_3) - F_{Bz} + L_A = 0
\end{aligned} \tag{24.16}
$$

$$
\begin{aligned}
&m'_4\ddot{\xi}'_4 - C'_{p2}(\dot{z} - \dot{\xi}'_4 + 0.5t_b\dot{\alpha}_x + l_2\dot{\alpha}_y) - K'_{p2}(z - \xi'_4 + 0.5t_b\alpha_x + l_2\alpha_y) \\
&+ K'_{t2}(\xi'_4 - q'_4) - F_{Bz} + L_A = 0
\end{aligned} \tag{25.17}
$$

where I_x and I_y are the moments of inertia of the sprung mass; $C_{pi}^{\prime('')}$ is the stiffness of the suspension at a wheel; $K_{pi}^{\prime('')}$ is the shock-absorber resistance factor reduced to a wheel, and K_{Rf} is the truck's antiroll bar torsional stiffness in front.

25.3 Stability of Motion

Stability of motion is the property of a vehicle to maintain specified differential parameters of motion or a specified course of variation in them, after the perturbing forces are removed [41]. A vehicle is caused to move by all kinds of forces, including applied forces, gravity forces, inertia forces, and finally, circumferential forces, the distribution of which among the wheels depends to a large extent on the driveline system. The time of action of the forces is finite but investigating the stability and their aftereffects should be considered even when some of them are no longer applied. Stability may be retained up to a certain velocity level that depends on road conditions, applied forces, and design parameters of the vehicle. When this velocity is reached, the motion of the vehicle becomes unstable, which also means that it is potentially dangerous for the subsequent level.

There are six generalized coordinates of a vehicle: three linear, along the x-longitudinal, y-lateral, and z-vertical axes, associated with the vehicle body and three angular turns about these axes. A vehicle may deviate from its specified path and lose stability in four of six generalized coordinates of the vehicle body: linear motion along the x-axis and along the z-axis of the vehicle are not included in this analysis.

25.3.1 Stability Assessment

The stability of motion for the autonomous utility truck with the boom equipment is accessed on the basis of following.

1. Grip Utilization Factor
2. Lateral roll angle of the sprung mass γ_1
3. Angular velocity r of vehicle about the vertical axis, known as the yaw rate.

Grip Utilization Factor

Lateral skid of a vehicle usually occurs before the vehicle rolls over. To assess tire-road interaction and the skid occurrence, the grip utilization factor K_μ is estimated for all the driving wheels:

$$K_\mu = \frac{\sqrt{F_x^2 + F_l^2}}{\mu_{px} R_z} \tag{25.18}$$

where μ_{px} is the peak friction coefficient in longitudinal direction of tire, F_x is the wheel circumferential force, and F_l is the wheel lateral reaction force acting perpendicular to the wheel plane (Fig. 25.5).

The wheel lateral reaction force is considered as a summation of the wind force acting in the lateral direction and a centrifugal force at the CG and can be represented in first approximation as

$$F_{li} = K_{yi}\alpha_i = \frac{1}{8} C_s \rho A V^2 + \frac{1}{2}\left(\frac{M_{ai} v_x^2}{R_a}\right) - \frac{1}{4} f_{By} \tag{25.19}$$

where

$$R_a = R_t + \left(\frac{M_{a1}}{K_{y1}} - \frac{M_{a2}}{K_{y2}}\right)\frac{v_x^2}{\delta} \tag{25.20}$$

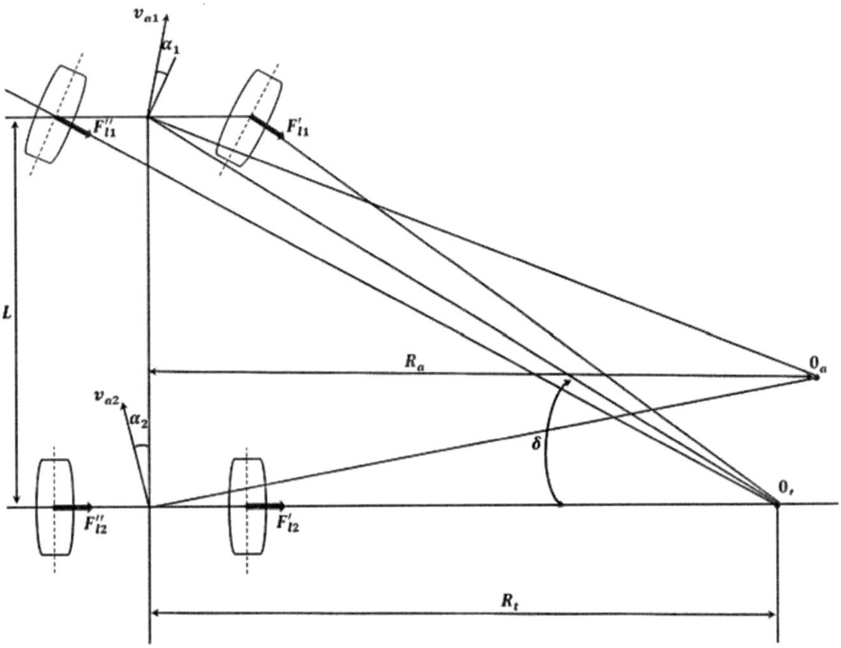

Fig. 25.5 Vehicle taking a turn

$$R_t = \frac{L}{\delta} \tag{25.21}$$

here, K_{yi} is the cornering stiffness of a tire, α_i is the wheel slip angle, δ is the wheel steering angle, and M_{a1} and M_{a2} are the masses obtained from the split of the truck's weight among the front and rear axles.

The lower value of K_μ, the higher the gripping potential of the wheels and the grater probability that the truck will move in the desired path set by the driver. It is usually assumed that lateral skid of a wheel occurs at $K_\mu > 0.5$ [41].

Lateral roll angle of the sprung mass γ_d

Among the dynamic maneuvers, a vehicle can experience rollover, considered one of the most severe threats to occupants. The rollover may be precipitated by one or a combination of factors. It may occur on flat and level surfaces when the lateral accelerations on a vehicle reach a level beyond that which can be compensated by lateral weight shift on the tires. The off-road surface may contribute along the disturbances to the lateral forces arising from curb impacts, soft ground, or other obstructions that may lead to the vehicle's rollover. Static roll angle is an index to estimate the capability of a vehicle to withstand the rollover and it is calculated as follows:

Equilibrium of sprung mass:

$$T_{st} = M_{Bx} + M_R - h\left(W_{av}\sin\gamma + f_{By}\cos\gamma\right) \tag{25.22}$$

where, $T_{st} = K_{st}\gamma$, is the moment acting from the tires and suspensions and γ is the static roll angle of the sprung mass. It's assumed that the static roll angle is less than 8° and hence, $\cos\gamma \approx 1$ and $\sin\gamma \approx \gamma$ (Fig. 25.6).

$$\gamma = \frac{M_{Bx} + M_R - f_{By}h}{K_{st} + W_{av}h} \tag{25.23}$$

The *Rollover Margin Factor* is another estimation index to estimate the rollover condition and defined as

$$K_{RM} = \frac{\gamma}{\gamma_d} \tag{25.24}$$

here, γ_d is the dynamic roll angle, which estimates the capability to withstand rollover on a bumpy road/terrain. The equations of sprung and unsprung masses of the utility trucks allow us to determine the dynamic roll angle. It is usually assumed that the value of K_{RM} should remain between 2 and 2.5 to prevent the rollover condition.

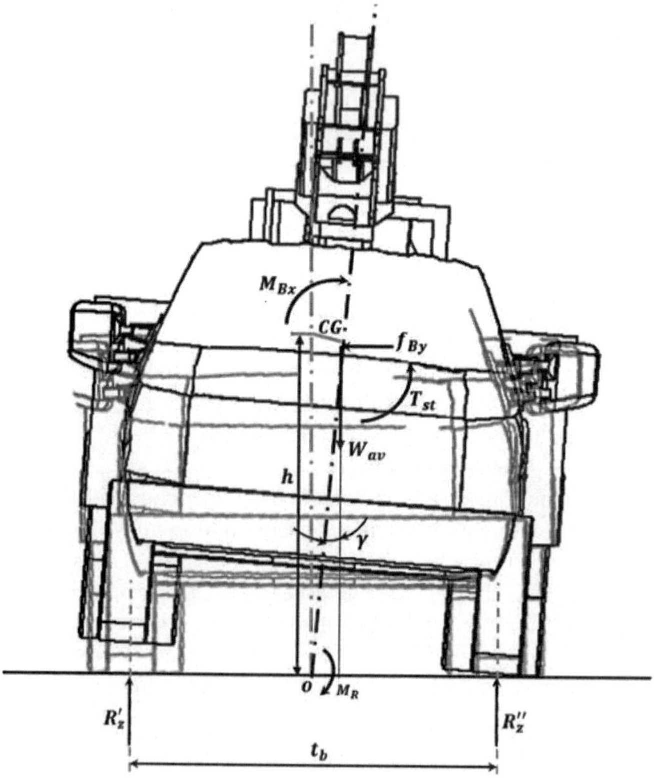

Fig. 25.6 Forces and moments acting to rollover a vehicle

Angular velocity r of vehicle about the vertical axis

Angular velocity is a critical parameter in vehicle dynamics because it affects the stability and handling of the vehicle. Yaw motions produce lateral accelerations causing roll motions, and roll motion, in turn, alters yaw response through the modification of tire cornering forces arising from lateral transfer and suspension action. As we assumed that the utility truck is subject to a uniform crosswind that moves in a steady, straight line, then the equation of the yaw rate in the first approximation from Fig. 25.5 can be written as follows

$$Yaw\ Rate\left(\frac{\text{rad}}{\text{s}}\right) = \frac{v_x}{R_a} \tag{25.25}$$

here, R_a is the actual turn radius and v_x is the utility truck velocity in the longitudinal direction.

25.4 DA-IMS Smart Morphing Devices

While driving, friction-like forces, including the normal force generated by the moving vehicle, considerably affect the vehicle's performance in terms of braking, cornering, and acceleration. These forces are related to each tire's vertical load and friction coefficients. Therefore, by modifying the external shape of the vehicle and managing the aerodynamic flow around the vehicle, the adhesion forces on the tires related to the normal force of the vehicle with the ground can be managed to improve the running stability. Furthermore, the morphing dynamics of the boom equipment majorly impact the wheel dynamic normal reactions. Therefore, such agile managing of wheel dynamic normal reactions is a crucial component of the DA-IMS systems that should stabilize the motion of the autonomous smart utility truck under hazardous environmental conditions. Thus, by determining the smart morphological properties of the entire autonomous utility truck 3D perimeter, it is achievable to reduce aero drag in extreme weather conditions, increase the tire-terrain grip/traction, and reduce the rolling resistance of the tires.

Two smart morphing devices, as shown in Fig. 25.7, (i) the morphable boom equipment and (ii) the morphing device, are used to manage the moment of inertia, tire-road forces, and the aerodynamics multi-phase forces, e.g., air wind, meteorology, and several multi-phase hazardous kinds of weather and roadway conditions. The boom equipment is considered a 5-DOF robotic manipulator which autonomously morphs its orientation. In contrast, the morphing device is designed with the eighteen NACA2412 morphable airfoils. Each airfoil can rotate from 0° to 180° (in a transverse plane) depending upon the required conditions to manage the aerodynamic flow.

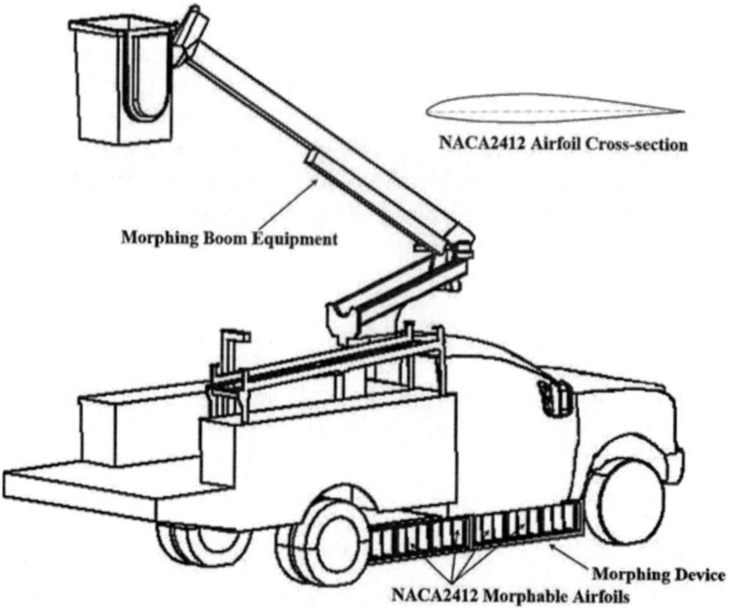

Fig. 25.7 Utility truck with the morphing boom equipment and morphing device

The autonomous system of the utility truck senses the weather conditions, analyzes the data, employs the stability criteria, makes an intelligent decision of "to go" or "not to go" into the in-coming weather conditions, and if it decides "to go", then subsequently adjust the utility truck's shape either using the morphable boom equipment or the morphing device, to enhance its driving conditions on the road and thus, reduces the risk of accidents and enhances the overall safety of the utility truck.

25.5 Results and Discussion

A numerical case is presented to analyze the static stability of the utility truck moving at a constant speed. In this case, we evaluated the utility truck moving in a steady, straight line at a speed of 40 mph on an asphalt road subject to a uniform crosswind. The absolute wind, $\boldsymbol{u}$, is acting at an angle of 25°.

The National Weather Service categorizes tornadoes and hurricanes based on severity using the Fujita scale [42] and Saffir-Simpson Hurricane Wind Scale [43]. According to their categorizations, a severe weather condition generated by either tornadoes or hurricanes will have a wind speed of at least above 50 m/s, which can cause severe damage. Thus, we gradually increased the absolute wind velocity, u, to 80 m/s to see the wind effect on the static stability of the utility truck in the crosswind.

In this considered case, the absolute wind is acting at an angle of 25°. As the velocity of the absolute wind increases, the change in normal static reactions at the front and rear right wheels, i.e., windward side wheels, are reducing. In contrast, the normal reaction force is increasing in leeward side wheels. It should be noted that there's a slide reduction in the front left wheel, which can be inferred by the uniform wind acting at an angle in the compass sense direction, as shown in Fig. 25.2. The grip utilization factor indicates that the utility truck will start skidding at the absolute wind speed of approximately 28 m/s, i.e., $K_\mu > 0.5$. However, it remains on the road, gripped by the other three wheels, generating the yaw rate. Actual skidding of the entire truck will start at the wind speed of 55 m/s when two other wheels (rear right and front left) lose lateral grip. It has been assumed that to increase truck safety, the value of static normal reaction forces should not decrease by more than 20% of its value on a flat surface in the transporting position. Thus, it can be seen from Fig. 25.8a that the reduction of 20% in the normal static reaction at the front left wheel occurs at the absolute wind speed of 30 m/s which also matches the grip utilization factor condition that happens at 28 m/s. The static roll angle at these values is 0.51°.

25.6 Conclusions

In conclusion, this paper presents a novel comprehensive study of the DA-IMS to enhance the safety and stability of an autonomous utility truck with boom equipment in extreme weather conditions. First, the mathematical formulations of utility truck's aero- and road-dynamics interactive models are presented, which include the severe weather conditions as an input in a two-domain interactive model-based design. Secondly, a utility truck's stability assessment is mathematically provided to investigate the stability of an autonomous utility truck with the boom equipment. Finally, a novel solution using two smart morphing devices is provided to manage the moment of inertia, tire-road forces, and the aerodynamic multi-phase forces. This novel formulation will provide a base for improving the rescuing team's safety and utility truck's productivity in extreme weather conditions.

25.7 Future Work

One of the promising directions of this research is to establish a safe zone for the utility truck. A safe zone is defined as one where the autonomous utility truck can slide under hazardous weather conditions without rollover. Computational and CFD numerical simulations with the validation of experimental results based on this novel formulation are work in progress that will be presented in a separate paper.

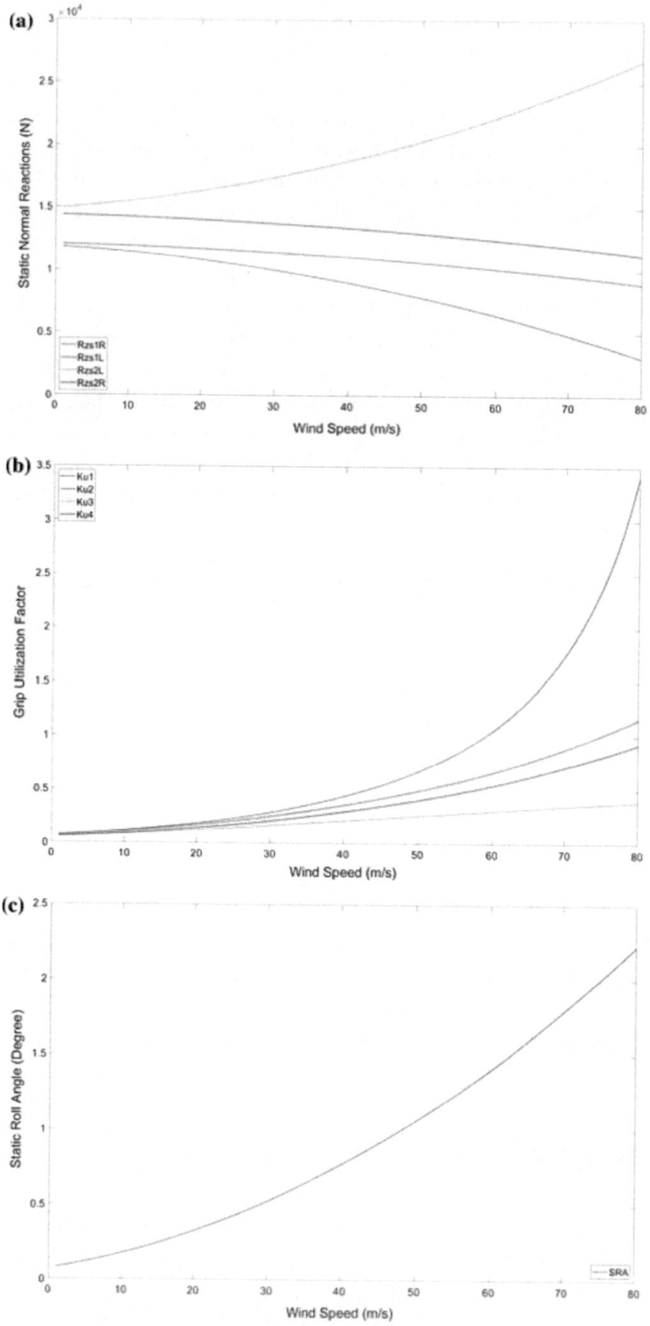

Fig. 25.8 Utility truck moving at a constant speed of 40 mph and the absolute wind angle of 25°. **a** Static normal reactions at the wheels. **b** Grip utilization factor. **c** Static roll angle (SRA)

Acknowledgements This research is supported by NSF award S&AS-1849264. We would like to express our gratefulness to Altec Inc., our research partner for providing utility truck and required data for the research analysis, which aided us to progress in our research.

References

1. Climate Science Special Report. https://science2017.globalchange.gov/chapter/7/
2. Freight facts and figures 2015, US department of transportation, Bureau of transportation statistics. http://www.princeton.edu/~alaink/Orf467F16/FHWA2201FreightFactsF_complete.pdf/
3. Burgin, K., Adey, P., Beatham, J.: Wind tunnel tests on road vehicle models using a moving belt simulation of ground effect. J. Wind Eng. Ind. Aerodyn. **22**(2–3), 227–236 (1986)
4. Hucho, W.-H.: Aerodynamics of Road Vehicles, 4th edn, p. 918. SAE
5. Kieselbach, R.J.: Aerodynamically Designed Commercial Vehicles, 1931–1961: Built on the Chassis of: Daimler Benz-Krupp-Opel-Ford, Kohlhammer (1983)
6. Patel, P.Y., Liu, J.H., Vantsevich, V.V., Koomullil, R.P.: Modelling effect of rain on aerodynamic performance of the Ahmed body. AIAA 2022-0335. AIAA SCITECH 2022 Forum, January 2022
7. Evgrafov, M., Vysotsky, A.N.: Aerodynamics of Wheeled Vehicles. Minsk, BelAutoTractor-Mashinostroenie (in Russian) (2001)
8. Bouferrouk, A.: On the applicability of trapped vortices to ground vehicles. In: The International Vehicle Aerodynamics Conference. Woodhead Publishing, p. 101 (2014)
9. Gilhaus, A.: The influence of cab shape on air drag of trucks. J. Wind Eng. Ind. Aerodyn. **9**(1–2), 77–87 (1981)
10. Mastinu, G., Plöchl, M.: Road and Off-road Vehicle System Dynamics Handbook. CRC Press (2014)
11. Wood, R.M., Bauer, S.X.: Simple and low-cost aerodynamic drag reduction devices for tractor trailer trucks. SAE Technical Paper, Technical Report (2003)
12. Chowdhury, H., Moria, H., Ali, A., Khan, I., Alam, F., Watkins, S.: A study on aerodynamic drag of a semi-trailer truck. Procedia Eng. **56**, 201–205 (2013)
13. Hyams, D.G., Sreenivas, K., Pankajakshan, R., Nichols, D.S., Briley, W.R., Whitfield, D.L.: Computational simulation of model and full-scale class 8 trucks with drag reduction devices. Comput. Fluids **41**(1), 27–40 (2011)
14. Qi, X., Liu, Y., Du, G.: Experimental and numerical studies of aerodynamic performance of trucks. J. Hydrodyn. Ser. B **23**(6), 752–758 (2011)
15. Fukuda, H., Yanagimoto, K., China, H., Nakagawa, K.: Improvement of vehicle aerodynamics by wake control. JSAE Rev. **16**(2), 151–155 (1995)
16. Han, M.-W., Rodrigue, H., Cho, S., Song, S.-H., Wang, W., Chu, W.-S., Ahn, S.-H.: Woven type smart soft composite for soft morphing car spoiler. Compos. B Eng. **86**, 285–298 (2016)
17. Kim, M.-H., Kuk, J.-Y., Chyun, I.-B.: A numerical simulation on the drag reduction of large-sized bus using rear-spoiler. SAE Technical Paper, Technical Report (2002)
18. Patten, J., McAuliffe, B., Mayda, W., Tanguay, B.: Review of aerodynamic drag reduction devices for heavy trucks and buses, vol. 205, p. 3. National Research Council Canada NRC Technical Report CSTT-HVC-TR (2012)
19. Bonnet, C., Fritz, H.: Fuel consumption reduction in a platoon: experimental results with two electronically coupled trucks at close spacing. SAE Technical Paper, Technical Report (2000)
20. Brinker, C.J.: Superhydrophobic coating. Sandia National Laboratories, Technical Report (2008)
21. Council, N.R., et al.: Technologies and Approaches to Reducing the Fuel Consumption of Medium- and Heavy-duty Vehicles. National Academies Press (2010)

22. McGowan, A.R.: NATO-perspectives on highly adaptive or morphing aircrafts. Technical Report, NATO, RTO-MP-AVT-168
23. Schweiger, J., Suleman, A.: The European research project active aeroelastic aircraft structures. In: Proceedings of the CEAS/NVVL/AIAA International Forum on Aeroelasticity and Structural Dynamics (2003)
24. Bowman, J.: Development of next-generation morphing aircraft structures. NATO, RTO-MP-AVT-168
25. Turnock, S., Keane, A., Bressloff, N., Nicholls-Lee, R., Boyd, S.: Morphing of 'flying' shapes for autonomous underwater and aerial vehicles (2009)
26. Miller, S.: Control of rotating spars for an adaptive aeroelastic wing. NATO, RTO-MP-AVT-168
27. Yuan, W.: Aerodynamics performance of UAV wings with morphing surfaces. NATO, RTO-MP-AVT-168
28. Nangia, R.: Morphing wings incorporating in-plane & folded-tips–, "Inverse" Aerodynamic Design Studies" Paper for RTO-AVT, vol. 168
29. Sainmont, C.: Multidisciplinary approach for the optimization of a laminar airfoil equipped with a morphing upper surface. NATO, RTO-MP-AVT-168
30. Patel, P., Vantsevich, V., Happawana, G., Harned, C., et al.: Dynamic Formulation of the Utility Truck with the Morphing Boom Equipment. SAE Technical Paper 2022-01-0917 (2022)
31. Patel, P.Y., Vantsevich, V., Koomullil, R.: Inverse dynamics of the utility truck's morphing robotic manipulator. In: Larochelle, P., McCarthy, J.M. (eds.) Proceedings of the 2022 USCToMM Symposium on Mechanical Systems and Robotics. USCToMM MSR 2022. Mechanisms and Machine Science, vol. 118. Springer, Cham (2022)
32. Patel, P.Y., Happawana, G., Vantsevich, V.V., Boger, D., Harned, C.: Morphing capabilities and safe operation zone of the utility truck boom equipment. In: Proceedings of the ASME 2020 International Mechanical Engineering Congress and Exposition. Volume 7B: Dynamics, Vibration, and Control. Virtual, Online. 16–19 Nov 2020. V07BT07A025. ASME
33. Patel, P.Y., Vantsevich, V.V., Whitson, J.: Utility truck: morphing boom equipment for terrain mobility. In: 20th International Conference and 9th Americas Conference of ISTVS 2021. ISBN 978-1-942112-52-5
34. Patel, P.Y., Jannoi, T., Zou, W., Vantsevich, V., Koomullil, R.: Aerodynamic analysis of the utility truck with the morphing boom equipment. In: Proceedings of the ASME 2022 Fluids Engineering Division Summer Meeting. Volume 2: Multiphase Flow (MFTC); Computational Fluid Dynamics (CFDTC); Micro and Nano Fluid Dynamics (MNFDTC). Toronto, Ontario, Canada. 3–5 Aug 2022. V002T05A033. ASME
35. Patel, P.Y., Krishnamurthy, C., Clausman, G., Vantsevich, V., Koomullil, R.: Modelling effect of rain on the external aerodynamics of the utility truck with the morphing boom equipment: computations and wind tunnel testing. AIAA 2023-1761. AIAA SCITECH 2023 Forum, January 2023
36. Baker, C.J.: A simplified analysis of various types of wind-induced road vehicle accidents. J. Wind Eng. Ind. Aerodyn. **22**, 69–85 (1986)
37. Lemay, J.: Revue de littérature portant sur la problématique de perte de contrôle des véhicules lourds causée par la présence de vents latéraux. Ministère des transports du Québec (2010)
38. Gillespie, T.D.: Fundamentals of Vehicle Dynamics. Premiere Series Books. SAE International, Warrendale (1992)
39. SAE J1939-21:2011: Surface vehicle recommended practice. The engineering society for advancing mobility land sea air and space
40. Vantsevich, V.V., Paldan, J.R., Gray, J.P: A hybrid-electric power transmitting unit for 4x4 vehicle applications: modeling and simulation. In: Proceedings of the ASME 2014 Dynamic Systems and Control Conference. San Antonio, Texas, USA, 22–24 Oct 2014. V003T44A002. ASME
41. Andreev, A.F., Kabanau, V.I., Vantsevich, V.V.: Driveline of Ground Vehicles: Theory and Design. In: Vantsevich, V.V., Scientific and Engineering Editor (eds.), p. 792. Taylor and Francis Group/CRC Press (2010)

42. National Weather Service, NOAA: Appendix: Fujita Scale (or F Scale) of Tornado Damage Intensity (2016). https://www.weather.gov/oun/tornadodata-okc-appendix
43. National Weather Service, NOAA: Saffir-Simpson Hurricane Wind Scale (2016). https://www.nhc.noaa.gov/aboutsshws.php. Accessed: 22 Nov 2016